D0561847

LABORATORY ....

SAFETY OFFICER..

DEPUTY SAFETY OFFICER OR PERSON RESPONSIBLE
FOR MAINTAINING FIRST-AID BOX ....

NEAREST HOSPITAL FOR CASUALTIES ....

.... Telephone..............

NEAREST AMBULANCE SERVICE ....

.... Telephone..............

NEAREST FIRE SERVICE ....

.... Telephone..............

OR DIAL 999 ASKING FOR AMBULANCE/FIRE SERVICE;
THEN GIVE EXACT LOCATION OF CASUALTY/FIRE

NEAREST DOCTORS ....

.... Telephone..............

....

.... Telephone..............

PERSONS ON PREMISES TRAINED IN FIRST AID.....................

....

....

PERSON TO WHOM ACCIDENTS MUST BE REPORTED

....

OTHER INFORMATION ....

....

....

HAZARDS
IN THE
CHEMICAL LABORATORY

HAZARDS
IN THE
CHEMICAL LABORATORY

edited by

G. D. MUIR, BSc, PhD, CChem, FRIC

*Consultant; formerly Technical Development Manager,
BDH Chemicals Ltd, Dorset*

LONDON
THE CHEMICAL SOCIETY

The information contained in this book has been compiled by recognised authorities from sources believed to be reliable and to represent the best opinion on the subject as of 1976. However, no warranty, guarantee, or representation is made by The Chemical Society as to the correctness or sufficiency of any information herein, and the Publisher assumes no responsibility in connection therewith; nor can it be assumed that all necessary warnings and precautionary measures are contained in this publication, or that other or additional information or measures may not be required or desirable because of particular or exceptional conditions or circumstances, or because of new or changed legislation.

ISBN 0 85186 699 9

© *The Chemical Society, 1977*

First published September 1971; second edition, March 1977

Published by The Chemical Society, Burlington House, London W1V 0BN, and distributed by The Chemical Society Publications Sales Office, Blackhorse Road, Letchworth, Herts, SG6 1HN.

Printed in Great Britain at the Alden Press, Oxford.

CONTENTS

D. W. NEILL, MSc, CChem, FRCPath, FRIC and
J. RUSSELL DOGGART, MSc, PhD, CChem, MRIC,
Clinical Biochemistry Laboratory, Royal Victoria Hospital, Belfast

S. B. OSBORN, BSc, PhD, FInstP, *Department of Medical Physics, King's College Hospital and Medical School, London ; Radiological Protection Adviser, South East Thames Regional Health Authority*

Preface to the Second Edition

Much has happened in the five years since the first edition was published to justify further changes in the scope and emphasis of *Hazards in the chemical laboratory*.

In the United Kingdom, the Health and Safety at Work etc Act 1974 has set new standards of safety and responsibility for the country which will soon permeate every practical activity in industry, teaching and research. In tune with this movement towards greater concern for good practice and the well-being of oneself and one's fellow worker, has been the publication by the Royal Institute of Chemistry of a *Code of practice for chemical laboratories* (London: RIC, 1976). This advises all who are involved in chemical laboratory work as to the current professional view of what constitutes good practice, at the same time leaving the responsible managers to devise and supervise the operation of their own laboratory rules and regulations in a manner that recognises the desirability and right of everyone in a laboratory team to play their part in promoting its smooth running.

The monumental, 10-year labour-of-concern, *Handbook of reactive chemical hazards* by Mr Bretherick, has had a major impact upon the laboratory scene. This is a unique publication of world-wide importance which tackles, in much greater depth than has been attempted hitherto, the very difficult job of gathering together the scattered information about the instability and dangerous reactivity of chemicals, both commonplace and in the 'research' category, that can catch the laboratory chemist or research worker unawares. Today's research chemical quite often becomes tomorrow's routine laboratory reagent and this handbook should do much to reduce accidents caused by violent reactions and explosions.

As Mr Luxon stresses in his 'Introduction', there is public as well as professional concern about the emergence of hazards arising from the unforeseen toxic properties of a number of chemicals that have been in regular use in industry, and sometimes in domestic life, for many years. As a sign of this anxiety, and the enactment of the Occupational Safety and Health Act of 1970, the US Department of Health, Education, and Welfare published the *Toxic substances list* in 1971; this now appears annually and the 1974 edition featured 13 000 different chemicals.

We are very fortunate to have Dr Magos joining our authors to present the current findings of that difficult but rapidly growing science, toxicology, in relation to chemicals. His chapter, 'Chemical hazards and toxicology', follows a similarly authoritative presentation by Mr Bretherick of 'Reactive chemical hazards'. These two new chapters require and deserve very careful study by those likely to be working with the types of materials described by their authors.

At a less mentally exacting, but equally important level are the two other new chapters on the Health and Safety at Work etc Act 1974 by Mr Corbett, and on Medical services and first aid by Dr Gilks. Mr Corbett spells out clearly how the new Act affects the chemist and it is worth noting that a useful new book *A guide to laboratory law*, which outlines the contents of the numerous earlier items of legislation that affect laboratory practice and now operating under the wing of the new Act, was published earlier this year. Dr Gilks outlines the role of the Employment Medical Advisory Service, which is also covered by the Act and stresses the importance of properly trained first aiders who are members of laboratory staffs.

The team that wrote the first edition is still with us, but these authors too have shown their awareness that we are in a period of reconsideration of old attitudes and practices, and have brought their contributions up to date and in line with the spirit of the legislative changes that are in progress.

It has been a great personal help and satisfaction to have Mr Bretherick's ready cooperation with fundamental changes in the book's largest section, Chapter 8 now renamed 'Hazardous chemicals'. The possibility of providing our readers with immediate references to a great many (but by no means all) of the entries in *Handbook of reactive chemical hazards* has greatly increased the scope of our book, particularly to those engaged in research who would like to go back to the accounts of original incidents. Apart from the new exclusively 'Bretherick' references, Chapter 8 now includes new monographs dealing with chemicals that have found their way more generally into chemical laboratories.

We have to thank old friends again for helping us to get up to date. In particular, BDH Chemicals Ltd records have again been valuable and we are indebted to Mrs E. M. Stubbings and Mr E. R. Sheppard of that company for their assistance in obtaining information. We are also grateful to Mr J. P.Moore of Shell Chemicals UK Ltd, and Mr J. A. A. Forbes of St Peter's Hospital, Chertsey, for valuable comments on parts of the text. An editor is a particular burden to the 'regulars' who tidy up behind him and in this category he places Miss Lynette K. Hamblin of the Chemical Society who 'translated' him and his fellow authors to the printers with great tact and efficiency; his sincere thanks to her must be recorded and to other members of the Society staff who have worked on the book with us.

G. D. MUIR
September 1976

Preface to the First Edition

The present volume is a successor to the Royal Institute of Chemistry's *Laboratory Handbook of toxic agents*, first published in 1960, and issued in a revised edition in 1966. Before the second edition went out of print, the future of the publication was considered by both the Institute's Publications Committee, and by the previous editor, Professor C. H. Gray, and myself. It was generally felt that, rather than merely revise the existing material, it would be preferable to alter the underlying philosophy of the book by changing its scope from toxic hazards to a consideration of all hazards likely to be encountered in the chemical laboratory.

The general format remains the same, with the major part of the book (printed on tinted paper) being an alphabetical guide to hazardous chemicals and measures to be taken in the event of accidents in their use. However, whereas previous editions have included details of measures to be taken against the toxic hazards of such chemicals, the present edition also includes methods for spillage disposal and extinguishing of fires where appropriate. Once more, an alphabetical listing is adopted to enable the use of the book as a speedy reference in the case of emergency.

Much assistance was required in preparing this chapter, now extended to over 430 hazardous chemicals, and the preceding one on first aid. I must record my special thanks to my colleague Mr W. G. Moss for his collaboration when we prepared these chapters for the first edition of the earlier book, and to BDH Chemicals Ltd for permission to use their extensive records on the hazards, handling and disposal of chemicals; also to my colleague Dr P. Mostyn Williams who has added considerably to the earlier medical advice of Dr W. B. Rhodes.

Dr D. P. Duffield and Dr K. P. Whitehead of Imperial Chemical Industries Ltd have also provided important medical advice on up-to-date first aid practice, particularly on the treatment of cyanide and phenol poisoning. The chemical world must always be grateful for the pioneer efforts of ICI in encouraging chemical factory and laboratory safety and we would record again our thanks to Dr A. J. Amor and Dr A. Lloyd Potter for their interest in the first book and to pay special tribute to the work of Dr L. J. Burrage who has contributed so much to promote laboratory safety in this country.

Many other firms have given us the benefit of special knowledge of certain of their products and we are also grateful to James North & Sons Ltd for permission to reproduce their chart advising on the types of glove to be used when handling different classes of chemicals.

Despite extensive practical experience of chemical hazards, the writers of a book such as this lean heavily on the authors of major works on industrial toxicology. Not many may have had the privilege of knowing the charm and intelligence of that great lady, Dr Ethel Browning, who guided the Institute and some of the authors when the first book was conceived and drafted, and wrote two renowned works upon which we have drawn extensively, *Toxicity and metabolism of industrial solvents* and *Toxicity of industrial metals*, as well as editing the important series of monographs on toxic agents in which they appeared. Her death last year ended a long life of devoted service to industrial safety.

The valuable publications of the Chemical Industries Association – *Marking containers of hazardous chemicals* and *Exposure to gases and vapours* – have been referred to frequently and we would like to thank the Association for the privilege of perusing the text of the latter at the proofing stage. The *Laboratory waste disposal manual* published by the Association's counterpart in the US, the Manufacturing Chemists' Association, was also consulted extensively in preparing chapter 6, as were the following works:

Dangerous properties of industrial materials by N. Irving Sax
Extra pharmacopoeia (Martindale) edited by R. G. Todd
Industrial hygiene and toxicology edited by Frank A. Patty
Industrial toxicology by L. S. Fairhall
Poisoning by drugs and chemicals by P. Cooper
Poisons by Brookes and Jacobs
Toxicology of drugs and chemicals by W. B. Deichmann and H. W. Gerarde.
Other acknowledgements appear in chapter 5.

Of the remainder of the book, new chapters have been provided by Mr Ackroyd, Dr Taylor and Mr Sheldon on fire protection and by Mr Neill and Dr Russell Doggart on the particular hazards facing chemical workers in hospital biochemistry laboratories. In addition, Mr Luxon, of the Department of Employment, has contributed an entirely new introduction to replace the one by the late Sir Roy Cameron which appeared in the earlier editions. Mr Beard and Dr Osborn have thoroughly revised chapters 2 and 7 respectively. To all these authors, I am extremely grateful for the time they have spent on and interest shown in this project.

Dr Farago and his staff in the Editorial Office of the Institute have my sincerest thanks for their painstaking help and I am particularly indebted to Dr Martin Sherwood for his close collaboration in co-ordinating our efforts, sharing in our proof reading, and carrying out all the necessary negotiations with the printers. With my co-authors he has made the task of editing not only light, but both stimulating and enjoyable.

Finally, I should like to pay tribute to the immense amount of work which Professor Gray put into the planning and production of both editions of the *Laboratory handbook of toxic agents.* Although pressure of work prevented Professor Gray from taking a full part in the editing of this volume, I have had the benefit of his wise advice throughout its preparation. Without this and the substantial contribution he made to the earlier book, it is unlikely that this volume would have been possible.

G. D. MUIR
April 1971

Chapter 1

Introduction

Since the publication of the first edition, there has been much public concern and debate on all matters relating to safety and health. This led to the publication of the Robens Report on Health and Safety at Work and in turn to the Health and Safety at Work etc Act 1974 which over the next decade will undoubtedly have a profound effect on the matters towards which this handbook is directed.

As indicated in Chapter 2, the general purpose of the Health and Safety at Work etc Act is to provide for one comprehensive and integrated system of law dealing with the health and safety of all persons engaged in work activities and the health and safety of others who may be affected by such activities. It is important that everyone concerned with work in the laboratory is aware of and accepts their new legal responsibilities. Until some specific subordinate legislation is enacted, guidance as to what may be required in laboratories may be drawn from existing parallel legislation in other areas such as factories, and from codes of practice such as that on the use of ionising radiation. More general guidance and the reasoning underlying such guidance is contained in this handbook.

Particular attention should be paid to systems of work and to the clear delegation of specific responsibilities to those organising units of laboratory activity. A further important aspect of the legislation lies in its requirement in respect of consultation and this places a particular responsibility on chemists who are best able to advise the layman on the hazardous nature and properties of chemical substances, and the precautions necessary to ensure their safe use. It is, therefore, important that everyone concerned makes an evaluation of all aspects of health and safety now and puts in train any necessary steps to ensure that their house is in order.

Regrettably, perhaps, in the past chemists, who habitually handle dangerous substances, have been inclined to disregard the hazards associated with them, particularly if such hazards are of a long term nature. Every human being, and chemists are no exception, tends towards the view that although an accident may happen to another, it will never happen to him personally because he is too wise and knowledgeable. Experience shows that nothing can be further from the truth. During work in the laboratory, many persons have suffered injury to their health which, because of the insidious symptoms, may never have been associated with their work activity. It is only when permanent injury has occurred that many persons come to realise that the observance of even elementary precautions could have prevented such injury or in extreme cases premature death.

It is against this background that one should look at this edition of the handbook. The contributors have attempted to indicate and discuss the dangers likely to arise in the laboratory and have offered practical advice on their avoidance. The inclusion of new chapters gives an authoritative background to many of the detailed recommendations in Chapter 8. This will, I believe, prove most useful in devising precautionary techniques in respect of the many reagents and substances which, for reasons of space, have not been included in this work.

At the same time this manual has, I believe, become a much more useful and complete work not only for chemists in the laboratory but also for all those who handle hazardous substances on a small scale *eg* in industry. Additionally the work will be useful in schools and higher education establishments where training in the correct use and handling of the substances should be considered an integral part of the curriculum of students in science subjects, and where the Health and Safety at Work etc Act has extended specific obligations in respect of safety and health matters.

The control of hazards of the laboratory is well known: the enforcement of safe work systems, the need for mechanical safety involving the guarding of dangerous parts of machines, even if driven by only fractional horse power motors, so that injury from contact with moving or trapping parts is prevented; the need to provide safe means of access to every place where anyone is at work even if the work is only undertaken on rare occasions; the need for good housekeeping to minimise the possibility of accidents occurring through persons striking or being struck by objects; the need for care in handling glassware; the need to protect electrical conductors and to provide or use low voltage supplies or sensitive earth trip switches; and of course, the matters with which this handbook is intimately concerned — the prevention of injury from fire, explosion or from exposure to toxic substances and pathogenic agents.

Identification

Perhaps the most important single step we can take in securing the safe handling of chemicals is to ensure that a proper system of labelling is used which will identify the substance, indicate the hazards involved, and set out the simple precautionary measures to be followed. There has, of recent years, been considerable international discussion concerning labelling systems and the Council of Europe has published proposals concerning the labelling of pure chemical substances. This system is described in the so-called *Yellow book, Dangerous chemical substances and proposals concerning their labelling*, which can be purchased from HMSO. The EEC has embodied these general proposals in specific directives on pure substances and solvents. Other directives dealing with formulated products are in the course of preparation. The Health and Safety Executive has recently consulted interested parties on the formulation of legislation to give effect to these proposals. Chapter 8 follows these proposals both in respect of substances covered by the directives and similarly in respect of other chemicals. Failure to give some such simple warning of the hazards is inexcusable, particularly in the laboratory where many chemicals may at some time be handled by inexperienced and unskilled persons who are not members of the lab staff *eg* cleaners and maintenance workers. Accordingly all chemicals should be labelled following the general guidelines used in these systems and set out elsewhere in this work.

Management's task

Safety and health is the responsibility of management and this must be appreciated from the director downwards. Not only must he and all members of the staff know the hazards involved but they most also be clearly seen to be directly interested and involved in the promotion of a safe and healthy environment. Strict procedures should be written into analytical and other methods or, where such methods do not exist, the work should be immediately supervised by a responsible person who is aware of the dangers, and precautions to be followed both during normal working and in any emergency that may arise. In larger laboratories a safety officer and hygienist should be appointed to provide advice and general supervision and, not least, to look critically at the procedures involved from outside the group undertaking the project. Experience shows that such a view dissociated from the actual scientific work is invaluable in bringing to light relatively simple hazards which may have been overlooked.

General management and supervision must be 'tight' to ensure that work is conducted in a predetermined and orderly manner, that unauthorised actions are checked and that proper care and attention is given to the minute by minute operation of process or experiments. In particular, at each meal or tea break and at the end of the day, a thorough check must be made to see that everything can be left safely and if there is any doubt arrangements should be made for the continued supervision of the operations still in progress.

The overall aim should be to engineer out hazards so that the whole system can operate in such a way that any possible human error is eliminated as far as is practicable. Chapter 3 deals with this general aspect of the problem.

Fire and explosion hazards

The dangers of fire are well known but again we must remember the maxim 'familiarity breeds contempt'. The very large number of fires in laboratories proves the seriousness of this problem. Chapter 4 gives detailed advice on such risks and the text under each substance in Chapter 8 indicates properties on which an assessment of the fire hazard can be based and makes suggestions as to the selection of fire fighting equipment.

It should be clearly understood that when a liquid is used having a flash point below the highest normal ambient temperature it can, in suitable circumstances, liberate a sufficient quantity of vapour to give rise to flammable mixtures with air. This can, given time and favourable conditions, accumulate in the workroom to such an extent as to give rise to the possibility of a serious explosion when an ignition source is introduced. More importantly and more simply a source of ignition already present may ignite the vapour-air mixture at some distance, causing a flash back to the original source, with the consequent possibility of disastrous fire.

A flammable gas or vapour must be present in a concentration of the order of 1 per cent or more by volume if its mixture with air is to be flammable, so it is a relatively simple matter to check whether or not a dangerous concentration is likely to be present in closed plants such as ovens *etc.* During normal working it is desirable in practice to ensure that one-quarter of the lower flammable limit is never exceeded. The amount of flammable vapour or gas in the workroom air should of course never approach this concentration during normal working procedures. Account must however be taken of possible leakages and spillages so that although it is perhaps unnecessary to provide special precautions such as flameproof electrical equipment in normal circumstances this may be very desirable to take care of possible emergencies arising from unexpected loss of flammable solvents having a flash point of less than 32 °C and particularly those having a flash point of less than 21 °C. Our aim therefore should be to restrict the use of such liquids to situations in which they are absolutely necessary and even then to reduce the quantity involved as far as possible. This is particularly important with solvents used for routine operations where in every case we should carefully consider whether or not it is possible to substitute an alternative having a higher flashpoint.

The quantity of all flammable materials and of solvents in particular should be kept to the absolute minimum. There is often a tendency to disregard this and allow large quantities of solvents which are used only occasionally to accumulate in the lab. When flammable substances are not in use there must be adequate supervision to ensure that they are kept in a properly constructed fireproof store.

Suitable fire fighting equipment should be readily available and adequate means of escape provided (*see* Chapter 4). All personnel should be trained and familiar with the use of the equipment so that a small fire can be quickly localised and prevented from spreading, while in the event of the fire getting out of hand everyone must know how to escape safely.

Reactive chemical hazards

Particular care must be exercised when using highly reactive or unstable substances which may be liable to cause an explosion. The quantities used should be kept to a minimum and if necessary several reactions carried out on a smaller scale. Consideration should be given to limiting the effect of the explosion—should one occur—by the provision of suitable reliefs venting to a safe place. In all cases protective screens should be provided or the experiment operated by remote control so as to ensure that the operator will not be injured. It should be remembered that such substances may be produced as a result of side reactions or in residues standing over a period of time.

Reactions which are exothermic should be carefully controlled and instrumented to ensure that there is no failure of the cooling or stirring systems. Again quantities should be kept to a minimum and suitable screening provided. No operation of this kind should be entrusted to anyone other than a highly skilled and competent chemist knowledgeable in the dangers involved and the precautions to be taken. All these matters are discussed in detail in Chapter 5.

Toxic hazards

Toxic substances can act in three ways causing poisoning by ingestion,

5

percutaneous absorption, and inhalation. Our first thought should always be: can a harmless or less hazardous substance be used instead of the substance under consideration? Such a step removes completely or reduces the danger in an infallible manner and this possibility should never be overlooked. The dangers of ingestion by contamination of the hands or food can be virtually eliminated if there is proper attention to personal cleanliness. Washing accommodation of a high standard should be provided together with means for drying which is always available. This precaution is of course equally applicable to other health risks and promotion of personal cleanliness should never be neglected. Another common but inexcusable danger in this category is the use of mouth pipettes. Such methods should never be used for pipetting hazardous liquids and even in schools we should train pupils to use a rubber bulb.

The contact of corrosive substances with the skin is generally obvious and so this is somewhat less dangerous than contact with percutaneous poisons. Nevertheless gloves (*see* Chapter 3, p 25) and, where necessary, protective clothing should always be worn where this hazard is present and, if contact with the skin occurs, the affected parts should be washed immediately with soap and water. Special attention is necessary to protect the eyes, and where corrosive substances are regularly used, eye protection should be worn as a routine precaution. It must always be remembered that it is often an innocent bystander and not the person actually carrying out the process who suffers injury.

Substances which can be absorbed through the skin present a more insidious hazard, particularly if the contact is repeated or prolonged. Great vigilance is necessary to ensure that the dangers are appreciated by everyone concerned. The aim should be to prevent contact. The use of protective gloves has been found in many cases to be of doubtful value as contamination can occur on the inner surfaces through pinholes or by careless removal and replacement unless strict working procedures are observed. Any splashes on the unprotected skin should be washed off immediately with soap and water. Where the more hazardous substances in this group are used regularly it is advisable to make arrangements for the periodic medical examination of the persons involved.

Inhalation of harmful vapours, gases, dusts and liquid aerosols is perhaps the most insidious and widespread danger in the laboratory. Many persons have at one time or another been exposed to excessive quantities of vapours such as mercury, benzene and carbon tetrachloride or to dusts such as lead and beryllium. The danger is the more serious because it is insidious, there often being no sensible perception of danger. Additionally, since it is all pervading, the contaminant must be breathed once it has entered the air of the workroom. Everyone is inclined to judge the danger by the short term effects whereas it is the long term effects which are the more serious and may give rise to permanent and irreversible injury. Unfortunately such effects may not be directly attributable to exposure to toxic chemicals. The affected person may have changed his employment or be no longer working with the hazardous materials. Therefore there is little statistical evidence as to the incidence of ill health brought about by such exposures and in consequence the very real and serious dangers tend to be disregarded. It is the duty of every responsible person to see that such substances are only used where it is absolutely necessary. International opinion is now coming around to this view and restrictions are being considered on the use of benzene and certain chlorinated hydrocarbons as solvents (*see* Chapter 3, p 27). If a hazardous substance must be used then adequate instructions must be given, proper supervision assured and the whole process carried out in such a way as to ensure that none of the hazardous material enters the air of the workroom where it may be breathed. A properly designed fume cupboard should always be used and all steps in the process or experiment carried out therein. Where the hazardous substance is used regularly, giving rise to long term effects, the possibility of monitoring the atmosphere at regular intervals should be considered. If the sample is taken in the breathing zone by means of a suitable personal sampler then confirmation can be obtained as to the efficiency of the precautionary measures and the worker reassured as to the absence of any long term health risk. The assessment of toxicity is a complex matter but general parameters are set out in Chapter 6 which, as indicated earlier, lock into the labelling system recommended in Chapter 8.

Two relatively harmless liquids or substances may, when in contact, liberate an unexpected or surprise poisonous gas. Perhaps the commonest example is bleaching powder and acid lavatory cleaners which liberate chlorine gas. Another common example is alkaline cyanides which, in contact with acid, liberate hydrocyanic acid gas. This phenomenon often occurs in sink traps or other parts of the drainage system or during the use of containers which have not been properly washed out from a previous operation.

It is necessary to consider whether any such groups of substances are to be used, particularly by unqualified persons, and whether or not chance contact between such groups could produce a more serious hazard. If this is possible, the operations should be segregated so as to avoid a possible chemical reaction. Working instructions should include the safe disposal of waste and require the routine cleaning of containers used to transfer such chemicals.

The subject of carcinogenicity is a difficult one. Many substances if implanted repeatedly in animals will produce a carcinogenic reaction in time, but they are clearly not hazardous when used in the normal way. On the other hand there are substances such as 2-naphthylamine which will almost certainly produce carcinogenic effects if ingested into the human body. Between these two extremes lie a large number of potentially hazardous materials. Where a known carcinogenic risk exists it is indicated in the text. Substitution of another reagent should always be considered but, if such substances must be used, all practical precautions should be taken as, unlike poisonous substances, it is often difficult to fix a safe level and as far as practicable it should be ensured that contact does not occur. The Carcinogenic Substances Regulations 1967 prohibit the use industrially and the importation of certain carcinogenic compounds namely, 2-naphthylamine, benzidine, 4-aminodiphenyl, 4-nitrodiphenyl and their salts and substances containing any of these compounds other than in very small concentrations.

General environment

As with all chemical hazards good housekeeping—that is the cleaning up and removal of spillages and lost material—is a cornerstone to safe working. In this context we must completely remove or render chemically inert all lost material which may become a hazard. Liquids may be removed with water if they are soluble. If insoluble detergents or solvents may be used. In the last resort it may be necessary to react the material chemically to an inert form as a method of decontamination.

Toxic dusts are particularly hazardous in this respect since if lost material is not removed it will become repeatedly airborne whenever it is disturbed either by movement of persons or materials or by strong air currents. Such dust if present must be removed at frequent intervals by an industrial vacuum cleaner fitted with a high efficiency filter to prevent recirculation of the fine particles in a breathing zone.

Past experience clearly indicates that many cases of poisoning in laboratories are due to background contamination brought about by a neglect of these principles.

Coupled with good housekeeping is the quality of the general working environment—in particular, lighting, ventilation and heating. Good lighting is important because hazards become immediately apparent. There is less need when carrying out intricate manipulations for the operative to bend close to the danger area to obtain an adequate view of the operation, and malfunction of the equipment or instrumentation is immediately apparent, permitting remedial measures to be taken before danger occurs.

Heating and ventilation go hand in hand. Good general ventilation is essential in laboratories where toxic materials are handled. A general standard of at least five air changes per hour should be the aim. Adequate heating must be provided during the winter months or such ventilation will not be used.

We may sum up all these general matters by saying that where hazardous materials are handled we should provide a pleasant, clean working environment properly planned with good access and means of escape and a high standard of lighting, ventilation and heating with regular cleaning to ensure removal of any spilt or lost material.

After a hazardous substance has been used there remains the problem of disposal. The first method to be considered is its return to stock *via* reprocessing. If this is not possible small quantities of combustible material may be burnt in a controlled manner so that a hazard will not arise from fire or toxic decomposition products. Small quantities of less hazardous chemicals can be flushed away to the drain with copious quantities of water but if larger quantities are involved the local authority should be consulted. Highly hazardous chemicals should be reacted in a fume cupboard to break them down into less hazardous compounds which are more readily disposable. It must be clearly understood that the responsibility for safe disposal rests with the user and under no circumstances should substances be disposed of in such a way as to constitute a hazard either inside or outside laboratories. In the last resort specialist companies are available who will undertake the task of disposal for a fee.

Where laboratories are situated in a built-up area or where the effluent from exhaust systems must be discharged at a low level, steps must be taken to ensure that means of trapping the dangerous substances are provided so that the exhausted air does not contain a dangerous quantity of harmful substance. Care should be taken to discharge the effluent at a safe height where it will not recirculate into the laboratory or cause a nuisance or danger to other persons in the vicinity.

Biochemistry hazards

The biological hazards which occur in biochemical and medical laboratories may be considered an extension of the dangers of inhalation, ingestion or skin contamination from toxic materials requiring, as they do in many instances, even more stringent precautions against infection. Chapter 9 sets out the working practices to be observed. It is particularly important that the precautions against the consumption of food and drink and the prohibition of smoking should be observed. Equally, the avoidance of contamination by the use of suitable protective clothing, the washing of contaminated skin and the prompt treatment of all injuries will further reduce the possibility of infection. As with other hazards an alert and well-informed safety officer can ensure that the precautions are observed and that any defects are remedied before danger arises.

Radiation hazards

These are dealt with very fully in Chapter 10. Reference should also be made to the Code of Practice on the use of Radiation in Research and Teaching Laboratories published by the Department of Health and Social Security. Sealed sources and generators of radiation such as x-ray equipment are usually well defined and their location well known. Adequate safety during their use can be achieved by restricting the radiation itself to as low a level as possible and certainly below the level set out by the International Commission of Radiological Protection.

Conclusions

It is clear that with the passage of time knowledge is advancing more and more rapidly in every field, not least in the toxicological effects of exposure to chemical substances and with each year we seen an improved understanding of the long term hazards associated with their use. A massive documentation now exists and for the great majority of substances some general assessment of the hazard can be made. Nevertheless our general policy should be to avoid exposure as far as possible in every case. This has become particularly important as the safe levels of substances are constantly being reduced and, in addition, new and unexpected toxic effects may become apparent with the passage of time.

It must be the aim, therefore, of every responsible person to ensure that all chemicals are handled safely without either immediate or long term dangers. A clean, healthy general working environment must be provided and each individual encouraged to become safety conscious. The subsequent parts of this book provide detailed information on particular types of hazards while in Chapter 8 under each individual substance details of its principal properties and dangers and the general procedures to be followed together with the threshold limit value and flashpoints, will enable the reader to decide for himself what should be done. At the same time it must be emphasised that, given a wise, commonsense approach to the problem, the general elementary precautions indicated above will enable any normal operation to be carried out safely.

Perhaps the most important single step is to pause before carrying out any new procedure—to stop and stand back and consider what hazards may arise, what precautions should be taken, and what emergency procedures may be necessary.

Chapter 2

Health and Safety at Work etc Act 1974

Background

Even though there are aspects of this recent legislation which are very pertinent to chemists, and others working in a professional capacity in a chemical environment, it is important to stress that the chapter is not intended to be a full explanation of the Act. For a legal review other works should be read and in specific cases expert (legal or otherwise) opinion should be sought.

The Act has been described as the most important and significant statute dealing with health and safety since the original Shaftesbury legislation of 1833 *et seq*. It arose from widespread consultation with organisations representing those likely to be affected, and after a studied review of all existing legislation and its implementation and effect. The committee undertaking this work (presided over by Lord Robens and to which the Royal Institute of Chemistry made significant submissions of evidence), made recommendations which in the main were acceptable over a wide range of political and other organised opinion, and became law in 1975.

The Act embraces and tidies up all existing legislation and brings into one legislative document one comprehensive system of law and law enforcement to cover the health, safety and welfare of people at work and the health and safety (but not the welfare) of those members of the general public likely to be affected by work activities of others.

As compared with previous legislation it extended 'factory law' to a further five to six million people, and introduced—universally—the 'duty of care' to third parties only previously recognised in some very specific types of plant and process. To ensure—in due time—a liberal, systematic and universally acceptable set of technical standards of performance and enforcement, the Act also established the machinery and administration for an integrated and nationally organised and controlled Executive and Inspectorate.

Chemists, of course, may be members of the Inspectorate and special instruction in its role will in this case be available to them. To a large extent also those chemists who have a direct and first hand functional role in health and safety or industrial hygiene will need a wider understanding of the subject than can be given here. For the majority of chemists, however, an appreciation of the philosophy behind this changed attitude to factory legislation, will suffice to ensure that their involvement is cooperative, intelligent and professional.

Main provisions

The general purposes of the Act are to maintain and improve standards for those at work or affected by work activities, to control the storage and use of dangerous substances and to control emissions into the air in prescribed circumstances. Employers have a general duty to ensure these standards, to consult with their employees and advise them and others (*via* general policy statements or arising from specific requests and their company's Annual Report) of ways in which they are discharging this obligation. They must draw attention to potential hazards and hazardous substances and describe the measures which must be taken for safe working and handling. All equipment or any modifications required to ensure a safe working environment must be freely provided by an employer, and employees must ensure that such facilities are not misused. Additionally employees must cooperate to the full with their employer and others in meeting statutory obligations: in most cases this will be by elected safety committees where consultation will have taken place. Third parties providing goods and services must also ensure that these are not harmful or unsafe if used or applied in accordance with accompanying instructions.

The technological limitations to a totally safe and healthy situation are recognised at many places in the Act, where phrases such as '... so far as is reasonably practicable', '... use the best practicable means ...' frequently occur. Among the new classes of 'workers' to whom the legislation has been extended are those in research and development, public utilities, teaching in schools, colleges and universities; medical and dental services, entertainment and the media, museums and art galleries, central and local government, fire and police and similar laboratories or ancillary services. Whereas before the Act chemists and others checked for individual *inclusions* in this type of legislation, the new Act is so all embracing that *exclusions* are those which need to be specifically checked.

Comprehensive application

The Act covers nearly all people and nearly all situations.

It does not define cases, but makes an overall assumption that an employer has a general duty of care to his employee, that an employee has this same duty to his co-workers and both have a duty to the public at large. This overall (or blanket) assumption is incorporated into a blanket law which in effect (although it deliberately includes, for the time being, relevant previous codified legislation) is an enabling instrument capable of being changed, extended or amended as circumstances change and experience is gained. It ensures that health and safety education and practice are on-going dynamic subjects capable of continuing and continuous debate and subsequent revision by consultation with and agreement by those most affected.

Except where specifically inoperative, legislation in these fields prior to the Act, is modified only as far as its administration and penal sanctions are concerned. In all cases, procedure and policy is determined by the Health and Safety Commission and administered and enforced by its Executive which includes all branches of the unified Inspectorate. Some of its functions have been and will be delegated to local authorities; some Inspectors, with wide experience and expertise will be given the wider powers of enforcement that the legislation allows.

Illegal situations can be corrected by the serving on an employer of an Improvement Notice (which defines the unsafe environment and procedure and allows time for its correction) or a Prohibition Notice either immediately or deferred (which can in fact shut down activities until the specified defect(s) has/have been corrected).

Involvement and consultation

Safe working procedures and environments are promoted, initially, by those employers anxious to ensure that their workers' productivity is not lowered by sickness or accident and their own commercial 'image' is not distorted by the reactions of critical neighbours and/or customers. In due time the best of all such working conditions becomes distilled into the 'reasonable norm' for that particular industry or situation. In fact the full spectrum of interests is almost invariably considered in the drawing up of the instructions or recommendations by discussions with some representative forum from the industry and/or the workers concerned.

Arising from such experience and consultation there can emerge, from the Health and Safety Commission, two types of authoritative document:

(a) Regulations   These are like Statutory Instruments; they are issued by the Minister in accordance with his authority under the Act; they are obligatory and have the full force of law. They are detailed as to time, temperature, size, weight, degree, and units *etc*: the quantities can be measured or accurately assessed and the letter of the law must be obeyed in every detail. Any infringement is an offence. Much of the older type of factory legislation will be brought up to date and reissued in this form. Many chemical plants and operations could be covered by these and chemists must be clear as to the precise directives and the part they themselves must play in monitoring or control procedures.

(b) Approved Codes of Practice   These are guidance rules only and they have limited legal effect. They are drawn up by experts from industry, trade unions, education, professional bodies *etc*, and they set out to define standard and acceptable good behaviour and practice in a particular set of circumstances. They are relevant and necessary where principles are reasonably clear but their application may be different from case to case due to widely differing circumstances.

Such codes are approved by the Commission but do not have the full force of law. Those affected by them however ignore them at their peril. Although it is not an offence to act differently from the Code, it is necessary, in defence of some infringement of the Act, to show that the adopted procedure or attitude was at least as good as, or better than, the one proposed in the Code.

Laboratory and some pilot plant operations, in particular, are difficult to classify and define in precise detail. They are most likely to be covered by a Code of Practice and in anticipation of this the Royal Institute of Chemistry has prepared a draft Code of Practice for Chemical Laboratories[1] and submitted it for consideration by the Health and Safety Commission.

The responsibility of the chemist

As has been indicated, the chemist is most likely to be working in a situation not now exempt from the provisions of the Act and he will therefore have responsibilities similar to those of other workers. These will be *inter alia* to wear his protective clothing and use such appliances as are necessary in prescribed procedures, to work in a manner which is not likely to injure the health and/or safety of his colleagues and to cooperate with his employer in the fulfilment of statutory obligations.

This is minimal and more often than not his responsibilities will be much greater: his knowledge, experience and training will dictate this. He may have a direct functional role in the health and safety organisation (using this in its widest sense), either at Executive or company level, as a consultant or as a teacher. He may be monitoring data, doing research work directed to health and safety matters, developing safer procedures, writing instructions for new plant, equipment or processes, contributing to the work of safety committees, working as a 'community politician' with his neighbours on environmental aspects of local industry—in fact in any way where his specialist contribution may lead to a better understanding of the health and safety aspects of life in an industrialised community.

His legal responsibility is that of any other person; some special responsibilities may arise from his employment—his terms of reference. In all cases these will be judged from the standard of a professional chemist and not the 'man on the Clapham omnibus'—the accepted legal norm for the general mass. He will find the wider aspects of professional conduct dealt with elsewhere[2] and his acceptance of this code will not conflict with any other interests he may have. At all times he must be fully up to date with his own branch of specialisation. Unless it can be proved otherwise, it will always be assumed that his knowledge and professionalism will be commensurate with his status, qualification and experience.

References

1. *Code of practice for chemical laboratories.* London: Royal Institute of Chemistry, 1976.
2. *Professional conduct—guidance for chemists.* London: Royal Institute of Chemistry (free to members only), 1975.

Chapter 3

Planning for Safety

In this chapter many aspects of safety that chemists usually take for granted are discussed. A laboratory, in addition to its being planned for the efficient conduct of whatever work has to be done in it, should also be designed with a view to eliminating accidents. In the context of this book, this applies with particular force to the more insidious dangers from chemicals. Each laboratory will have its own special problems. Here, therefore, we are concerned, in the first place, with such general features of laboratory planning as will restrict danger of contact with harmful chemicals to a minimum and then with suggesting means by which harmful chemicals can be handled with relative safety. Though some remarks on specific chemicals will be included, the discussion will, on the whole, be of a general nature. The hazards of handling radioactive and biochemical test materials are considered separately (*see* Chapters 10 and 9). Thus the account is not to be taken as an exhaustive study of the subject, and much of it reflects the personal views of the authors. It aims at giving assistance rather than definite instructions, for it is thought that each laboratory or organisation should build up its own code of practice to suit local conditions. The Royal Institute of Chemistry has published a *Code of practice for chemical laboratories* which could be taken as a basis.

Safety planning involves the selection of the least hazardous equipment, methods and materials and if necessary the substitution of existing equipment and techniques by newer, safer ones. It also includes the containment or isolation of the hazard so as to avoid or minimise the exposure of personnel by foreseeing and anticipating circumstances which might arise. Adherence to the underlying concept of good laboratory safety planning *ie* the avoidance of unnecessary exposure to chemicals and physical agents, will do much in the long run to ensure safe laboratory operations.

Laboratory planning

Hard and fast rules for the design of chemistry laboratories cannot possibly be laid down, since the requirements in any laboratory will depend on the type of work to be carried out there. Nevertheless, certain general principles should always be observed if the necessary protection against toxic hazards is to be obtained. Ample light and adequate working space, together with careful attention to the disposition of work benches and other facilities, are perhaps the first essentials. Overcrowding and insufficient storage space, which can result in confusion of unused equipment and chemicals, always represent potential hazards.

Handling of hazardous materials

Where a large volume of work with a particular toxic chemical is involved, then consideration should be given to the provision of small rooms apart from the main laboratory in which these operations can be carried out. Thus, where larger than normal volumes of solvents are handled, as for example in banks of solvent extraction equipment, a separate room should be set aside—preferably flameproof with additional ventilation and adequate emergency exits in case of fire—so as to reduce the vapour concentration; this in turn will reduce the toxic hazards as well as the flammability hazard (*see* Chapter 4). Hydrogen sulphide is a reagent often used in considerable quantities in analytical laboratories and this may be isolated in a separate room. It is extremely hazardous as well as having an objectionable smell and the use of a manifold exhaust system to carry away the excess gas is recommended. This excess gas can then be burnt by leading it into the air supply of a gas jet so situated that the products of combustion are discharged outside the building.

Where separate rooms are used for hazardous processes, it should be realised that personnel may be working in them unobserved and may be overcome by fumes without being noticed. Suitable routines to prevent this happening should be adopted.

Water supply and drainage

Water supplies are essential in all chemistry laboratories but it is surprising how often they are inadequate or inconveniently placed. Each working position should have a readily accessible sink with sufficient taps so that one is always available in emergency. The drains should be designed to carry away the maximum water flow and should be so disposed as to have a continuous fall to the outflow so that no stagnant pockets of effluent can be left. The type of trap to be adopted depends very much on personal choice and the type of work in the laboratory. The common domestic U-trap has little to commend it. The use of plastic piping for waste systems is becoming more common and, while the design of plastic trap is superior to the conventional U-trap and the material is on the whole more resistant to laboratory effluents, some plastics can be softened or degraded by solvents so that there is the need for additional care in the disposal of solvents down a sink waste—a practice which should in any case be discouraged. A straight-down pipe with the lower end drowned in a gulley-trap is probably more satisfactory; the discharge from each trap can lead into a common gulley at floor level and thus be passed outside the building. The gulley should be sealed (not open) to prevent vapours from re-entering the work area. This arrangement may not, however, be practicable where laboratories are made by reconstruction of existing buildings. Gullies or gutters of chemical stoneware or glazed earthenware should be inspected frequently to see that the jointing is in good condition and should be readily accessible so that they can be cleaned when necessary.

Ventilation and fume cupboards

Good general ventilation is very necessary for all chemistry laboratories. In the past, this was often achieved by the use of a high, louvred roof, but this is not very satisfactory as it gives rise to heating problems in winter and does not make it easy to maintain adequate cleanliness in the laboratory. Where forced ventilation is required this should be by a positive pressure supply so as to assist rather than compete with the forced draught in fume cupboards. For normal laboratory purposes 6 to 10 changes of air per hour is satisfactory, depending on the function of the laboratory. Three types of fume cupboard ventilation system are commonly employed: hot-air-assisted natural draught (which is not very efficient), air ejector systems from either a compressed air supply or a separate low-pressure blower, and direct extraction fans. Air ejector systems can be designed to work satisfactorily but they are in general very noisy and wasteful in using quantities of compressed air. They have been recommended where highly corrosive gases are likely to be present but modern methods of protecting fan blades from corrosion have made them unnecessary even for these purposes. Direct-extraction fans are undoubtedly the most satisfactory. They should preferably be mounted outside the laboratory so as to reduce noise and reduce the chance of leakage back into the laboratory. The ducting to them should be as short and straight as possible without any sharp bends. The fan discharge should be carefully sited to avoid any danger that the fumes will be blown back into the building or adjacent buildings. Separate small fans for each cupboard, so that fumes from one cupboard cannot leak back into another, are to be preferred to one large one into which a group of cupboards are manifolded. In this latter system a second danger is that the draught in any particular cupboard may suddenly be reduced below the safe minimum if too many of the other cupboard fronts are simultaneously opened.

It is advisable that a minimum air flow of 0.5m/sec (100 ft/min) air velocity at a 0.3m (1 ft) opening should be maintainable under all conditions. Services into the cupboard should not obstruct the opening so that the cupboard front cannot be properly closed when the cupboard is being used. The cupboards themselves should be so designed as to encourage their use and, where preparative processes are involved, should have adequate head room to allow complex apparatus such as reflux and fractionating columns to be erected in them with ease. Hoods are very commonly used for the removal of less noxious fumes, but they are only really effective where a direct vertical uptake is fitted and a good thermal draught can be secured.

When handling highly toxic materials, the use of a glove box as an alternative to the use of a fume cupboard is to be preferred whenever practicable.

Fume cupboards are often used for the storage of highly flammable or other volatile chemicals which can evolve toxic vapours. It is dangerous to mix the storage of oxidative, corrosive chemicals such as fuming nitric acid with that of organic materials such as petroleum. The provision of separate, well ventilated storage cupboards for each is essential and experiments should not be carried out in fume cupboards used for storage.

Benches and fittings
The choice of teak for bench tops, though traditional, has certain disadvantages unless very carefully seasoned wood is used. Any shrinkage of the timber in the warm and often dry atmosphere of the laboratory can give rise to cracks in the bench surface which are undesirable traps for spilled chemicals, especially mercury, and represent a potential toxic hazard. Laminates faced with melamine, such as Formica, make very good surfaces for bench tops and can be easily kept very clean. They are hard-wearing and can be obtained in sheets large enough to avoid any joints. A dark colour is preferred because of the possibility of staining with organic reagents. Acid-resisting tiles set in a minimum of acid-resisting cement also make a very suitable bench-top. Where special cleanliness is essential, then Vitreolite is probably the most suitable, provided its poor heat resistance is acceptable.

Reference may be made here to the British Standards Institution's *Decorative laminated plastics sheet* (BS 3794: 1973) and *Recommendations on laboratory furniture and fittings* (BS 3202: 1959).

Bench fittings should be as uncomplicated as possible. Much will depend upon the function of the bench itself, but the classical idea of a bench with a raised stand in the middle carrying a standard set of reagent bottles is very rarely an ideal arrangement. Many of these bottles contain toxic or corrosive chemicals which are not necessarily used with any frequency. There is much to be said for a flat-topped bench with no fittings other than the normal services at the back of the bench, the controls being at the front below the level of the bench top. Electrical outlets should be of the separately switched type and sufficient should be provided on any length of bench for any experimental work which may be conducted on that bench so that distribution boards or long lengths of trailing wires are avoided. When outlets are provided at other than normal mains voltage—*eg* a low voltage AC or a DC supply—these should be different in form and shape from the normal mains outlet so that no possibility exists of apparatus being incorrectly connected to the wrong voltage outlet.

Reagent bottles can, with advantage, be in trays, designed to contain spillages, which can be moved when not required (the selection of reagents in each tray depending on the work in progress). Reagents that have to be stored on shelves should be so arranged as to be accessible to the operator, who should not have to reach across the bench for them.

Floors and floor coverings
Laboratory flooring always presents something of a problem. Wooden floors are not really satisfactory as they are neither resistant to spillages nor completely impervious. Concrete or tile floorings are ideal but very tiring to work on continuously. Vinyl tiles are very commonly used in modern laboratories and are reasonably satisfactory although they have poor resistance to some solvents and to strong caustic alkali solution. An impervious rubber flooring or properly laid linoleum is probably the most satisfactory as well as easy to keep clean. Polishing of laboratory floors is not normally advisable. For areas which are likely to be susceptible to frequent spillages of acid or alkali, a filled epoxide resin finish laid on top of concrete is both hard wearing and chemically resistant.

Fire hazards
Although fire protection is considered in the next chapter, it is worth remembering that consideration of fire as well as toxic hazards during the planning stage of a laboratory or its function is paramount. Adequate means of escape must be provided from laboratories in which flammable solvents are handled and the positioning of fire extinguishers and other fire fighting

equipment must be such as to be readily accessible in the event of emergency. Remote operation of all extraction fans so that they can be turned off or on from the outside of a laboratory in the event of a serious fire which necessitates the evacuation of the laboratory, should always be installed.

Safety organisation

It is not possible to lay down any hard and fast rules as to the safety organisation to be adopted in all circumstances. So much depends on the size of the company or laboratory; whether the chemical laboratory is part of a larger department or establishment and whether it is a teaching, research or other type of laboratory. The RIC *Code of practice* discusses in detail some of the requirements of a safety organisation. In larger organisations where the Safety Officer may not be a professional chemist, it may be advisable that in each chemical laboratory, a competent chemist should be appointed as a Laboratory Safety Officer. In a school laboratory, the LSO would be a member of the teaching staff and the laboratory steward might be trained as Deputy.

Laboratory Safety Officer
The functions of the Safety Officer will depend very much on the laboratory or organisation concerned, but he should at all times act as focal point for all safety matters including the provisioning of safety equipment and training individuals in their proper use, the compilation of safety instructions and other standing orders covering safety and the maintenance of records on accidents to name but a few. Whether he should be assisted by a safety committee is for local management to decide but it should be borne in mind that such a committee may be a requirement under regulations made under the Health and Safety at Work etc Act 1974.

Many large organisations have systems for reporting accidents and most industrial laboratories are required by statute to keep records of accidents involving injury to personnel. Such records are of value in that when examined and analysed at intervals, they may disclose aspects of safety precautions which are defective and require either reconsideration or more strict enforcement. The Chemical Society and *Chemistry in Britain* are always interested to receive accounts of more unusual accidents and occurrences so that they can be published. By this means, a wider public can be made aware of unexpected hazards and thus a repetition can be avoided.

General safety precautions

It is generally recognised today that every laboratory or group of laboratories, however small, should have a set of safety rules which are appropriate to their particular type of work. These should take into account the hazards of the materials handled, and in this section some of the more important points which should be included are considered.

No set of rules is of any value unless they are fully understood by those concerned and are recognised as being designed for their protection. Safety training must therefore play a major part in ensuring freedom from serious accidents in any laboratory. This should take place at an early stage in the individual's career, be it at school, college or in industry, and should aim to explain the basis for the rules so that the benefit of working within them can be appreciated. Provision of a list of 'do's' and 'don'ts' is not training. Everyone working in a laboratory needs to be made aware that he has responsibility for the safety of others working alongside him as well as for his own safety.

Gas cylinders

The use of a wide range of compressed gases in cylinders is now common practice in chemical laboratories. These may range from the flammable such as hydrogen, propane and acetylene to the toxic such as carbon monoxide and chlorine. Wherever possible cylinders of any gases should be installed outside the laboratory building and a permanent pipe system arranged to deliver gas, at the required operating pressure, to the point in the laboratory where it is needed. This avoids collections of cylinders in the laboratory where they all too frequently block the gangways between benches and constitute a major hazard in the event of a fire in the laboratory. All cylinders should be securely shut-off at the valve on the cylinders when not in use and always at the end of the working day.

Although not common reagents, toxic gases such as chlorine or carbon monoxide are occasionally used as compressed gas in cylinders. When using such gas cylinders, the normal precautions for use of compressed gases should be observed, but on account of their toxicity it is inadvisable to leave them overnight in a laboratory, as a slight leakage from the valve could very quickly build up a highly dangerous concentration in an unventilated room. Care should be taken to check the valve out of doors before a cylinder is first taken into use to ensure that it can be operated easily and can be shut off tightly without leakage. Such cylinders should be used whenever possible in a fume cupboard, so that any leakage of gas due to malfunctioning of the apparatus can be rendered harmless to the operator.

Vacuum and pressure equipment

Apparatus which involves the use of vacuum or high pressure must always be both carefully designed and constructed. Any equipment subject to high pressure must be regularly maintained and tested to a suitable pressure in excess of its normal working pressure and provision must be made in the design for the safe consequences of accidental over-pressurisation during use. Glass equipment which is subject to reduced pressures such as filter flasks, vacuum handling lines and desiccators should be enclosed by shatterproof safety screens to retain any glass fragments in the event of an accidental implosion.

Discipline

The consumption of food or drink in chemistry laboratories should be absolutely forbidden. The danger of accidental contamination of food should be perfectly obvious to anyone working in a laboratory, and the practice of drinking tea, which takes place in many laboratories, though less objectionable, should not be encouraged. It is far safer to set apart for this purpose an office or mess room where there are no chemicals.

It is far more difficult to generalise on the question of smoking in laboratories. There are clearly many chemical processes, such as those in analytical work, where the presence of tobacco smoke is undesirable on general principles. It may not always be realised that comparatively harmless materials can be converted into dangerous toxic ones on passage through the hot zone of a cigarette. For example, chlorinated hydrocarbons, such as trichloroethylene, produce phosgene on partial oxidation. Very small quantities of some fluorine compounds, such as polytetrafluoroethylene, have been known to get mixed with tobacco and on subsequent smoking give rise to symptoms similar to those of acute influenza within two or three hours. It would thus seem advisable to limit smoking in laboratories, especially those in which organic chemicals and solvents are handled.

Students and junior staff should have impressed upon them the dangers of carrying out unauthorised experiments. While it is clearly not desirable to discourage legitimate initiative, the dangers of forming toxic gases or vapours should be emphasised. Even the more commonplace chemicals can react together in unfavourable circumstances to produce hazardous, even explosive, products, as instanced by the report (*Chemy Ind.* 1970, 185) of an explosion following the mixing of waste acetone and chloroform used for chromatography. The gases evolved in qualitative analysis should not be inhaled deeply; a cautious sniff is safer and just as informative.

Chemicals used for laboratory experiments are not subject to the provisions of the Poisons Acts. Nevertheless, scheduled poisons are usually labelled as such and these are the only agents described as poisons in Chapter 8. It is essential that a close control should be kept on the most dangerous, such as cyanides. These should be in the charge of a responsible person and should be issued only when required for specific experimental purposes and then only in the quantities immediately required. Any material transferred to a secondary container should be clearly labelled so as to prevent mistakes. Everyone handling such scheduled poisons has a moral obligation to ensure that they are not improperly used and do not go astray, with possible serious consequences.

Isolated and unattended experiments

When processes involving hazardous chemicals are to be carried out, it is often inadvisable for the operator to be working in a laboratory by himself. Depending on the scale of operations, or the extent of the hazard involved, it may be sufficient for a second person to look in at intervals to see that all is well. Where the potential danger is high, a second person in the room should be ready on the spot to take immediate action in event of an emergency. In either case it is essential that the nature of the materials being used should be clearly understood by someone other than the operator so that the appropriate emergency action may be taken. These points should be borne in mind when arranging for the working of overtime.

Where equipment has to be left running unattended during silent hours, either overnight or at weekends, it is essential that those who are responsible for the laboratory or building during such times or in emergencies are fully informed of the processes and materials involved and the appropriate action to be taken in the event of an accident or doubt, as well as the quickest method of making contact with the individual responsible for the experiment. This can be conveniently summarised on a safety notice board placed adjacent to the experiment or process, so that those not directly concerned will know what precautions should be taken.

Protective equipment and its use

To discuss in detail the precautions necessary for each type of toxic agent would be both lengthy and repetitive. The information given in Chapter 8 defines clearly the potential dangers, and the types of protection and preventive action to be taken can be deduced from them. It will be noted that a distinction is made in the level of risk by the use of the words 'avoid' and 'prevent'. For example, 'avoid' is used where contact with the skin is clearly undesirable but because of the slow rate of absorption or attack on the tissue, the wearing of gloves would not be essential when handling small quantities, provided that any splashes were immediately washed off. If, however, the same chemical were handled in quantity, then clearly gloves would be advisable. Where 'prevent' is used, then protection should be considered as obligatory even when handling small quantities of the chemical if there is the slightest possibility of splashing. It must be for the operator or supervisor to decide what is the extent of this possibility. Even though contact between the skin and hydrofluoric acid must be prevented, it is daily handled without gloves in analytical laboratories and with no serious consequences, because the quantity used for any one operation is rarely more than a few millilitres.

Nevertheless, if any large-scale preparative use were under consideration the wearing of both gloves and a face-mask would then be essential. The wearing of safety glasses or equivalent eye protection in certain instances, is now covered by the Protection of Eyes Regulation 1974. The requirement for adequate protection applies not only to the operators involved in handling hazardous materials but also to any other personnel present in the same area. This emphasises the point that others can be involved in accidents arising from mishandling hazardous materials or the unforeseen failure of a piece of equipment. The eyes are liable to damage not only from corrosive chemicals but also from high energy light sources such as lasers and ultraviolet lamps (*see* p 465). Adequate protection must be provided against both of these.

It is probably a sound principle to limit the compulsory use of protective clothing to those operations where it is absolutely essential. Its use is, in general, not popular, because it is always somewhat restrictive and hampers freedom of movement; instructions for its use are more likely to be obeyed when it is realised that it is done for the good of the operator and not as a pedantic precaution. Nevertheless, no-one should ever be discouraged from using protective equipment if he feels more confident when doing so. There is often a tendency for senior staff to claim that experience can replace the necessity for personal protection. This is most regrettable, as the example of senior staff can be the major factor in impressing juniors that regulations for its use are sensible and designed for the protection of the operators.

The main point to remember about protective clothing is that the easier and more comfortable it is to wear the greater the likelihood of its being used. Cumbersome and unnecessarily restrictive clothing should be avoided unless absolutely essential. Eyeshields and face-masks which stand well clear of the face and are adequately ventilated are to be preferred to close-fitting goggles. Gloves which are easy to slip on and off are more likely to be used for such operations as handling strong acids and alkalis than close-fitting surgical ones. These, however, should be available where careful manipulation of apparatus has to be carried out with protected hands. The choice of material for protective gloves must depend on the material to be handled. No glove is resistant to all organic materials and when in doubt preliminary permeability tests using a selection of glove materials against the compound to be handled should be carried out. As a guide, Table 3.1 will be of value. When carrying large quantities of strong acids and other potentially dangerous materials, the use of suitable baskets is recommended. Specially designed polythene carriers are now available for holding Winchester quart bottles.

The provision of breathing apparatus will depend on the type of work being carried out. In normal laboratories, where only small quantities of toxic gases are being handled, emphasis should be placed on fume cupboards and good ventilation for protecting personnel. The use of respirators to give protection against toxic gases or vapours can give rise to difficulties because of their specificity which can involve the provision of a wide range of different respiratory canisters and the chance of incorrect selection. The use of self-contained compressed air breathing apparatus is to be preferred for dealing with emergency situations where it is necessary to work in a toxic concentration of gases or vapours.

Dust respirators should always be available when handling finely-powdered chemicals in bulk or when conducting grinding or sieving operations. The use of asbestos materials can introduce severe respiratory hazards. Its use should be avoided whenever possible but when unavoidable amosite or chrysotile, which are reputed to be less hazardous, must be employed. The use of crocidolite is controlled by the Asbestos Regulations 1969 which also lay down the maximum permitted limit for the number of asbestos fibres per unit volume of air. Gloves and aprons made from asbestos cloth should be avoided unless they are made from specially resin-impregnated cloths which are non-dusting.

All protective equipment should be readily available in each laboratory and should be kept in a specific place set apart for the purpose—such as a glass-fronted cupboard painted green—which is known to all persons working in the laboratory. First aid equipment should also be provided nearby (*see* p 98). After use the equipment should be washed and dried and returned to the cupboard ready for the next user. Any defects should be reported immediately so that they can be made good by the issue of new equipment.

Since, in many instances, it is not only a question of protecting the individual operator but also other people working in the same room, fume cupboards or other types of protection around the equipment should be used whenever possible. It is probably true to say that, apart from incidents arising from careless handling in transferring chemicals from one container to another, the greatest hazard arises from the inhalation of toxic vapours and gases or the splashing of chemicals over the laboratory by excessively violent behaviour of the contents of a reaction vessel. An efficient fume cupboard affords complete protection against the first of these, and transparent portable safety screens that can be easily erected round a piece of apparatus and dismantled on completion of the process will go far to reduce the second of these hazards.

Special operations

Electroplating

Electroplating baths give rise, not to vapours but to aerosol dispersions of the plating solution produced by the gases evolved at the electrodes. With cyanide baths this danger is obvious, and with chromium plating baths the danger is well recognised. For other plating baths, however, such as nickel and the fluoroborate baths now used for lead and other metals, this aspect may not be so frequently borne in mind. It is advisable that all plating baths, unless very small and infrequently used, should be provided with extra ventilation ducts, which can be conveniently placed behind the bath so as to draw air across the surface of the liquid and away from the operator.

Pipetting

Toxic chemicals and their solutions should never be pipetted by mouth. When small quantities only are involved, a Pasteur pipette should be used, but for larger quantities or precise work several types of safety pipette are now available from laboratory suppliers. One type resembles a hypodermic syringe, in that the suction is created by the action of a ground-glass piston running in a closely fitting cylinder. A simpler but equally effective one consists of a conventional pipette, to the top of which is fitted a large rubber bulb having three valves, which are operated by the fingers to control the liquid flow; solutions can be pipetted safely with the usual accuracy.

Solvents

The dangers of many laboratory processes are not always realised, even by experienced chemists. It is a common practice in almost all laboratories to remove solvents by evaporation in the open. Even if the fire risk is ignored the toxic hazards can be quite considerable, as the chronic effects of inhalation of even small quantities of, say, benzene vapour over a long period are very serious. The extreme toxicity of benzene, for example, is very often not appreciated by those who may use it as a matter of course in the laboratory. Its use as a solvent should be completely prohibited and this follows a Council of Europe recommendation that the use of benzene should be banned. This applies in addition to carbon tetrachloride as well as pentachlor- and tetrachlor-ethane. If evaporation rather than distillation is essential it should be carried out in a forced draught, either in a fume cupboard or in a special solvent room as discussed above (*see* p 17).

Handling of mercury

Mercury is perhaps one of the most casually handled chemicals in many laboratories, yet its vapour can give rise to serious effects both chronic and acute (*see* Chapter 6). Unless absolutely essential for the operation in hand it should never be left around the laboratory in open vessels. Spillage of mercury can result in small droplets collecting in cracks and crevices in floors or benches and it is essential that all spills of mercury are cleaned up without delay and that dirty mercury is kept under dilute sulphuric acid while awaiting cleaning and distillation. When mercury-vapour diffusion pumps are used in high-vacuum equipment, cold traps should be installed on the discharge side of the pumps to prevent mercury vapour being pumped into the atmosphere. In addition the pump discharge can be led outside the laboratory. If pumps are left running unattended, a safety switch should be incorporated to cut off the heating to the pump when the cooling water fails. This will also prevent excessive discharge of mercury vapour into the atmosphere. It should be remembered that mercury is a cumulative poison and can give rise to serious effects after exposure to small concentrations over a very long period.

Monitoring of laboratory atmospheres

Where laboratory conditions are such that the continued use of a specific chemical may be such as to suggest that the Threshold Limit Value (*see* p 112) may be exceeded, it is necessary to provide some means of monitoring the atmosphere for this particular component. This can often be most conveniently carried out by the use of chemical detector tubes which are manufactured by several companies. These are specific for a single chemical compound or class of compounds and will usually give an indication of the atmospheric concentration at the TLV which is accurate to about ± 25 per cent. The tubes operate on colour reactions which occur with reagents absorbed onto granules packed into a narrow glass tube. The length of the granule bed over which the colour has changed is a direct measure of the concentration according to the calibration provided with the tube. Some interference from other compounds may occur, but these are clearly stated by the maker and allowance will have to be made for such side reactions.

For other gases and vapour, such as mercury, isocyanates, styrene *etc*, continuously recording apparatus is available which can be used to monitor the atmosphere and protect personnel who are likely to be working for prolonged periods in such atmospheres where the TLV may be approached.

Handling of cryogenic fluids

The use of these materials is now commonplace in laboratories and can present hazards which are not only associated with their low temperatures. Because of the low temperatures, materials can very markedly change their properties; metals can become brittle and sensitive to fracture by shock. Hence all the materials used for handling cryogenic fluids must be carefully selected not only for their compatibility with the fluid but for their low temperature properties. Adequate ventilation must be provided if the gases generated by evaporation of the fluids are discharged into the laboratory. Because they are cold they tend to layer at floor level and can create hazards. Liquid oxygen can produce an atmosphere in which very vigorous combustion can occur and which can be particularly dangerous if it is trapped in an individual's clothing due to spillage on transfer or handling. Clothing should always be well ventilated if spillages have occurred, otherwise the injudicious use of matches or a cigarette lighter could cause severe burns.

Liquid nitrogen though in many ways safer than liquid air or liquid oxygen could produce an irrespirable atmosphere at low levels in trapped pockets which could have fatal consequences as the result of spillages in confined spaces. Because of its lower boiling point, liquid nitrogen can condense oxygen from the atmosphere in cold traps and this could subsequently give rise to an enrichment of the oxygen concentration.

Even brief contact with metallic surfaces at cryogenic temperatures can result in severe 'burns' or torn flesh and the use of thick protective gloves should be obligatory in such circumstances. Eyes are particularly vulnerable to splashes and goggles should be worn at all times when handling cryogenic fluids. Operators who may be transferring cryogenic fluids from bulk storage to ready use containers should also be provided with an impervious coat or apron.

Because the hazards of using cryogenic fluids may be more real than apparent, instructions for their safe use should be laid down in detail in the laboratory code of practice. This should be supplemented by an initial course of instruction to all staff when first handling cryogenic fluids.

Spillages and waste disposal

Finally it is advisable to consider what steps should be taken to dispose of waste hazardous chemicals. Recommendations for the disposal of spillages are given in Chapter 8 and these can usually be adapted for the disposal of small quantities of waste. In most laboratories the quantities involved should not be more than can be washed down the normal service drain with a copious flow of water. Care should be taken that no untoward reaction between waste materials occurs in the drains with the generation of noxious or toxic gases. The disposal of flammable wastes is discussed in Chapter 4.

Table 3.1 Glove resistance ratings (by courtesy of James North & Sons Ltd).

Resistance rating code: E—Excellent; F—Fair; G—Good; NR—Not recommended.

Chemical	Natural Rubber	Neoprene	Nitrile	Normal pvc	High-grade pvc	Chemical	Natural Rubber	Neoprene	Nitrile	Normal pvc	High-grade pvc
Organic acids						Ammonium nitrate	E	E	E	E	E
Acetic acid	E	E	E	E	E	Ammonium nitrite	E	E	E	E	E
Citric acid	E	E	E	E	E	Ammonium phosphate	E	E	E	E	E
Formic acid	E	E	E	E	E	Calcium hypochlorite	NR	G	G	E	E
Lactic acid	E	E	E	E	E	Ferric chloride	E	E	E	E	E
Lauric acid	E	E	E	E	E	Magnesium chloride	E	E	E	E	E
Maleic acid	E	E	E	E	E	Mercuric chloride	G	G	G	E	E
Oleic acid	E	E	E	E	E	Potassium chromate	E	E	E	E	E
Oxalic acid	E	E	E	E	E	Potassium cyanide	E	E	E	E	E
Palmitic acid	E	E	E	E	E	Potassium dichromate	E	E	E	E	E
Phenol	E	E	G	E	E	Potassium halides	E	E	E	E	E
Propionic acid	E	E	E	E	E	Potassium permanganate	E	E	E	E	E
Stearic acid	E	E	E	E	E	Sodium carbonate	E	E	E	E	E
Tannic acid	E	E	E	E	E	Sodium chloride	E	E	E	E	E
						Sodium hypochlorite	NR	F	F	E	E
Inorganic acids						Sodium nitrate	E	E	E	E	E
Arsenic acid	G	G	G	E	E	Solutions of copper salts	G	G	G	E	E
Carbonic acid	G	G	G	E	E	Stannous chloride	E	E	E	E	E
Chromic acid (up to 50%)	G	F	F	E*	G	Zinc chloride	E	E	E	E	E
Fluorosilicic acid	G	G	G	E	G						
Hydrochloric acid (up to 40%)	G	G	G	E	G	**Alkalis**					
Hydrofluoric acid	G	G	G	E*	G	Ammonium hydroxide	E	E	E	E	E
Hydrogen sulphide (acid)	F	F	G	E	E	Calcium hydroxide	E	E	E	E	E
Hydrogen peroxide	G	G	G	E	E	Potassium hydroxide	E	G	G	E	E
Nitric acid (up to 50%)	NR	NR	NR	G*	F*	Sodium hydroxide	E	G	G	E	E
Perchloric acid	F	G	F	E*	G						
Phosphoric acid	G	G	G	E	G	**Aliphatic hydrocarbons**					
Sulphuric acid (up to 50%)	G	G	F	E*	G	Hydraulic oil	F	G	F	G	E
Sulphurous acid	G	G	G	E	E	Paraffins	F	G	E	G	E
						Petroleum ether	F	G	E	F	G
Saturated salt solutions						Pine oil	G	G	E	G	E
Ammonium acetate	E	E	E	E	E						
Ammonium carbonate	E	E	E	E	E						
Ammonium lactate	E	E	E	E	E						

* Resistance not absolute, but the best available.

Chemical	Natural Rubber	Neoprene	Nitrile	Normal pvc	High-grade pvc
Aromatic hydrocarbons†					
Benzene	NR	F	G	F	G
Naphtha	NR	F	F	F	G
Naphthalene	G	G	E	G	E
Toluene	NR	F	G	F	G
Turpentine	F	G	E	F	G
Xylene	NR	F	G	F	G*
Halogenated hydrocarbons†					
Benzyl chloride	F	F	G	F	G
Carbon tetrachloride	F	F	G	F	G
Chloroform	F	F	G	F	G
Ethylene dichloride	F	F	G	F	G
Methylene chloride	F	F	G	F	G
Perchloroethylene	F	F	G	F	G
Trichloroethylene	F	F	G	F	G
Esters					
Amyl acetate	F	G	G	F	G
Butyl acetate	F	G	G	F	G
Ethyl acetate	F	G	G	F	G
Ethyl butyrate	F	G	G	F	G
Methyl butyrate	F	G	G	F	G
Ethers					
Diethyl ether	F	G	E	F	G
Aldehydes					
Acetaldehyde	G	E	E	E	E
Benzaldehyde	F	F	E	G	E
Formaldehyde	G	E	E	E	E
Ketones					
Acetone	G	G	G	F	G
Diethyl ketone	G	G	G	F	G
Methyl ethyl ketone	G	G	G	F	G

Chemical	Natural Rubber	Neoprene	Nitrile	Normal pvc	High-grade pvc
Alcohols					
Amyl alcohol	E	E	E	E	E
Butyl alcohol	E	E	E	E	E
Ethyl alcohol	E	E	E	E	E
Ethylene glycol	G	G	E	E	E
Glycerol	G	G	E	E	E
Isopropyl alcohol	E	E	E	E	E
Methyl alcohol	E	E	E	E	E
Amines					
Aniline	F	G	E	E	E
Butylamine	G	G	E	E	E
Ethylamine	G	G	E	E	E
Ethylaniline	F	G	E	E	E
Methylamine	G	G	E	E	E
Methylaniline	F	G	E	E	E
Triethanolamine	G	E	E	E	E
Miscellaneous					
Animal fats	F	G	G	G	E
Bleaches	NR	G	G	G	E
Carbon disulphide	NR	F	G	F	G
Degreasing solution	F	F	G	F	G
Diesel fuel	NR	F	G	F	E
Hydraulic fluids	F	G	G	G	E
Mineral oils	F	G	E	G	E
Ozone resistance	F	E	G	E	E
Paint and varnish removers	F	G	G	F	G
Petrol	NR	G	G	F	G
Photographic solutions	G	E	E	G	E
Plasticizers	F	G	E	G	E
Printing inks	G	G	E	G	E
Refrigerant solutions	G	G	E	F	G
Resin oil	F	G	G	G	E
Vegetable oils	F	G	G	G	E
Weed killers	G	E	E	G	E
White spirit	F	G	G	F	G
Wood preservatives	NR	G	G	F	G

† Aromatic and halogenated hydrocarbons will attack all types of natural and synthetic gloves. Should swelling occur, switch to another pair, allowing swollen gloves to dry and return to normal.

Chapter 4

Fire Protection

Researchers and laboratory staff, although highly trained in their particular fields, are often unacquainted with the elements of fire prevention. Laboratory and research managers must accept responsibility for providing a safe working environment, while those working in laboratories should appraise the fire hazards of their own particular work. This involves considering the flammability or explosibility of the materials which are to be handled, examining all the sources of ignition which may be present, and understanding how fire spreads and the probable outcome of a fire, should one occur. It is recommended that consultation should take place with the local fire authority and the fire insurers at the design stage of the laboratory (*see* Insurer's approach to fire protection).

Chemical laboratories can be the site of very hazardous processes and materials but the knowledge that a chemist can bring to the problem of how substances react when they are mixed or subjected to heat, pressure or other physical changes should enable him to control the materials with which he is working.

If fire should occur it may spread beyond the control of any one individual. Laboratory staff should, therefore, know how to call the fire brigade and summon other help.

The nature of the risk

Flammable liquids

The most common of the hazardous materials to be found in laboratories are flammable liquids and it is essential to know their fire properties, such as flash point and ignition temperature. If the flash point is below room temperature then these liquids will always constitute a fire hazard and careful control should be maintained over them. Liquids with flash points well above room temperature will support a fire if heated to a temperature exceeding their flash point.

Before ignition can occur, a flammable vapour has to be heated, at least locally, to a temperature exceeding its ignition temperature. Almost invariably the ignition temperatures are well below those of common igniting sources such as flames, sparks and incandescent surfaces. For some materials ignition temperatures can be extremely low. Carbon disulphide, for example, will be ignited by a source whose temperature just exceeds that of boiling water, while ethers and aldehydes usually have low ignition temperatures. Other properties of liquids have to be taken into account such as their propensity to form more hazardous substances after long periods. Ethers are one such class of materials, and form highly explosive peroxides after long standing. The possibility of mutual interaction between flammable liquids should be considered when keeping them in laboratories or stores. Only the minimal amount of flammable liquids should be kept in laboratories. One day's supply is often recommended but this may be inconvenient or impracticable. However, the requirements of the Highly Flammable Liquids and Liquefied Petroleum Gases Regulations 1972 may limit the total quantity of highly flammable liquids allowed to be stored in the laboratory to 50 litres within suitable containment. Nevertheless, the quantity of flammable liquid should be kept to a minimum.

Bottles containing flammable liquids should be positively identified by labels which can withstand the deleterious effects of any atmosphere likely to be present in the laboratory. Sand-blasted labels would make loss of labelling impossible. Large containers of flammable liquids should never be carried in the hand and in particular Winchester bottles should not be carried by the neck. Suitable Winchester carriers are available and should be used for the transfer of liquids from one place to another. Only trained staff should refill bottles with flammable liquids in a special room or in a laboratory where all ignition sources have been removed.

Damaged glassware should not be employed in experiments as it may crack and spill the contents during the experiment. Poorly assembled or unsuitable apparatus may introduce serious fire risks. Badly fitting corks and bungs will introduce a fire risk and ground glass joints are usually to be preferred. The breakage of equipment by localised overheating using direct gas flames is a hazard which can be easily avoided by using water baths, hot plates, heating mantles or sand baths.

Compressed and liquefied gases

Compressed or liquefied gases present hazards in the event of fire since heating will cause the pressure to rise and may rupture the container. Leakage or escape of flammable gases can produce an explosive atmosphere within the laboratory which can be ignited and result in a devastating explosion. Cylinders of gases should be provided with a suitable pressure regulating valve. The pressure at which the gas is to be used should be determined by this valve and not by operating the needle valve, as this leads to erratic control and the application of full pressure on any tubing should the exit of the tubing become blocked. Gas cylinders should preferably be placed outside and the gas piped into the laboratory. Properly designed compartments and reinforced walls should be used to protect personnel against possible explosion.

Hazardous solids

Most combustible solids will not present a great fire risk unless they are ground into powder. Powders of combustible solids can be explosive when dispersed in air and if large quantities are ground to a fine state then precautions may have to be taken against dust explosion. The main hazard will be those solids which are unstable and are likely to decompose explosively if they are heated or subjected to friction or even excessive light. Other solid materials may be hazardous due to their oxidising properties so that a hazardous situation can result if they should become contaminated with combustible material.

Another class of hazardous solids is those which will react spontaneously and exothermally with water or air. Obvious examples are alkali metals, metal hydrides and certain organometallic compounds. Even aluminium powder can react with water and although the reaction is not very vigorous hydrogen—which can be exploded by a small spark—may accumulate.

Special methods of storage of these materials are called for and periodical inspections are necessary to ensure that these conditions are being maintained. For example, some unstable materials are kept damped down with water or a high flash point liquid. During a long period of storage there may be evaporation of the liquid and drying out so that the unprotected solid is exposed. When attempting to scoop out some of the unstable material friction with the scoop may then ignite it, leading to fire or explosion.

Working practice

It is inevitable that flammable vapour or gases will be produced in the laboratory atmosphere from time to time. The best general precaution is to ensure that the laboratory is well ventilated. Windows and doors cannot be relied upon to provide adequate ventilation as they may well be kept closed much of the time and mechanical ventilation should be used. Any chemical operation which involves the possible production of flammable vapours should be carried out in a fume cupboard. The apparatus should be set up over a large metal tray to catch any flammable liquids which may escape due to accidents, such as breakage of equipment. The recommended air-flow through the face of a fume cupboard is 30 metres per minute. The air-flow should be maintained by means of a fan and the fan motor should be placed outside the duct serving the fume cupboard, driving the fan by means of a shaft. The extraction duct should be made of non-combustible material and it is desirable that it should pass directly to the outside of the building without passing through ceilings. If, however, it must pass through one or more floors, or there is intercommunication with other fume cupboards, then careful consideration must be given to the control of spread of fire and toxic gases within the building. Several alternatives are available, *viz*:

1. The extraction duct-work should be of a fire-resisting construction.

2. The extraction duct-work should be enclosed in a fire-resisting structure.

3. Fire dampers may be fitted at positions where the duct passes through ceilings or partition-walls. However, the complexity of the problems associated with ducts intercommunicating and/or passing through building require expert advice.

Considerable amounts of heat are released during some reactions. This is obvious in acid–base neutralisation or oxidation reactions and can often be predicted when the mechanisms of the expected reaction are worked out beforehand. Such considerations may be second nature to experienced researchers but may not be so apparent to students and inexperienced laboratory assistants who should only undertake such work under the supervision of an experienced person.

When new work is being attempted, full consideration should be given to the possible dangers it presents. In these cases, before proceeding on the scale desired a very small scale experiment should be carried out to determine whether a large amount of heat is liberated or gases are produced or an unusually vigorous reaction occurs. These factors, although not serious on a small scale, may become extremely important in a large scale experiment. On scaling up a reaction heat loss will only increase according to a square law while the volume of reactants and heat produced will accord to a cube law. Additional means of cooling, such as cooling coils, may be required to take care of the extra heat produced.

The rate at which one reagent is added to another affects the rate at which heat is produced. In the case of a known exothermic reaction the reactants or reagents should be added as slowly as possible with adequate stirring to make sure that the heat is liberated slowly. One way in which too much reactant may be added at one time is when it is added to an overcooled mixture and accumulates in the reaction mix where the rate of reaction is artificially low. It then only requires a small increase in temperature, either due to the mixture warming up normally or to the gradual accumulation of heat of reaction, for a run-away condition to occur, leading to disastrous results. A better course is to allow the reaction to take place at a temperature where it can proceed at a suitable speed and add the reactant slowly with adequate stirring to make sure that it is consumed at the same rate at which it is added.

Flammable vapours can often escape into the atmosphere because they are not condensed as rapidly as they are produced. A balance must be struck between the rate at which they are heated and the cooling provided for in a water-cooled condenser.

Experiments should not be left unattended and if the person in charge of an experiment is called away he should ensure that whoever takes over is fully aware of the dangers involved and the precautions to be taken. No experiment should be left running at night unattended unless it can be ensured that somebody will come round and inspect it at regular intervals.

The disposal of waste materials calls for special attention. There have been fatal accidents during the disposal of waste materials and the disposer should be fully aware of the risks involved. The overriding consideration should be not to try to dispose of too much material at any one time. This is particularly important in the case of heavy metals such as silver and mercury which can form explosive compounds of which fulminates are a well-known example. Such by-products can be produced accidentally without the intentions of experimenters (*see* Chapter 3, p 23) and full knowledge of the conditions under which explosive compounds can be produced should be available to such persons.

Untreated flammable liquids should not be disposed of down the drain, apart from *very* small quantities of water-soluble miscible solvents. Limited spillage of flammable water insoluble solvents may be disposed of by the dispersion method described in Chapter 8. Larger quantities of flammable liquids should preferably be collected in sealed metal containers and recovered or disposed of in a safe fashion, for example by burning in a shallow metal tray in the open air. Drains should be properly trapped and vented and they should preferably discharge into an industrial waste sewer rather than a sanitary sewer. There are specially designed incinerators for burning flammable liquids and if one attempts to employ this method a full investigation should be made of the types of material it can handle.

Materials which are known to decompose spontaneously and explosively should be kept in a safe manner until they can be disposed of safely. Waste material should be removed daily and destroyed or disposed of in a safe manner. Filter papers, residues and wiping cloths which have been in contact with unsaturated oils should be kept in covered receptacles of limited size to ensure frequent disposal of the contents. Bins used for disposal should be labelled appropriately and clearly so that no mistake is made in introducing the wrong type of material into the bin. Benches and glass apparatus should not be cleaned with flammable solvents after experiments.

Control over sources of ignition

Electrical equipment, open flames, static electricity, burning tobacco, lighted matches and hot surfaces can all cause ignition of flammable material.

Gas supplies to laboratory outlets should be by means of rigid permanent piping. A laboratory may have many outlets for gas supply and in this case a control valve should be placed just outside the laboratory so as to be able to cut supplies off in the case of an emergency.

Electric wiring should, as far as possible, be of a permanent nature and installed in accordance with the latest edition of the IEE regulations. Switches, sockets and terminals should be placed where they are easily accessible and are safe from accidental wetting by water or other liquids. Temporary wiring should be installed by a competent electrician or at least inspected by one before use. Where it is likely that large amounts of flammable gases or vapours may be released into the atmosphere then it is necessary to install electrical equipment designed for use in explosible atmospheres.

Flameproof equipment is not manufactured for all types of flammable vapours and gases and it may be necessary to use pressurised equipment. Care should be taken when using flameproof equipment that additions are not made which are below flameproof standard. A common mistake is to use domestic refrigerators for storage of materials and explosions have occurred when low flash point liquids have been placed in such refrigerators. Drying ovens are also a likely source of ignition and it should be remembered that, although the oven may be flameproof the surface temperature may exceed the ignition temperature of some vapours. Thermostats on oil baths have often failed with consequent overheating and fire. Independent excess-temperature manual-reset cutouts should be incorporated to disconnect the supply if overheating should occur. The use of organosilicon fluids in oil baths has much to commend it as they are highly stable and can have boiling points in excess of 400 °C. There are a number of general precautions which can be taken. Hotplates, furnaces and ovens should stand on heat resisting surfaces. Where the heated unit is on a wooden bench top there should be an air space between it and the bench to prevent charring. Gas jets not in use should be turned off or adjusted to give a small luminous flame as the pale blue flame cannot be seen easily in bright sunlight.

The discharge of accumulated static electricity can provide a spark which will ignite flammable vapours. Static electricity is created by the relative movement of two materials. Non-polar materials such as hydrocarbon solvents accumulate static charges readily as they have high insulating values and do not allow the charge to leak away. It should be noted that a dispersion of an immiscible polar liquid in a non-polar liquid can generate static charges even more rapidly than a non-polar charge alone. Some improvement can be obtained by the use of a small proportion of a conductive additive where this would not affect the chemical properties of the fluid. Even crystallisation can produce static electricity and the discharge of carbon dioxide has been known to produce static sparking. Laboratory coats and clothing made of certain synthetic fibres are prone to generate static electricity and should be avoided. Where possible all metalwork should be bonded and earthed when there is a static hazard. When no solution can be found to the static problem then all processes must be carried out as slowly as possible to give the accumulated charge time to disperse.

Space heating should be safe and adequate. Hot water or low pressure steam radiators are desirable and installation and maintenance should be first class so as to reduce the likelihood of the introduction of uncontrolled and dangerous forms of space heating. Radiators should not be used for drying materials and sloping wire mesh over them should prevent this abuse.

Good housekeeping will help to reduce the number of fires in laboratories in addition to reducing other types of accidents. Cleaning up as soon as possible after an experiment has been completed will help and the provision of adequate storage space beneath the bench is an advantage. A clear bench will enable the experimenter to see any dangerous procedure without being distracted by unnecessary equipment. The provision of suitable waste bins, preferably of non-combustible material such as sheet iron, will help to encourage a positive attitude to housekeeping.

Structural protection and segregation

The floor of the laboratory should be impervious to chemicals. It may be considered necessary to place a sill at the door which should be provided with a ramp which is not too steep. There should be at least two means of escape from the laboratory and benches should be laid out so that people can escape easily if a fire occurs on the bench. This means that blind aisles should be avoided and the bench should more or less run parallel with the line of the exits. A space of at least four feet should be provided between benches for passage of personnel and equipment. Reagent shelves should be situated on the bench so that it is unnecessary to lean across experimental apparatus in order to reach reagent bottles. Similar consideration should apply to services such as gas, electricity supply and water supplies and drainage.

If laboratories are situated in single storey buildings, then it may be sufficient to construct the buildings of noncombustible material. If the laboratories are situated in multi-storey buildings then walls, floors and ceilings should be of fire resisting construction. Fire resistance should be of at least two hours but will vary according to the risk in adjoining compartments. Columns and beams should also be protected to the same standard of fire resistance. Openings made in walls for the passage of ducts, pipes and cables should be fire-stopped. Stairways should be enclosed by fire-resisting walls. Exit doors should be hung to swing outwards.

The quantities of flammable materials in the laboratory shall be kept as small as possible. Where it is possible a separate building should be used for the storage of flammable liquids and those laboratories subject to the HFL and LPG Regulations 1972 will have certain legal obligations regarding quantities stored. This should be a single storey building constructed from non-combustible materials. The roof should be of light construction and easily shattered or blown off in the event of an explosion. If a separate building cannot be provided the store should be on the ground floor and should be of fire-resisting construction. It should be well ventilated and unheated and have doors which open outwards. There should be a sill on the doorway in such a storage compartment so as to contain any spillages which may occur accidentally. As far as possible separate stores should also be used for materials of differing hazards, and materials which react vigorously with one another should not be stored together.

Pilot plant experiments are carried out principally because operation on this scale presents the possibility of unforeseen results. Laboratories which are used for pilot plant work are particularly hazardous as larger quantities of materials will be used and special buildings or areas should be set aside for these purposes.

Types and quantities of extinguishers

Portable extinguishers

For use on ordinary combustible materials. Water is the best extinguishing agent for fire involving ordinary solid combustible material such as wood, paper and textiles. The cooling power of water is unsurpassed by other media and is especially useful on materials which are likely to re-ignite and for penetration into deep-seated fires. Portable water extinguishers are usually of two gallon capacity but smaller (1–1½ gal) extinguishers which are particularly suitable for use by women are available. Small hose reels which should be at least ¾ in inside diameter and capable of delivering at least five gallons per minute give a practically inexhaustible supply.

For use on flammable liquids. Flammable liquid fires cannot normally be extinguished with water. Dry powder, foam, carbon dioxide and vaporising liquids can be effective. Dry powder rapidly extinguishes the flames over burning liquid and gives a quick 'knock down'. It is particularly effective in the case of spill fires, but gives little protection against re-ignition. It can be safely used on electrical equipment as dry powder is non-conductive. The capacity of portable dry powder extinguishers ranges from about 2 lb to 20 lb.

Foam extinguishes a fire by forming a blanket which floats on the surface of the liquid preventing the access of air or the escape of vapour. The foam blanket remains in position for sufficient time to allow the liquid and surroundings to cool, so preventing re-ignition. Foam is particularly effective in dealing with fires in containers which have become overheated and is more effective than dry powder. It is impossible to form a foam blanket over liquids flowing down a vertical surface and difficult when liquids are flowing freely over a horizontal surface. Polar liquids break down the normal protein-based foam. An alcohol resistant foam has to be employed in fighting fires involving alcohol, acetone *etc*. Foam is electrically conductive and should not be used when electrical equipment is involved in the fire. Foam extinguishers are available in 1–2 gal capacities which will produce approximately eight times this volume of foam.

Carbon dioxide smothers flames mainly by excluding oxygen. It is effective in situations where it is not easily dispersed, and so can be used on fires in refrigerators and ovens. It can also be used safely where there is danger of electric shock and where delicate equipment is involved since it leaves no residue. The extinguishers come in various sizes ranging from $2\frac{1}{2}$ to 15 lb in capacity.

Vaporising liquids are halogen containing substances which act by inhibiting the flame reactions. They work very effectively and are safe to use on electrical equipment. Some are very toxic. The principal agents going into service at the present time are bromochlorodifluoromethane (BCF or Halon 1211) and bromotrifluoromethane (BTM or Halon 1301). All the vaporising liquids produce toxic products on pyrolysis and they should not be used or kept in confined spaces or any place where there is a risk that the vapours or their products could be inhaled. The capacity of these extinguishers is normally one pint or one quart.

For use on special risks. Some of the substances used in laboratories such as sodium, potassium and metal hydrides will react with extinguishing media and it would be ineffectual or even dangerous to attempt to use these agents. Special dry powders are available to deal with such hazards, and manufacturers and suppliers of special risk chemicals will be able to assist in the selection of the appropriate dry powder for these substances.

A multiplicity of extinguishers is undesirable and probably all risks can be covered by two or three types. It is valuable to obtain extinguishers which are uniform in their method of operation; all personnel should be instructed in the use of these appliances and which type to use. Extinguishers should conform with British Standards Specifications and be approved by the Fire Offices' Committee. Further advice on choice of extinguishers may be obtained from the local fire authority or insurer.

Distribution of extinguishers
The number and size of the extinguishers required will vary according to the risk but there should be at least one of each required type in every laboratory. In large laboratories there should be a minimum of one water-type extinguisher for every 200m² of area and the total should be not less than four gallons capacity on every floor. Further information regarding the use of water and other types of extinguishers can be found in the Fire Protection Association's Fire Safety Data Sheets Nos 6001, 6002 and 6003.

Fixed fire extinguishing systems
Large laboratories and pilot plant areas can be protected by means of fixed installations. The system used can be automatic or manually operated and any type of agent can be employed in the system. Hydrants, hose reels or automatic sprinklers can provide general protection. Hydrants are advisable for premises covering a large area, remotely situated areas and tall buildings. Hose reels have already been mentioned. They are permanently connected to the water supply, simple to use and very effective for tackling fire at an early stage. Further details are given in the Fire Protection Association's Fire Safety Data Sheets Nos 6006 and 6007.

Although water is the extinguishing agent, automatic sprinklers are acceptable in many laboratories. It is advisable to link the automatic sprinkler alarm to the local fire brigade. As the sprinkler heads open automatically in response to elevated temperature the detection and operation is completely automatic. Automatic sprinklers are the most effective form of automatic protection (*see Rules of the Fire Offices' Committee for Automatic Sprinkler Installations*, 29th edn).[1]

Protection for special risks can be given by water spray, foam, inert gas and dry powder. Water spray systems consist of pipes with high or medium pressure outlets projecting sprays for a predetermined droplet size. This system is suitable for protection of large scale flammable liquid and liquefied gas risks. Such a system is not only used to extinguish a fire but to protect plant, by cooling, from a nearby fire. Foam can be applied to a given risk by fixed pipework connected either to a self-contained foam generator or an inlet to which the fire brigade can connect their foam-producing equipment. Carbon dioxide or other inerting gases can be directed into plant or rooms by pipework. Most of the gases used are primarily suitable for flammable liquids, electrical equipment, valuable water-sensitive equipment and water reactive chemicals. The gas is delivered through pipes and the system involves automatically closing doors and ventilation ducts. Warning bells have to be sounded so as to warn occupants to make their escape as inert gases will poison or asphyxiate them.

A dry powder system consists of a dry powder container which is coupled to a gas cylinder and pipework and outlets. The pressurised gas drives the powder to the outlets. The system is suitable for flammable liquids, electrical equipment and materials where it is essential to avoid water contamination.

Action in the event of fire

Action to be taken in the event of a fire or explosion should be to a prearranged plan. All personnel must be made aware of the procedure and fire drills should be carried out at least twice a year in order to familiarise staff with these procedures. The essential elements which should be covered by instruction are *(i)* raising the alarm, *(ii)* summoning the local fire brigade, *(iii)* first aid fire fighting practice, and *(iv)* evacuation.

Raising the alarm

Personnel should be trained to recognise the severity of an outbreak of fire and the immediate danger presented by it. They should be able to report accurately on the fire situation, being able to discern whether a normal procedure or emergency procedure should be brought into operation. In the normal procedure the person discovering the fire may decide that it is relatively minor. He should report the fire immediately to the switchboard or instruct some other person to do so. A person in a position of responsibility should decide if it is necessary to evacuate the building. Meanwhile the switchboard operator should call the fire brigade and notify all persons who have been allocated special responsibility.

Emergency procedure is necessary when the fire has been found to have spread over a wide area or hazardous materials and processes are threatened. In such circumstances the premises must be evacuated immediately and the manual fire alarm system should be operated by the person discovering the fire. Then, if possible, the switchboard operator should be notified. The switchboard operator should have standing instructions to call the fire brigade on hearing the fire alarm unless the alarm is directly linked to the fire station.

Summoning the fire brigade

When the fire brigade is called by the switchboard operator it should be informed at the time which entrance is nearest to the fire. In all cases responsible persons should be sent to the entrance to give the brigade information it may need on the location and extent of the fire, water supplies, the nature of special risks and details of casualties and trapped persons. If any particularly hazardous materials or processes have been brought into the laboratory the fire brigade should be forewarned. Ideally the fire brigade should have prior knowledge of the hazards through a liaison or safety officer who should be responsible for informing the local fire authority on hazards at the time they are introduced into the laboratory (*see* Chapter 3, p 21).

First aid fire fighting
Provided that no danger is involved, the fire should be attacked with first aid fire fighting equipment as soon as it is discovered. It must be decided on relative merit whether attacking the fire takes precedence over reporting it. It can be dangerous to waste time trying to tackle a fire which cannot be controlled instead of immediately reporting the incident; on the other hand it is undesirable to allow a small fire to obtain a hold through spending time reporting it. In most circumstances there is no conflict, for there is normally more than one person near the scene of the outbreak, and one person can report the fire leaving the others to try to extinguish or contain it. On returning that person can aid the other personnel.

Fire fighters should withdraw from the scene if the heat and smoke threatens to overcome them or if the fire endangers their escape route. They should also retreat if the fire spreads towards explosive materials or gas cylinders. On withdrawing, windows and doors should be closed in an effort to contain the fire.

Evacuation of personnel
The essential features of evacuation are that it should proceed by a pre-arranged plan and that all personnel should be familiar with the escape routes to be used. It should be impressed on everybody that they should leave, without panic, immediately they have received instructions to evacuate the building. The assembly point should be fixed and known to all personnel, and the head of each department (or his deputy) should be responsible for ascertaining that all persons in his charge, including visitors, have been accounted for. To ensure that this is thoroughly carried out, one person must be delegated the responsibility of searching the department, including the lavatories and cloakrooms to ensure that nobody has been left behind. No person should attempt to re-enter the building without the expressed permission of the person taking the roll call. Escape routes should be clearly marked as such.

All responsibilities must have been allocated in advance in order to avoid delay and doubt. Moreover, each of the persons bearing responsibilities should designate deputies who could take over in the event of absence, illness or accident. This delegation of responsibilities could easily be made to coincide with that of the delegation of normal administrative duties.

Insurers' approach to fire protection

As previously advocated, consultation should take place with the fire insurers at the design stage of the laboratory. Whilst appreciating that safety of life is the prime object of fire protection, fire insurers are principally concerned with material and consequential loss which may arise from damage to buildings and contents. The fire authority approach will be to ensure that legal requirements in relation to the safety of life are complied with *eg* that there are sufficient means of escape provided and that the construction is capable of containing the fire long enough to allow the personnel to escape.

In some instances, insurers and brigades may put forward requirements in which the emphasis differs. In general, however, there is close liaison and difficulties are usually resolved and a solution is found acceptable to both. This removes the dilemma of management having to make a decision between two differing requirements. In most cases the fire insurer's requirements will be the more stringent, since they start from the basis that there is a statistical probability that a fire will occur at a particular risk some time. Their philosophy is to apply compartmentation by having the structure suitably divided by a fire resisting construction to contain a fire within a compartment to allow additional fire fighting equipment to be brought into use if necessary. By this means the premises at risk are sub-divided and they accept the possible loss of a portion of the premises, but seek to prevent any extension of the damage. In recognition of the fact that the extent of fire damage may be limited by the presence of suitably approved automatic sprinkler systems,[1] fire alarm systems,[2] or portable fire extinguishing equipment, the insurer may grant an allowance in the form of a reduced premium. The value fire insurers attach to automatic sprinkler systems is illustrated by the generous allowances which are given for this type of protection.

Legislation

All statutory requirements cannot be covered in a survey of this nature, but it can show the range of legislation. The original legislation should be consulted by interested parties since no attempt is made here to interpret the requirements. An extensive guide to relevant legislation is contained in Chapter 15 of reference 4 (*see* Bibliography, p 49).

The legislation concerned with fire and explosion in laboratories is mainly contained in Factories Act 1961 when laboratories are concerned with process control, and Petroleum (Consolidation) Act, 1928 and Regulations and Byelaws made under these Acts. The Highly Flammable Liquids and Liquefied Petroleum Gases Regulations 1972 (SI 1972: No. 917) made under the Factories Act regulate the use and storage of such materials. The recently enacted Health and Safety at Work etc Act 1974 has far-reaching implications in that it enables regulations to be made to cover premises previously exempted from other legislation.

In addition to statutory requirements, consideration has to be given to Common Law duties of occupier and employer.

There is a strict duty in Common Law in respect of injury to one's neighbour's property by fire and to employees in respect of the need to provide safe means of access, a safe place of work, a safe system of work and competent supervision; an obligation which almost certainly extends to fire protection measures. Common Law affects laboratories which are outside the scope of the Factories Act.

The Factories Act, 1961

The Factories Act deals with many matters having a direct bearing on fire protection. These are fire drill, fire warning and means of escape as well as fire extinguishing equipment and fire prevention matters. The following sections of the Act are of special interest:

Provisions for plant employing flammable gases, vapours and dusts (Section 31)

Provision and maintenance of means of escape (Sections 40–47 inclusive)

Safety in case of fire (Section 48)

Escape instruction (Section 49)

Enabling special regulations for fire prevention to be made (Section 50)

Provision, maintenance and testing of fire fighting equipment (Sections 51 and 52)

Dangerous conditions and practices (Section 54)

Notification of accidents and dangerous occurrences (Sections 80 and 81)

Persons empowered to inspect premises (Section 148)

Persons held responsible for compliance with the Act (Sections 120, 155, 160 and Second Schedule)

Regulations made or deemed to have been made under the Factories Act, 1961

Certain processes and materials are considered to be extra hazardous and extra legislation has been introduced to deal with these. The requirements of such legislation can be considered supplementary to the requirements of the Factories Act.

The Chemical Works Regulations (SR & O 1922: No. 731)

In these comprehensive Regulations there are requirements for electrical installations and the prohibition of naked lights, matches *etc*, and special precautions for heating installations in any place where there is danger of an explosion from or ignition of flammable gas, vapour or dust. Notices prohibiting smoking and the presence of naked lights, matches *etc*, must be fixed at the entrance to any room or place where there is risk of explosion from flammable gas, vapour or dust.

The Electricity Regulations, 1908 (SR & O 1908: no. 1312) *as amended by the Electricity (Factories Act) Special Regulations* (SR & O 1944: No. 739)

The principal Regulation relating to fire is Regulation 27 of the Electricity Regulations, 1908, which requires that all conductors and apparatus exposed to the weather, wet, corrosion, flammable surroundings or explosive atmosphere, or used in any special process shall be so constructed or protected, and such precautions shall be taken as may be necessary adequately to prevent danger in view of such exposure or use.

The Highly Flammable Liquids and Liquefied Petroleum Gas Regulations 1972 (SI 1972: No. 917)

These Regulations impose requirements for the protection of persons employed in factories and other places to which the Factories Act 1961 applies, in which any highly flammable liquid or liquefied petroleum gas is present for the purpose of, or in connection with, any undertaking, trade or business.

Regarding highly flammable liquids, the Regulations contain requirements as to the manner of their storage, the marking of storage accommodation and vessels, the precautions to be taken for the prevention of fire and explosion, the provision in certain cases of fire-fighting apparatus and the securing in certain cases of means of escape in case of fire. Regarding liquefied petroleum gases, the Regulations contain requirements as to the manner of their storage and the marking of storage accommodation and vessels.

The Petroleum (Consolidation) Act, 1928

The Act requires premises used for the bulk storage of petroleum spirit, defined as petroleum which evolves a flammable vapour at less than 73 °F tested as prescribed, to be licensed and empowers licensing authorities to attach to licences such conditions as are necessary to ensure the safe keeping of the petroleum spirit.

The Secretary of State is empowered to make Regulations governing the conveyance of petroleum spirit by road, and the keeping, use and supply of petroleum for use in vehicles, motor boats and aircraft. In fact for the latter use certain exemptions are made by the Petroleum Spirit (Motor Vehicles) Regulations (SR & O 1929: No. 952). Provision is made to enable Regulations to be made under the Act which can extend the Act or portions of the Act to substances other than petroleum spirit.

Petroleum (Transfer of Licences) Act, 1936
This Act empowers local authorities to transfer petroleum spirit licences granted under the Act of 1928.

Orders and Regulations made under the Petroleum (Consolidation) Act, 1928. The Petroleum (Mixtures) Order, 1929 (SR & O 1929: No. 993)
This applies the whole of the Petroleum (Consolidation) Act to all mixtures of petroleum with any other substances, which possess a flash point below 73 °F.

The Petroleum (Carbon Disulphide) Order, 1958 (SI 1958: No. 257)
This applies the requirements of the 1928 Act in respect to labelling and conveyance by road to carbon disulphide. This Order is modified by The Petroleum (Carbon Disulphide) Order, 1968 (SI 1968: No. 571).

The Petroleum (Compressed Gases) Order, 1930 (SR & O 1930: No. 34)
Air, argon, carbon monoxide, coal gas, hydrogen, methane, neon, nitrogen and oxygen when compressed into metal cylinders are brought under certain sections of the 1928 Act.

The Petroleum (Organic Peroxides) Order, 1973 (SI 1973: No. 1897)
Specified organic peroxides and certain mixtures or solutions of them are brought under some sections of the 1928 Act.
Other relevant regulations are:
   The Gas Cylinder (Conveyance) Regulations, 1931 (SR & O: No. 679)
   The Gas Cylinder (Conveyance) Regulations, 1947 (SR & O: No. 1549)
   The Gas Cylinder (Conveyance) Regulations, 1959 (SI 1959: No. 1919)
However, compressed gases used for powering motor vehicles are subject to The Compressed Gas Cylinders (Fuel for Motor Vehicles) Regulations, 1940 (SR & O 1940: No. 2009).

The Petroleum (Inflammable Liquids) Order, 1971 (SI 1971: No. 1040)
This applies certain sections of the 1928 Act to a large number of flammable liquids, other than petroleum derivatives with a flash point below 23 °C (73 °F), and also to certain other substances which are considered hazardous. The liquids and other substances are set out in Parts I and II of the Schedule attached to the Order.

Explosives Acts, 1875 and 1923
Explosive Substances Act, 1883
These Acts, and Orders and Regulations made under them, govern the manufacture, sale, importation and conveyance by road of all explosives.

Acetylene

Order of the Secretary of State No. 5 dated 28.3.1898

Order of the Secretary of State No. 5A dated 29.9.1905

These provide that under certain conditions compressed acetylene in admixture with oil–gas is not deemed to be an explosive within the meaning of the Act.

Order of the Secretary of State No. 9 dated 23.6.1919

This provides that when acetylene is contained in a homogeneous porous substance with or without acetone or some other solvent and provided certain prescribed conditions are fulfilled shall not be deemed to be an explosive within the meaning of the Act.

Order in Council No. 30 dated 2.2.1937 as amended by

The Compressed Acetylene Order, 1947 (SR & O 1947: No. 805)

This prohibits the keeping, importation, conveyance and sale of acetylene compressed to over 9 psig. Certain exceptions are made among which is that acetylene at pressures not over 22 psig may be manufactured and kept under conditions approved by the Secretary of State. Subject to certain conditions, acetylene at pressures not exceeding 300 psig and not mixed with air or oxygen can be used in the production of organic compounds.

Other Acts and Regulations which may be applicable to laboratories in chemical works

Fire Precautions Act 1971

Premises to which this Act can apply include those used for purposes of teaching, training or research.

The Public Health Act, 1961

Under this Act there have been made the very important Building Regulations 1972 (SI 1972: No. 317). These Regulations have far reaching provisions determining height, floor area, cubic capacity and siting, and fire resistance of structural elements of new buildings. Scotland and Inner London Area have their own legal requirements.

Radioactive Substances Act, 1948

Radioactive Substances Act, 1960

These Acts regulate the acquisition, storage, transport and disposal of any radioactive substances.

The Ionizing Radiations (Sealed Sources) Regulations, 1969. (SI 1969: No. 808)

These regulations impose requirements for the protection of persons employed in factories against ionising radiation arising from sealed sources, and machines or apparatus intended to produce ionising radiations. Northern Ireland has its own version of these Regulations

The Ionizing Radiations (Unsealed Radioactive Substances) Regulations, 1968 (SI 1968: No. 780)

Bibliography

Fire and related properties of industrial chemicals. London: Fire Protection Association 1974.

N.I. Sax, *Dangerous properties of industrial materials*, 4th edn. New York: Reinhold, 1975.

R.R. Young and P.J. Harrington, *Design and construction of laboratories.* (Lecture series 1962, no.3). London: Royal Institute of Chemistry, 1962.

Guide to fire prevention in the chemical industry. London: Chemical Industries Association, 1968.

D.D. Libman, 'Safety in the chemical laboratory', *Laboratory Equipment Digest*, October, 1967.

Mathew M. Braidech, 'Fire and explosion problems in laboratories and pilot plants', *J. chem. Educ.* 1967, **44** A319.

N.V. Steere (ed.), *Safety in the chemical laboratory*, Vol. 1 1967, Vol. 2 1971, Vol. 3 1970. Easton, Penn: Chemical Education Publishing Co., 1967.

P.J. Gaston, *The care, handling and disposal of dangerous chemicals.* Aberdeen: Northern Publishers, 1970.

Safety measures in chemical laboratories, 4th edn. London: HMSO, 1971.

H.A.J. Pieters and J.W. Creyghton, *Safety in the chemical laboratory*. London: Butterworth, 2nd edn, 1957.

The General Safety Committee of the Manufacturing Chemists' Association Inc. *Guide for safety in the chemical laboratory.* New York: Van Nostrand, 1962.

List of approved portable fire extinguishing appliances, London: Fire Offices' Committee, 1970.

Fire Boklist. London: Fire Protection Association, 1975.

References

1. *Rules for automatic sprinkler installations*, 29th edn. London: Fire Offices' Committee, 1968.
2. *Rules of the Fire Offices' Committee for Automatic Fire Alarm Installations*, 11th edn. London: Fire Offices' Committee, 1973.

Chapter 5

Reactive Chemical Hazards

All chemical reactions involve changes in energy, usually evident as heat. This is normally released during exothermic reactions, but may be occasionally absorbed into the products in endothermic reactions, which are relatively few in number.

Reactive chemical hazards invariably involve the release of energy in a quantity or at a rate too great to be dissipated by the immediate environment of the reacting system, so that destructive effects appear. To try to eliminate such hazards from chemical laboratory operations, attempt to assess the likely degree of risk involved in a particular operation, and then plan and execute the operation in a way which will minimise the risks foreseen. It may, of course, be necessary to accept that a certain degree of risk is likely to be attached to a particular course of action, but in this case, personal protection appropriate to the risk will be necessary.

This chapter is concerned with various aspects of the recognition and assessment of reactive hazards, of practical techniques for reaction control, and of personal protection where this is deemed necessary. Several references to sources of more detailed information are given by superscript numbers in the text, and where appropriate, bracketted page references to a monograph on reactive hazards[1] are also given.

Physico-chemical factors

Many of the underlying causes of incidents and accidents in laboratories which have involved unexpected violent chemical reactions are related to a lack of appreciation of the effects of simple physico-chemical factors upon the kinetics of practical reaction systems. Probably the two most important of these factors are those governing the relationship of rate of reaction with concentration, and with the rise in temperature during the reaction.

Concentration of reagents

It follows from the law of mass action that the concentration of each reactant will directly influence the velocity of reaction and the rate of heat release.

It is, therefore, important not to use too-concentrated solutions of reagents, particularly when attempting previously untried reactions. In many preparations, 10 per cent is a commonly used level of concentration where solubility and other considerations will allow, but when using reagents known to be vigorous in their action, 5 per cent or 2 per cent may be more appropriate. Catalysts are commonly employed at these or even lower concentrations.

Of the many cases where increasing the concentration of a reagent, either accidentally or deliberately, has transformed a safe procedure into a hazardous event, three examples will suffice. When concentrated ammonia solution was used instead of the diluted solution to destroy dimethyl sulphate, explosive reaction occurred (327). Omission of most of the methanol during preparation of a warm mixture of nitrobenzene and sodium methoxide led to rupture of the containing vessel. Catalytic hydrogenation of *o*-nitroanisole at 34 bar (34×10^5 Pa) under excessively vigorous conditions (250 °C, 12 per cent catalyst, no solvent) ruptured the hydrogenation autoclave (532).

Reaction temperature
According to the Arrhenius equation, the rate of a reaction will increase exponentially with increase in temperature, and in practical terms an increase of 10 °C roughly doubles the reaction rate in many cases. This has often been the main contributory factor in cases where inadequate temperature control had caused exothermic reactions (normal, polymerisation, or decomposition) to run out of control.

An example relevant to the first two reaction types is the explosive decomposition which occurred when sulphuric acid was added to 2-cyano-2-propanol with inadequate cooling. Here, exothermic dehydration of the alcohol produced methacrylonitrile, acid-catalysed polymerisation of which accelerated to explosion, rupturing the vessel (793). An example of an exothermic decomposition reaction is the violent explosion which occurred during storage of *m*-nitrobenzenesulphonic acid in sulphuric acid at ~150 °C under virtually adiabatic conditions. It was subsequently found that exothermic decomposition of the solution set in at 145 °C, and the pure acid decomposed vigorously at 200 °C (481).

Many other examples of various types of hazardous and unexpected reactions have been collected and classified.[1, 2]

Operational considerations

Effective control is essential to minimise possible hazards associated with a particular reaction system, and to allow you to achieve such control, relevant knowledge is necessary for you to assess potential hazards in the system.

Try to find what is already known about the particular procedure or reaction system (or a related one) from colleagues, or from existing literature.[1-4]

If no relevant information can be found, or the work proposed is known to be original, it will be necessary to conduct cautiously a very small scale preliminary experiment to assess the exothermic character and physical properties of the reaction system and its products.

When subsequently planning and setting up larger-scale reactions or preparations, attention to many practical details may be required to ensure safe working as far as possible. Relevant factors include:

- adequate control of temperature, with sufficient capacity for heating, and particularly cooling for both liquid and vapour phases;
- proportions of reactants and concentrations of reaction components or mixtures;
- purity of materials, absence of catalytic impurities;
- presence of solvents or diluents, viscosity of reaction medium;
- control of rates of addition (allowing for any induction period);
- degree of agitation;
- control of reaction atmosphere;
- control of reaction or distillation pressure;
- shielding from actinic radiation;
- avoiding mechanical friction or shock upon unstable or sensitive solids, and adequate personal protection if such materials will be isolated or dried (without heating).

Further details of specialised equipment, techniques and safety aspects are to be found in the publications devoted to preparative methods.[5-7]

Pressure systems
Some of the above considerations assume greater significance when conducting reactions in closed systems at relatively high pressures and/or

temperatures. In high pressure autoclaves, for example, the thick vessel walls and generally heavy construction necessary to withstand the internal pressures implies high thermal capacity of the equipment, and really rapid cooling of such vessels to attempt to check an accelerating reaction is impracticable. This is why bursting discs or other devices must be fitted as pressure reliefs to high pressure equipment. A brief account of autoclave techniques in high-pressure hydrogenation is readily available.[8]

A further important point specific to closed systems which will be heated is to make adequate allowance for expansion of liquid contents. Several cases are recorded in which cylinders or pressure vessels partially filled with liquids at ambient temperature have burst under the hydraulic pressure generated when heating caused expansion of the liquid to fill completely the closed vessel. This is also likely to happen when cylinders of liquefied gases are exposed to fire conditions.

Adiabatic systems
If exothermic reactions proceed under conditions where heat cannot be lost (*ie* in adiabatic systems) they may accelerate out of control.

Although, in general, few laboratory situations will approach adiabatic conditions, occasionally the combination of a uniform heating system which is unusually well-insulated (such as a thick heating mantle with top jacket to surround completely a flask), or which is of unusually high thermal capacity and inertia (such as a deep well-lagged oil-bath), coupled with a strongly exothermic reaction may approximate to an adiabatic system.

If it is really necessary for technical reasons to use such systems, provision must also be made for application of rapid cooling in the event of an untoward rise in internal temperature. This may be effected by lowering the heating mantle or bath, and application of an air blast, or of cooling to an internal coil.

Types of decomposition
There are three distinct types of chemical decomposition (fast reactions) and the violent effects manifest during deflagration, explosion or detonation are directly related to the rate of energy release, usually derived from combustion or similar processes. The two former types are most likely to be experienced in the laboratory, and usually arise from combustion of an inflammable vapour or gas mixed with enough air to give a composition within the flammable limits.

If the fuel-air mixture is relatively small in volume and virtually unconfined when ignition occurs, a deflagration or 'soft' explosion will occur. Persons in the close vicinity may suffer flash burns from such an incident, but material damage other than scorching by the moving flame-front will be minimal. This is because the unconfined rapid combustion will give no significant pressure effects. Such a situation might arise from a small release of gas or vapour and its ignition in a relatively large room, or from a spill of flammable liquid under an open-fronted hood or lean-to building. There is, however, a volume effect, and larger-scale incidents of this type can produce significant destructive effects.

The effects of explosions under conditions of confinement are invariably more serious, and if the confinement is relatively close, and the fuel-air mixture is nearly stoicheiometric, instantaneous pressures several times that of the normal atmosphere may readily be produced. Such pressures are sufficient to demolish a laboratory building of normal construction. It is for this reason that operations of high potential hazard (involving highly energetic substances, and/or high temperature and pressure) are conducted in isolation cells of reinforced construction designed to withstand possible pressure effects.[9]

Explosion situations may also arise if a reaction accelerates out of control and to the point where the containing vessel fails and/or the vaporised contents reach their auto-ignition temperature. Fire will then definitely occur, but explosion may not.

Detonation is the name applied to a particularly severe form of explosion where the velocity of explosive propagation and the associated decomposition temperature and pressure are much higher (by up to two or three orders of magnitude) than in deflagrations. Under some circumstances, a gaseous deflagration can accelerate into detonation, but the necessary conditions (physically long vessels or pipelines) are seldom present in laboratories. However, explosive decomposition of unstable solids or liquids may occasionally involve detonation (during a few μs and with propagation velocities up to 8 km/s) with the associated violent shattering effects. A more detailed analysis of explosive phenomena is available.[10]

Chemical composition in relation to reactivity

It may eventually be possible to calculate the hazards related to the stability and reactivity of chemical compounds and reaction mixtures in advance of experience. However, at the moment and in most cases, quantitative or qualitative assessments based on known examples represent the best practicable means of assessment of such hazards. Assessment may be based either on overall composition or on detailed structure of the materials involved.

Overall composition and oxygen balance

One of the fundamental factors which may determine the course of a reaction system is that of the overall elemental composition of the system. It is a fact that the majority of reactive chemical accidents or incidents have involved oxidation systems and, especially in organic systems, the oxygen balance is an important criterion.

Mixtures. Oxygen balance is the difference between the oxygen content of a system (a compound or a mixture) and that required to oxidise fully the carbon, hydrogen, and other oxidisable elements present to carbon dioxide, water, *etc.* If there is a deficiency of oxygen, the balance is negative, and if a surplus, positive. Oxygen balance is often expressed as a weight percentage with appropriate sign.

In laboratory oxidation reaction systems, one should plan the operations to keep the negative oxygen balance at a maximum to minimise the potential energy release. This consideration will dictate, therefore, that wherever possible an oxidant will be added slowly (and with appropriate control of cooling, mixing, *etc*) to the other reaction components, to maintain the minimal effective concentration of oxidant throughout the reaction. It is important to establish as early as possible, from the physical appearance or thermal behaviour of the system, that the desired exothermic reaction has become established. If this does not happen, relatively high concentrations of oxidant may accumulate before onset of reaction, which may then become uncontrollable.

Two relevant examples may be quoted. Mixtures of several water-soluble organic compounds (ethanol, acetaldehyde, acetic acid, acetone, *etc*) with aqueous hydrogen peroxide show clearly defined limits within which the mixtures are detonable (790–791). Oxidation of 2,4,6-trimethyltrioxane ('paraldehyde') to glyoxal with nitric acid is subject to an induction period, and the reaction may become violent if addition is too fast. Presence of nitrous acid eliminates the induction period (773).

In other cases it may be necessary for practical reasons to add one or more of the reaction components to the whole (or preferably part) of the oxidant, but the other considerations will still apply.

Compounds. The concept of oxygen balance has more usually been applied to isolated compounds rather than to reaction mixtures as mentioned above. A fairly rapid appreciation of any potential tendency towards explosive decomposition may be gained by inspection of the empirical formula of a particular compound.

If the oxygen content of a compound approaches that necessary to oxidise the other elements present (with the exceptions noted below) to their lowest state of valency, then the stability of that compound is doubtful. The exceptions are that nitrogen is excluded (it is usually liberated as the gaseous element), and halogen will go to halide if a metal or hydrogen is present. Sulphur, if present, counts as two atoms of oxygen.

This generalisation is related to the fact that most industrial high-explosives are well below zero oxygen balance, and some examples follow.

- Compounds of negative balance include:
    trinitrotoluene (525)
    $C_7H_5N_3O_6 + 10.5O \rightarrow 7CO_2 + 2.5H_2O + 1.5N_2$

    peracetic acid (311)
    $C_2H_4O_3 + 3O \rightarrow 2CO_2 + 2H_2O$
    Presence of an oxidant will decrease the negative balance.
- Zero balance:
    performic acid (237)
    $CH_2O_3 \rightarrow CO_2 + H_2O$

    ammonium dichromate (714–715)
    $Cr_2H_8N_2O_7 \rightarrow Cr_2O_3 + 4H_2O + N_2$
- Energy release is maximal at zero balance.
  Compounds of positive balance include:
    ammonium nitrate (812)
    $H_4N_2O_3 \rightarrow 2H_2O + N_2 + O$

    glyceryl nitrate
    $C_3H_5N_3O_9 \rightarrow 3CO_2 + 2.5H_2O + 1.5N_2 + 0.5O$

    dimanganese heptoxide (864–865)
    $Mn_2O_7 \rightarrow Mn_2O_3 + 4O$
Presence of a fuel or reductant will increase the potential energy release.

Compounds with unusually high proportions of nitrogen and N–N bonds are also suspect (66). Hydrazine (87.4 per cent nitrogen), hydrogen azide (97.6 per cent) are both explosively unstable, but not ammonia (82.2 per cent).

In practical terms, the margin between potential and actual hazard of explosive decomposition may be very narrow or quite wide, depending on the energy of activation necessary to initiate the decomposition. Performic acid is treacherously unstable (low energy of activation), and a sample at $-10\ °C$ exploded when moved. TNT, on the other hand is relatively stable and will not detonate when burned, or under impact from incendiary bullets, but requires a powerful initiating explosive ('detonator') to trigger explosive decomposition.

Molecular structure
Instability and/or unusual reactivity in single compounds is often associated with a number of molecular structural features, which may include the specific bond systems given below.

$C\equiv C$	Acetylenes (3), haloacetylenes (57), metal acetylides (76)
CN_2	Diazo compounds (37)
$C-NO$	Nitroso compounds (104)

$C-NO_2$	Nitro compounds (102)
$C-(NO_2)_n$	*gem*-Polynitroalkyl compounds (131)
$C-O-NO$	Alkyl or acyl nitrites (6)
$C-O-NO_2$	Alkyl or acyl nitrates (12, 5)
$C=N-O$	Oximes, metal fulminates (83)
$N-NO$	*N*-Nitroso compounds (104)
$N-NO_2$	*N*-Nitro compounds (103)
$C-N=N-C$	Azo compounds (24)
$C-N=N-O$	Arenediazoates (21), bisarenediazo oxides (25)
$C-N=N-S$	Arenediazo sulphides (21), xanthates (40), bisarenediazo sulphides (25, 40)
$C-N=N-N-C$	Triazenes (144)
N_3	Azides (24)
$N=N-NH-N$	Tetrazoles (141)
$C-N_2{}^+$	Diazonium salts (38–40)
$N-C(N^+H_2)-N$	Guanidinium oxosalts (254–256)
N^+-OH	Hydroxylammonium salts (69)
$N-Metal$	*N*-Metal derivatives (heavy metals) (81)
$N-X$	*N*-Halogen compounds (61), difluoroamino compounds (45)
$O-X$	Hypohalites (69)
$O-X-O$	Halites (29), halogen oxides (63)
$O-X-O_2$	Halates (79, 91)
$O-X-O_3$	Perhalates (15, 20, 121), halogen oxides (63)
$C-Cl-O_3$	Perchloryl compounds (122)
$N-Cl-O_3$	Perchlorylamide salts (122)
$Xe-O_n$	Xenon-oxygen compounds (145)
$O-O$	Peroxides (123)
O_3	Ozone (917–918)

A further large group of compounds which cannot readily be represented by line formulae, and which contains a large number of unstable members is the amminemetal oxosalts (17). These are compounds containing ammonia or an organic base coordinated to a metal, with coordinated or ionic chlorate, nitrate, nitrite, nitro, perchlorate, permanganate or other oxidising groups also present. Such compounds as dipyridinesilver perchlorate (577), tetraamminecadmium permanganate (627), or bis-1,2-diaminoethanedinitrocobalt(III) iodate (431) will decompose violently under various forms of initiation, such as heating, friction or impact.

An interesting application of computers to stability considerations is a program[11] which calculates the maximum possible energy release for a compound or mixture of compounds containing up to 23 elements. No information other than the chemical structure is necessary, and the result, which is semi-quantitative, is used as a screening guide to decide which reaction systems need more detailed and/or experimental investigation.

Redox compounds

When the coordinated base in the last described compounds is a reductant (hydrazine, hydroxylamine), decomposition is extremely violent. Examples are bishydrazinenickel(II) perchlorate (676) and hexahydroxylamine-cobalt(III) nitrate (706), both of which have exploded when wet during preparation.

Other examples of highly energetic and potentially unstable redox compounds, in which reductant and oxidant functions are in close proximity in the same molecule, are salts of reductant bases with oxidant acids, such as hydroxylammonium nitrate (814) hydrazinium chlorite (650) or chlorate (651); or double salts such as potassium cyanide–potassium nitrite (reductant and oxidant respectively).

Pyrophoric compounds

Materials which are so reactive that contact with air (and its moisture) causes oxidation and/or hydrolysis at a sufficiently high rate to cause ignition are termed pyrophoric compounds. These are found in many different classes of compounds, but a few types of structure are notable for this behaviour.

- Finely divided metals: calcium (618), titanium (948)
- Metal hydrides: potassium hydride (756), germane (753)
- Partially- or fully-alkylated metal hydrides: diethylaluminium hydride (7), triethylbismuth (513)
- Alkylmetal derivatives: ethoxydiethylaluminium (512), dimethylbismuth chloride (319)
- Analogous derivatives of non-metals: diborane (188), dimethylphosphine (329), triethylarsine (513), dichloro(methyl)silane (249)
- Carbonylmetals: pentacarbonyliron (434), octacarbonyldicobalt (543)

Many hydrogenation catalysts containing adsorbed hydrogen (before and after use) will also ignite on exposure to air.

Where such materials are to be used, an inert atmosphere and appropriate handling techniques and equipment are essential to avoid the distinct probability of fire or explosion.[12]

Peroxidisable compounds

A group of materials which react with air much more slowly and less spectacularly than pyrophoric compounds, but which give longer-term hazards, may now conveniently be described.

Peroxidation usually takes place slowly when the liquid materials are stored with limited access to air and exposure to light, and the hydroperoxides initially formed may subsequently react to form polymeric peroxides, many of which are dangerously unstable when concentrated and heated by distillation procedures.

The common structural feature in organic peroxidisable compounds is the presence of a hydrogen atom which is susceptible to autoxidative conversion to the hydroperoxy group –OOH. Some of the typical structures susceptible to peroxidation are:

O–C–H	in ethers, cyclic ethers, acetals
$\begin{array}{c} H_2C \\ \\ H_2C \end{array}\!\!\!\!\!\!\!>\!C\text{–H}$	in isopropyl compounds, decahydronaphthalenes
C=C–C–H	in allyl compounds
C=C–H	in vinyl compounds, dienes (*ie* monomers)

C–C–Ar
  |
  H
	in cumene, tetrahydronaphthalenes, styrenes

Several commonly-used organic solvents including diethyl ether, tetrahydrofuran, dioxan, 1,2-dimethoxyethane ('glyme'), bis-2-methoxyethyl ether ('diglyme') are often stored without initiators being present, and are therefore susceptible to peroxidation, and many accidents involving distillation in use of the peroxide-containing solvents have been reported. It is essential to test these solvents for peroxide (with acidified potassium iodide) before use, and if present, peroxides must be eliminated by suitable means[13] before proceeding.

Di-isopropyl ether must be mentioned as particularly dangerous for two reasons. Its structure is ideal for rapid peroxidation, and the peroxide separates from solution in the ether as a readily detonative crystalline solid. Several fatal accidents have occurred, and it should not be used.

If allyl, and particularly vinyl monomers become peroxidised, they are potentially dangerous for two related reasons. The peroxides of some vinyl monomers, such as 1,1-dichloroethylene (288) or butadiene (389) separate from solution and are extremely explosive. Even when this does not happen, the peroxide present may initiate the exothermic and sometimes violent polymerisation of any vinyl monomer during storage. Reactive monomers must therefore be inhibited against oxidation and stored cool, with regular checks for presence of peroxide.[13]

A few inorganic compounds, such as potassium and the higher alkali metals, and sodium amide are subject to autoxidation and production of hazardous peroxidic products. Many organometallic compounds are also subject to autoxidation and require handling in the same way as pyrophoric compounds.[12]

Water-reactive compounds

Second to air (oxygen), water is the most common reagent likely to come into contact, deliberately or accidentally, with reactive chemical compounds.

Some of the classes of compounds which may react violently, particularly with a limited amount of water, are:

- alkali and alkaline-earth metals, (potassium, 841; calcium, 619)
- anhydrous metal halides, (aluminium tribromide, 167; germanium tetrachloride, 698)
- anhydrous metal oxides, (calcium oxide, 623–624)
- non-metal halides, (boron tribromide, 182; phosphorus pentachloride, 702)
- non-metal halide oxides (*ie* inorganic acid halides, phosphoryl chloride, 694; sulphuryl chloride; chlorosulphuric acid, 641)
- non-metal oxides (acid anhydrides, sulphur trioxide, 922)

Concentrated solutions of some acids and bases also give an exotherm when diluted with water but this is a physical effect.

Endothermic compounds

Most chemical reactions are exothermic, but in the relatively few endothermic reactions heat is absorbed into the reaction product(s) which are thus endothermic (and energy-rich) compounds. These are thermodynamically unstable, because no energy would be required to decompose them into their elements, and heat would, in fact, be released.

There are a few endothermic compounds with moderately positive values of standard heat of formation (benzene, toluene, 1-octene) which are not usually considered to be unstable, but the majority of endothermic compounds do possess a tendency towards instability and possibly explosive decomposition under various circumstances.

Often the structure of endothermic compounds involves multiple bonding, as for example in acetylene, vinylacetylene, hydrogen cyanide, mercury(II) cyanide, dicyanogen, silver fulminate, cadmium azide or chlorine dioxide, and all these compounds have been involved in violent decompositions or explosions. Examples of explosively unstable endothermic compounds without multiple bonding are hydrazine and dichlorine monoxide.

In general terms, endothermic compounds may be considered suspect with regard to stability considerations.

Hazardous reaction mixtures

Although the number of combinations of chemical compounds which may interact is virtually unlimited, the combinations which have been involved in hazardous incidents are limited to those which led to an exotherm too large or too fast for effective dissipation under the particular experimental conditions.

The exotherm may have arisen directly from the primary reaction, but a two-stage exotherm is also possible. This will arise when the undissipated primary exotherm leads to instability and subsequent further exothermic decomposition of a reaction intermediate or product.

The great majority of incidents of these types have involved a recognisable oxidant admixed with one or more oxidisable components (or fuels). Examples are the vigorous combustion of glycerol in contact with solid potassium permanganate (844, the high viscosity of glycerol prevents effective heat transfer), and the violent or explosive oxidation of ethanol by excess concentrated nitric acid (involving formation of unstable fulminic acid, 760–761). Some further common examples are given in the partial list of incompatible chemicals, and many others are available in classified form.[1,2]

A smaller group of incidents has involved recognisable reductants admixed with materials capable of oxidation. Examples here are the explosive decomposition of a heated mixture of aluminium powder and a metal sulphate (164), the shock-sensitivity of sodium in contact with chlorinated solvents (888–889), or the violent interaction of powdered magnesium with moist silica when heated (861).

As might be expected, interaction of obvious oxidants and reductants is always potentially hazardous and must be conducted under closely controlled conditions with ample cooling capacity available. Attempted reduction of dibenzoyl peroxide by lithium tetrahydroaluminate led to a fairly violent explosion (602), and hydrazine is decomposed explosively in contact with chromium trioxide (811). Rocket technology furnishes further examples of the extreme energy release possible in undiluted redox systems.

Potential storage hazards

Most of this chapter has been devoted to the possible outcome of the deliberate interaction of chemicals, but it must not be forgotten that hazardous reactions may occasionally arise from accidental contact of chemicals due to breakage, spillage, or more seriously, from fire in a chemical store. This possibility will dictate that a good deal of thought must be given to segregation of different materials in storage to minimise the effects of accidental contact.

The lists below give a selection of materials which need to be segregated on the grounds of potential reactive or toxic hazards.

Partial list of incompatible chemicals (reactive hazards)

Substances in the left hand column should be stored and handled so they cannot possibly accidentally contact corresponding substances in the right hand column under uncontrolled conditions, when violent reactions may occur.

Acetic acid	Chromic acid, nitric acid, hydroxyl-containing compounds, ethylene glycol, perchloric acid, peroxides, and permanganates.
Acetone	Concentrated nitric and sulphuric acid mixtures.
Acetylene	Chlorine, bromine, copper, silver, fluorine, and mercury.
Alkali and alkaline earth metals, such as sodium, potassium, lithium, magnesium, calcium, powdered aluminium	Carbon dioxide, carbon tetrachloride, and other chlorinated hydrocarbons. (Also prohibit water, foam, and dry chemical on fires involving these metals—dry sand should be available).
Ammonia (anhyd.)	Mercury, chlorine, calcium hypochlorite, iodine, bromine and hydrogen fluoride.
Ammonium nitrate	Acids, metal powders, flammable liquids, chlorates, nitrites, sulphur, finely divided organics or combustibles.
Aniline	Nitric acid, hydrogen peroxide.
Bromine	Ammonia, acetylene, butadiene, butane and other petroleum gases, sodium carbide, turpentine, benzene, and finely divided metals.
Calcium oxide	Water.
Carbon, activated	Calcium hypochlorite.
Chlorates	Ammonium salts, acids, metal powders, sulphur, finely divided organics or combustibles.
Chromic acid and chromium trioxide	Acetic acid, naphthalene, camphor, glycerol, turpentine, alcohol, and other flammable liquids.
Chlorine	Ammonia, acetylene, butadiene, butane and other petroleum gases, hydrogen, sodium carbide, turpentine, benzene, and finely divided metals.
Chlorine dioxide	Ammonia, methane, phosphine, and hydrogen sulphide.
Copper	Acetylene, hydrogen peroxide.
Fluorine	Isolate from everything.
Hydrazine	Hydrogen peroxide, nitric acid, any other oxidant.
Hydrocarbons (benzene, butane, propane, gasoline, turpentine, *etc*)	Fluorine, chlorine, bromine, chromic acid, peroxide.
Hydrocyanic acid	Nitric acid, alkalies.
Hydrofluoric acid, anhyd. (hydrogen fluoride)	Ammonia, aqueous or anhydrous.
Hydrogen peroxide	Copper, chromium, iron, most metals or their salts, any flammable liquid, combustible materials, aniline, nitromethane.
Hydrogen sulphide	Fuming nitric acid, oxidising gases.
Iodine	Acetylene, ammonia (anhydr. or aqueous).
Mercury	Acetylene, fulminic acid,* ammonia
Nitric acid (conc.)	Acetic acid, acetone, alcohol, aniline, chromic acid, hydrocyanic acid, hydrogen sulphide, flammable liquids, flammable gases, and nitratable substances.
Nitroparaffins	Inorganic bases, amines.
Oxalic acid	Silver, mercury.
Oxygen	Oils, grease, hydrogen, flammable liquids, solids, or gases.
Perchloric acid	Acetic anhydride, bismuth and its alloys, alcohol, paper, wood, grease, oils.
Peroxides, organic	Acids (organic or mineral), avoid friction, store cold.

Phosphorus (white)	Air, oxygen.
Potassium chlorate	Acids (see also chlorates).
Potassium perchlorate	Acids (see also perchloric acid).
Potassium permanganate	Glycerol, ethylene glycol, benzaldehyde, sulphuric acid.
Silver	Acetylene, oxalic acid, tartaric acid, fulminic acid*, ammonium compounds.
Sodium	See alkali metals (above).
Sodium nitrite	Ammonium nitrate and other ammonium salts.
Sodium peroxide	Any oxidisable substance, such as ethanol, methanol, glacial acetic acid, acetic anhydride, benzaldehyde, carbon disulfide, glycerol, ethylene glycol, ethyl acetate, methyl acetate, and furfural.
Sulphuric acid	Chlorates, perchlorates, permanganates.

* Produced in nitric acid–ethanol mixtures.

Partial list of incompatible chemicals (toxic hazards)

Substances in the left hand column should be stored and handled so that they cannot possibly accidentally contact corresponding substances in the centre column, because toxic materials (right hand column) would be produced.

Arsenical Materials	Any reducing agent*	Arsine
Azides	Acids	Hydrogen azide
Cyanides	Acids	Hydrogen cyanide
Hypochlorites	Acids	Chlorine or hypochlorous acid
Nitrates	Sulphuric acid	Nitrogen dioxide
Nitric acid	Copper, brass any heavy metals	Nitrogen dioxide (nitrous fumes)
Nitrites	Acids	Nitrous fumes
Phosphorus	Caustic alkalies or reducing agents	Phosphine
Selenides	Reducing agents	Hydrogen selenide
Sulphides	Acids	Hydrogen sulphide
Tellurides	Reducing agents	Hydrogen telluride

* Arsine has been produced by putting an arsenical alloy into a wet galvanised bucket.

Protection from reactive hazards

Where it has been decided after assessment of the various factors discussed in this chapter that a potential reactive hazard may exist in work being planned, consideration must also be given to the level of personal protection which may be required to allow the work to be executed safely or with minimum risk.

There is a considerable range of possibilities which may be involved, depending on the scale of operations and the type of assessed hazard, but in all cases eye-protection will be mandatory, and that from approved safety spectacles may be supplemented with a visor or full face mask.

A small-scale reaction not involving toxic hazards could be run on a wall-facing laboratory bench behind a portable safety screen, but a larger scale reaction, or one also involving toxic materials would best be run with the greater degree of protection afforded by a good fume cupboard. Where exceptionally reactive materials are involved, provision of a specific fire extinguisher additional to the general purpose laboratory extinguishers may be necessary. Analogously, where materials of high toxicity (especially cylinder gases) are involved, a specific respirator, or air breathing set may be required.

Where the possibility of explosive decomposition has been assessed, and in any case where high-pressure reactions are being used, an isolation cell of suitable design is appropriate.[9]

In the absence of laboratory facilities appropriate to the degree of hazard assessed, operations must be deferred until such facilities can be made available.

References

1. L. Bretherick, *Handbook of reactive chemical hazards*. London: Butterworth, 1975.
2. *Manual of hazardous chemical reactions*, 491M, 5th Edn. Boston: National Fire Protection Association, 1975.
3. A.I. Vogel, *A text-book of quantitative inorganic analysis*, 3rd Edn. London: Longmans, 1962.
4. A.I. Vogel, *A text-book of practical organic chemistry*, 3rd Edn. London: Longmans, 1957.
5. *Organic syntheses*, various eds, New York: Wiley, Coll. Vols 1–5, 1944–1973, annual volumes thereafter.
6. *Inorganic syntheses*, various eds, London: McGraw-Hill, Vols 1–15, 1939–1974, annual volumes thereafter.
7. G. Brauer, *Handbook of preparative inorganic chemistry*, 2nd Edn. London: Academic, Vols 1 and 2, 1963–1965.
8. Ref. 4, pp. 866–870.
9. W.G. High, 'The design of a cubicle for oxidation or high-pressure equipment', *Chemy Ind.*, 1967, 899–910. *Safety in the study of chemical reactions at high pressure*, Tech. Bull. No. 100, A.L. Glazebrook, Erie, Autoclave Engineers.
10. V.J. Clancey, 'Explosion hazards', *Protection*, 1971, **8**, (9) 6.
11. *CHETAH, ASTM Chemical thermodynamics and energy release evaluation program.* Philadelphia: American Society for Testing and Materials, 1975.
12. D.F. Shriver, *Manipulation of air-sensitive compounds*. London: McGraw-Hill, 1969.
13. H.L. Jackson, W.B. McCormack *et al,* 'Control of peroxidisable compounds', *J. chem. Educ.*, 1970, **47**, A175–188.

Appeal for further information

The author of this chapter will be pleased to receive information on reactive chemical hazards from relatively inaccessible sources for a revised version of reference 1. Contributions will be acknowledged in print.

Chapter 6

Chemical Hazards and Toxicology

How toxic is 'toxic' or the basic rules of toxicology

In everyday language terms like poison, toxic, toxicity, imply some absolute quality. The actual meaning of these terms is that a compound (or an element) under controlled circumstances, which includes its quantity, is able to produce a definite harmful effect in a biological system, *eg* in the human body. To say that water is non-toxic is true only in the sense that the usual water consumption up to a certain limit has no adverse effect, but in deserts or in conditions of hot work uncontrolled water consumption of non-salted water can cause severe muscle cramps. Drugs used for the treatment of diseases exert beneficial effects up to a certain dose level, but toxic effects become apparent in the case of overdoses. Thus every chemical is a potential hazard to health, and the degree of this potential hazard can be measured under controlled circumstances by the quantification of toxicity. Consequently, to say that a compound is toxic or non-toxic does not convey more information than to say that a chemical is soluble without defining its physical state, the solvent, temperature and pressure. It sounds even more absurd to make a general statement of the solubility of a group of compounds. Unfortunately this happens often when people who are not aware of the basic rules of toxicology speak about toxic hazards. An example taken from life will help to pinpoint the most frequent mistakes and to summarise the guidelines.

A few years ago it was reported that many of the great lakes of North America had been contaminated with mercury derived from the effluent of chloralkali and pulp factories, and fish caught in these lakes frequently contained mercury in concentrations more than 1 ppm (1 mg/kg) and sometimes as high as 6–12 ppm. Mercury in fish is mainly in the form of methylmercury, the causative agent of irreversible nervous system damage, which is called Minamata disease after the location of the first methylmercury epidemic. Against the suggested, and sometimes exaggerated preventive measures which followed this discovery an eminent professor of chemistry sent a protest letter to a scientific journal with the following statements:

1. Neither had he suffered any adverse effects with the exception of loose teeth after a 3 h exposure to butylmercury vapour or distilling dimethylmercury for two days, nor his technician who was exposed to an unidentified concentration of dimethylmercury for 200 days;

2. 25 rats exposed to dimethylmercury vapour for 20 days when 25 g dimethylmercury was evaporated in a 4 m³ static inhalation chamber showed no visible signs of intoxication;

3. Neither he nor his students were affected with the exception of skin rashes when they carelessly manipulated alkylmercury salts;

4. During a 'mercury scare' which occurred 40 years ago tests carried out on himself showed that he eliminated per day as much mercury as he absorbed;

5. Organomercurial diuretics with a maximum dose of 1 g per day were essentially the only diuretics used prior to 1952.

There are some basic rules in toxicology which cannot be ignored when one makes a statement about toxic chemicals and their effects. The communication of knowledge assumes that a certain word means the same thing to different people. A list of rules with explanatory notes presented below aims to give an introduction to the understanding of the language of toxicology.

The potential health hazard depends on the physical and chemical properties of a compound, and these properties are usually changed with alteration in the molecular structure.

The first basic rule in toxicology is that, without proof one should never state that different compounds of an element have the same potential to be harmful or hazardous and their toxicities are identical. In every statement concerning toxicity the exact molecular structure of a chemical must be identified. Changes in the molecular structure caused by oxidation, reduction, cleavage of a bond, introduction of a new ligand or replacement of one ligand with another can have a profound effect on the potential of a compound to be hazardous. Some of these changes might occur during storage or as a result, intentional or otherwise, of working with a chemical. In these cases consideration must be given to the properties of the possible derivatives.

A change in the molecular structure can alter the physical properties. Thus dimethylmercury has a higher vapour pressure than methylmercury chloride. The volatility of methylmercury depends on the anion: methylmercury chloride is 350 times more volatile than the corresponding dicyandiamide salt. The second important physical property is solubility. The solubility of the mercury metal in water and biological fluids is very low which explains why 2 lb of mercury taken orally in four divided doses caused no ill effect in a young man. The solubility of mercurous chloride is also low compared with mercuric chloride, and in the case of oral administration the more soluble mercuric salt is 60 times more toxic than the mercurous salt. Other examples are: lipid solubility favours absorption through the upper layer of the skin, water solubility helps the transport through the lower layers. Consequently, of two lipid soluble compounds, like aniline and nitrobenzene, the absorption of aniline is somewhat faster due to its higher water solubility. The third physical property which can influence the potential hazard of a chemical is viscosity. Increasing viscosity diminishes the danger not only of mouth contamination but also aspiration when pipetting.

Change in the molecular structure can affect the reactivity of a compound. Thus dimethylmercury having no charge, behaves like a neutral gas and only a part of it is decomposed to methylmercury in the body. In mice within $2\frac{1}{2}$ hr after the iv administration of dimethylmercury 85 per cent of the dose was exhaled and only 10 per cent was converted to methylmercury, that is, to the neurotoxic agent. Dibutylmercury behaves similarly as dimethylmercury but the metabolic product, butylmercury, is less toxic than methylmercury and it seems to lack neurotoxic properties. Thus one cannot take generalisations even for the toxicity of alkylmercury compounds, not to mention organomercurials or the entire group of mercury compounds. Organomercurial diuretics are rapidly metabolised in the body and the role of the organic part of the molecule is to carry mercury and release it as mercuric ion at receptor sites which are involved in diuresis. This effect is qualitatively different from the neurotoxic effect of methylmercury which affects the sensory nerves or from the effect of the vapour of the mercury metal which produces psychic disturbances, the extreme form of which was not infrequent in the hat industry when mercury was used for processing fur ('mad as a hatter').

The biological system responding to a chemical must be defined. The definition should include species, sex, age and must consider individual variations even within a well defined group.
Not every species reacts with the same degree or type of response to equal

doses of a chemical. Dimethylmercury might be quite innocuous to rodents as they are able to convert only a small proportion of it to methylmercury, but man might be more sensitive. The experiences of the professor and his technician with dimethylmercury are not the only ones. In 1863 two laboratory technicians who worked with dimethylmercury died of methylmercury poisoning. In 1971 a chemist died of methylmercury poisoning in Czechoslovakia after synthesising 600 g dimethylmercury. As the exact exposure is not known in these cases, it is open for argument whether those who died were more sensitive to dimethylmercury or had higher exposure, but it is most likely that the 'carelessness' of the professor was neutralised by very efficient ventilation in his laboratory which diminished the actual hazard. However even if individual sensitivity contributed to the tragic outcome in the other cases, preventive measures must aim to protect every person against any irreversible damage which could handicap the affected persons for life.

The main difficulties in interpreting toxicological results are the extrapolation of data from one species to another and the biological variation in sensitivity within one species. Age and sex are the most important factors which influence sensitivity within one species, but there are more subtle causes such as nutritional or genetic. Thus male rats are more sensitive to the renotoxic effect of sublimate ($HgCl_2$) than females, young rats are more sensitive to methylmercury than older ones, a selenium rich diet gives some protection against methylmercury in rats and susceptibility to lead depends on a wide variety of dietary factors like calcium, iron, protein, vitamin D, ascorbic acid, nicotinic acid, other metals and alcohol consumption. In rats methylmercury is liable to damage the kidney, in human cases of methylmercury intoxication renotoxicity is always absent. Different strains of rats have different sensitivity to the hepatotoxic effect of CCl_4: Fischer rats are more sensitive than Wistar rats. The best known examples for the significance of genetic differences in humans are glucose-6-phosphatase deficiency resulting in hypersusceptibility to haemolitic agents like aniline, and serum antitrypsin deficiency which results in increased sensitivity to respiratory irritants. But even in a homogenous inbred group, which is homogenous in age, sex and nutrition, there will be some variation in the response to the same dose of the chemical. Thus individual observations though they may be clinically interesting, as they reveal some qualitative aspect of the toxic effect, do not satisfy the requirements of a responsible statement on the toxicity of a chemical. This needs greater numbers for suitable statistical analysis.

Information on the dose or exposure is the prerequisite for the evaluation of toxic hazards
How toxic is a compound, compared with another, cannot be answered

without exact data on the dose, or on the exposure which depends on the concentration of the compound in the breathing zone of the atmosphere or on the amount in contact with the skin. Toxicity is measured by relating exposure or dose to observed biological reactions. To give an exact amount of dimethylmercury to a closed chamber seems to satisfy this demand, but only the repeated analysis of the toxicant in the atmosphere can prove that there was no leakage or other loss and the tested compound was actually in the air inhaled by the experimental animals for a definite period of time. In the laboratory, carelessness is not a measure of exposure, as other factors might diminish the exposure to below the level to which a careful chemist is exposed in a not so well constructed laboratory. The only conclusion which can be drawn from the quoted statements is that the exposure to dimethylmercury was not high enough to cause obvious neurological damage. The exposure to other mercury compounds was high enough to cause inflammation of the gum, and skin rashes indicated contact with the skin. These observations, unrelated to dose or exposure, do not help either to compare the toxicity of different mercury compounds or to judge their potential hazard.

Evaluation of a chemical hazard is based on two relationships, one is the dose effect relationship, the other is the dose response relationship. The methylmercury epidemic in Iraq which involved many thousands of people proved that the daily consumption of certain amounts of methylmercury contaminated bread caused paraesthesia [sensory disturbances], higher level of exposure caused ataxy [uncoordinated movements] and even higher daily uptake resulted in the loss of vision, deafness, or even death. Thus methylmercury is able to cause many effects depending on the exposure. Another example is the narcotic and hepatotoxic effect of carbon tetrachloride and many other solvents. In the case of lead, porphyrin in the urine, anaemia, abdominal pain, palsy and, mainly in children, encephalopathy are the main constituents of the dose effects relationship. Rats given 1–2 mg/kg amphetamine run around the walls in the cage, when treated with 6 mg/kg amphetamine they soon become stationary, but they move their head repetitively and higher doses produce cardiovascular disorders. Listing effects against the corresponding dose gives the dose effects relationship. If one effect, like anaemia, or loss of nerve function can be measured on a graded scale of severity, the gradation of this effect in relation to dose is also used as dose effect relationship. If one effect is selected as a response and the percentage of animals giving this response in every dose group is plotted against the dose, a dose response curve is produced. From the dose response curve one calculates what is the dose which is able to produce a response in 50 per cent of the animals and this dose is called ED 50 (ED for effective dose). If the response is death, the dose which killed 50 per cent of the animals is called LD 50 (LD for lethal dose). Both ED 50 and

LD 50 can be calculated from single or multiple administration experiments, but in every case the route of administration, dose, treatment and observation periods must be stated. In the case of inhalation exposure, the atmospheric concentration of the test compound and the length of the exposure are the essential data. Both ED 50 and LD 50 are valid only for the species tested with the qualification of age, sex and nutrition. Even when men or animals can be killed by a single dose, this does not necessarily mean that by lowering the dose, a dose-effect relationship which includes all the possible effects can be established. In the case of methylmercury, the typical neurological symptoms in rats can be produced only after repeated administrations. Many of the chemical carcinogens must be given for an extended period of time and the observation period might cover nearly the whole lifetime of the animal.

Naturally LD 50 values obtained in experimental animals can give only a rough estimate of the degree of toxicity, but even this can be helpful to classify toxic compounds into broad categories. The categorisation showed in

Table 6.1. Categories in relation to relative acute toxicities.

Toxicity rating	term of toxicity	probable human lethal dose for a 70 kg man	compound belonging to the group with oral LD 50 for rats in mg/kg	
1	practically non toxic	>15 g/kg	propylene glycol:	26 000
2	slightly toxic	5–15 g/kg	sorbic acid:	7400
3	moderately toxic	0.5–5 g/kg	isopropanol:	5800
4	very toxic	50–500 mg/kg	hydroquinone:	320
5	extremely toxic	5–50 mg/kg	lead arsenate:	100
6	supertoxic	<5 mg/kg	nicotine:	50

Table 6.1 is widely used and indicates some relationship between a single oral dose which may cause death in man and the single oral LD 50 for rats.

LD 50 values can be used as a first estimate of acute toxicity, but naturally if the toxic compound accumulates in the body or there is an accumulated effect, or the compound is carcinogenic or can produce sensitisation, the potential hazard has no relationship to the above toxicity rating. Thus, based on a single dose, methylmercury is less toxic than mercuric chloride, but in the case of repeated administration animals develop a resistance to the renotoxic effect of mercuric chloride but not to the neurotoxic effect of methylmercury. Di-isocyanates are only slightly toxic according to the toxicity rating but in people exposed to them it is not uncommon that following acute symptoms, an asthma-like syndrome is precipitated even by exposure to a minute amount of di-isocyanate. One of the difficulties when testing carcinogens is the absence of definite early signs before the occurrence of the tumour cells.

There is a term in toxicology, the knowledge of which is extremely important when the development of a toxic effect depends on the accumulation of a toxic compound in the body. This term called half time denotes the time which is needed to excrete half of the total amount of the compound from the body, that is half of the body burden when there is no further exposure. Accumulation of a toxic compound in the body also depends on the half time as, in every day, a fixed proportion of the body burden (and not of the daily dose) is eliminated. If the daily uptake (dose or exposure) remains constant, the body burden for this compound increases to the point at which absorption and elimination are equal. That condition is called the steady state. For example methylmercury has a 70 day half time and if a person is ingesting 2 mg methylmercury per week of which 95 per cent is absorbed from the gastrointestinal tract, one can calculate that after 6 months the body burden for this type of mercury will be 24 mg, after 1 year 28 mg, and 28.9 mg at steady state when he will excrete exactly 2 mg of mercury a week. From the observations that our professor of chemistry excreted exactly the same amount of mercury as he absorbed (the form of mercury was not given) one can conclude that for this type of mercury he was in a steady state condition, but an increase in his exposure should have resulted in a positive, and a decrease a negative balance until the body burden reached the steady state corresponding to the new level of exposure.

As far as half time is concerned, methylmercury belongs to the group of compounds which have only one half time, and daily body burden plotted on a semilog paper against time gives a straight line. There are other compounds, like lead or mercury vapour which have more than one half time, a shorter half time followed by one or more longer half times. This indicates that the half time for different organs or tissues is different, and as time goes on, tissues with a longer half time dominate the elimination curve. In the case of lead, bone has the longest half time. Another complicating factor is the metabolism of the chemical yielding derivatives which might have a longer or shorter half time than the parent compound. Thus the half time of dimethylmercury is very short and the half time of methylmercury is very long. In this case the derivative is the toxic component and not the parent compound. Phenylmercury, the alkoxyalkyl mercury salts or the organomercurial diuretics are rapidly metabolised to inorganic mercury and afterwards the half time of their mercury component is identical with the half time of mercuric chloride.

One of the most important problems in toxicology is whether a compound has a real borderline between safe and hazardous exposure or does any level of exposure represent a certain risk?
The main aim of a toxicologist is to establish whether the compound can be used safely and what are the limits and conditions of safe use. For

occupational exposure to toxic gases, vapours or dusts the threshold limit values (TLV) represent a time weighted average concentration to which all workers may be repeatedly exposed day after day without any adverse effects. The maximum allowable concentrations (MAC) are threshold limits which should not be exceeded any time during a normal working day. Industrial experience and experimental studies have proved for a wide variety of compounds that up to a certain exposure every individual is safe from toxic effects. In other words if the mildest effect is selected as response, and the percentage of people with a positive response is plotted against the logarithm of the dose, the dose response curve intercepts the abscissa at zero response level and the intercept gives the highest dose (or exposure) with no harmful effect (upper right hand part of *Fig. 6.1*, solid circles). Individual variation in susceptibility, genetic or acquired, is responsible for a plus-minus variation from the intercept, but this does not change the validity of the concepts: there is exposure without danger to health and preventive measures achieve their aim if exposure is below the TLV or MAC.

The most reliable TLVs are those which are based on decades of industrial experience. Experiments on human volunteers are limited for the observation of reversible acute effects. However, the main bulk of data on the toxicity of chemicals have derived from animal experiments and present the problem of translating animal data to man. This extrapolation is relatively easy if a reference point is established both for the experimental animal and man. For example if the dose required to elicit a single critical response is established both for man and the experimental species, there is a scientific base for extrapolation from the experimental dose response curve to safe exposure level in man. Data on the half time and the metabolism of the toxic chemical in man and in the experimental animal also helps extrapolation. Without parallel data extrapolation from animal experiments there is only guesswork—approximation based on analogies and probabilities with bias on the side of safety. Thus the threshold limit for uranium established on the basis of the urinary catalase activity of rabbits exposed to this metal was reduced four-fold when the same response had been estimated in hospitalised volunteers.

The situation is quite different for chemicals which are able to cause some irreversible damage either after a long term exposure or a long time after a short term exposure. Though observations on humans can give invaluable information on their toxicity and safety, cases of human intoxication—like malignant tumours caused by vinylchloride or the thalidomide babies—are the signs of failure in the toxicological evaluation or preventive measures. The toxicological evaluation of toxic compounds which might cause irreversible damage—like cancer—is a complex procedure which needs not only carefully controlled animal experiments covering nearly the whole lifetime of the species, but a complicated system of extrapolation. Before any acceptable exposure is extrapolated to man, an extrapolation is needed to compute the exposure to the acceptable risk in the experimental species, which can be as low as 1 in 10 thousand or 100 millions. For this extrapolation a simple dose-response curve with a range between zero and 100 per cent response is unsuitable: firstly because it is not sensitive enough in the lower range of response; secondly because of the assumption that for certain types of toxic effects there is no threshold dose or exposure and with decreasing exposure the risk is lowered but never eliminated. The difficulty created by the assumed absence of zero response and the actual observation that below a certain dose the experimental animals do not respond (*eg* with cancer) is overcome by the upward correction of the per cent responses, resulting in the elimination of zero responses. An experiment in which none of the 10 mice treated with a certain dose of a carcinogen developed cancer actually indicates that if the experiment is repeated 100 times, in 99 of the 100 similar experimental groups the frequency of cancer will not be higher than 37 per cent and thus based on a single experiment the upper limit of tumour risk with a 99 per cent assurance is 37 per cent. If the number of animals tested is increased to 100 and none of them develop cancer, the risk is decreased to 4.5 per cent or in the case of 1000 mice to 0.45 per cent and so on, but though the upper limit of risk decreases with the increase in the number of animals it never reaches the zero level. Similar correction based on binominal distribution gives the upper limit of risk as 87 per cent when actually 13 of 20 animals treated with the same dose developed cancer.

One of the graphically expressed transpositions of a dose response relationship with no zero response level is based on normal distribution giving a bell shaped curve. The middle, that is, the highest point of the curve corresponds to the ED 50 and from right (in the direction of higher doses) and to the left (in the direction of lower doses) the first standard deviation (SD) corresponds to 50 ± 34 per cent responses, the second SD to 50 ± 47.7 per cent responses, the third to 50 ± 49.86 per cent responses and so on. Units on the ordinate are given in SD units or normal deviates which in the corresponding per cent probabilities never reach the 100 or zero per cent, but can indicate by extrapolation a risk as low as 1 in 100 millions (left hand part of *Fig. 6.1*, empty squares).

Without going into the details of the use of this or any other model this simplified outline might help one to appreciate the difficulties in the safety evaluation of carcinogens. The reason for these difficulties lies mainly in the deficiency of our knowledge on the mechanism and dynamics of cancer formation. One must ask whether it is possible to cause cancer with one molecule if this one molecule hits at the right moment a sensitive target cell in the sensitive individual, or this degree of sensitivity does not exist at all and the use of mathematical models with no zero frequency level is theoretically not justified. The same considerations apply to non-carcinogenic compounds

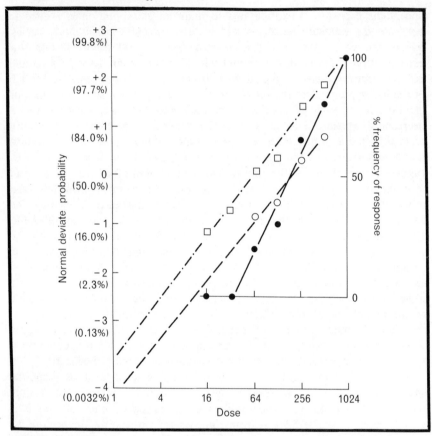

Fig. 6.1. Dose-response curves in log dose/frequency of response (●); in log dose/normal deviates of 0 ≺ frequency of response < 100 per cent (○); in log dose/normal deviates of responses corrected to the upper limit of risk with 99 per cent assurance (□). The number in brackets under the number of normal deviates gives the risk in per cent. The ordinate can be extended to the infinite.

if they are able to cause chronic irreversible damage. If nerve cells can tolerate a certain concentration of methylmercury without being damaged, extrapolation by using mathematical models without zero response level is nonsensical. However if there is not such a threshold limit it must be assumed that there is a certain risk at any level of exposure to methylmercury.

The problem of risk evaluation is not made easier if the chemical is a physiological one which is continuously formed in the body and at the same time it is a very common environmental toxic agent like carbon monoxide. For a long time it was believed that CO has only acute effects which are present if at least 20 per cent of the haemoglobin is made unsuitable for oxygen transport. Recently it has become apparent that exposure resulting in less than this level is able to produce an irreversible chronic effect by accelerating the ageing process of the cardiovascular system. A similar effect—though probably by a different mechanism—is exerted by another simple compound, carbon disulphide. The acceleration of the ageing process presents similar problems for risk evaluation as carcinogenic chemicals. Ageing is a normal physiological process but due to exposure to the toxic chemical it becomes evident earlier than usual. Depending on the size of the population and their life span, some individuals would develop cancer without the administration of a carcinogen, but the frequency of malignancies is increased by the exposure.

No model is available to predict the safe level of exposure if, as a result of the formation of antigens from the chemical and physiological proteins, antibodies are produced against the chemical. The reaction between the chemical and the antibodies in the sensitised individual can produce rashes (urticaria), allergic dermatitis or asthma (see also Chapter 7 p 91). Sensitisation is quite different from hypersensitivity. In the case of sensitisation the response differs from the normal toxic effect of the chemical and its severity depends much less on the level of exposure than on the degree of sensitisation. In the case of hypersensitivity the reaction is the expected effect, only for a hypersensitive person the dose response curve is shifted to the direction of lower doses.

In the case of sensitisation the removal of the person from the exposure solves the problem. When a chronic irreversible damage has developed, the pathological process has already reached the stage of no return. That is why the work with chemicals which can cause irreversible damage, like cancer, demands the utmost care in the control of exposure.

Exposure as a measure of relative hazard

Exposure and the route of uptake

Occupational exposure to toxic substances differs considerably from the intentional administration of a drug or a poison. In the latter case a known dose is selected to give the desired effect, in the case of occupational exposure the exact dose is not known and sometimes it needs a thorough examination even to identify the route of uptake. In occupational circumstances a toxic chemical can enter the body by any or by a combination of the following routes: inhalation, orally or through the skin. The respiratory tract is by far the most important route by which toxic chemicals enter the body. The inhaled air in the lungs can diffuse into the blood through a surface as large as 50 to 100 m^2, while the skin surface is only 1.8 m^2.

If atmospheric contamination is responsible for the exposure, the difference in the concentration of the chemical in the inhaled and exhaled air multiplied with the exposure time gives the dose. However, without any attempt to calculate the exact dose, the atmospheric concentration can be related to one or more responses. This is the base for TLV or MAC values.

It is more difficult to obtain the measure of exposure when oral or percutaneous absorption is the route of uptake or at least contributes to the total exposure. If the chemical is a gas which is absorbed through the skin or dissolved in the mouth, air concentration is a measure of total exposure. Hydrogen cyanide—though a gas—can penetrate the skin to such an extent that without inhalation exposure the body can take up enough cyanide from the atmosphere to cause severe or even lethal intoxication.

In occupational circumstances oral or percutaneous exposures are usually interlinked. Contamination on the hand is the usual source of oral uptake through contamination of food and cigarettes. Absorption through the skin depends on the physical and chemical qualities of the toxic substance, but the surface area contaminated and time allowed between contamination and decontamination are equally important factors. The skin has a protective layer against those chemicals which are not solvents for lipids. Lipid soluble chemicals can penetrate the skin quite easily, though further transport is facilitated by water solubility. Solvents, aromatic amino or nitro compounds are the main group of compounds which can easily be absorbed through the skin. Contamination within a rubber glove presents the most favourable conditions for absorption: increased temperature, vapour pressure, sweating and softening of the epidermis. Skin and the mucous membranes can be directly affected by the chemical. Chemical burns are usually caused by liquids.

In laboratories where the working procedures are alternating, it is unlikely that the occasional measurement of the atmospheric concentration of a chemical will give a truthful picture about the level of exposure. That is why it is so important to judge whether the conditions are such that appreciable exposure exists at all. The awareness of the presence of toxic chemicals and of the conditions which might transform a potential hazard to an actual hazard is the best safeguard against harmful exposure. Nevertheless, in experimental plants or when a work is carried out for extended periods, *eg* synthesis, distillation, *etc* it might be necessary to analyse the working environment for possible contamination.

There is no similar method to measure absorption through the skin or through the gastrointestinal tract. If the chemical has a colour, discolouration of the skin can indicate the contact. Similarly the analysis of the skin wash may show how carefully the toxicants are handled. A very useful indicator of exposure is the concentration of a chemical or its metabolite in blood or urine samples. These biological tests give an overall picture of exposure independently of the route of entry.

Physical conditions and other factors in relation to exposure

The physical conditions (solid, liquid or gas), vapour pressure and solubility of a chemical are the factors which determine the type of exposure. The volume of the chemical, dilution, temperature and ventilation are the external factors which influence the degree of hazard. The knowledge of each of these factors is important in the assessment of possible danger to health.

If the toxic chemical is a gas, like carbon monoxide or a vapour like carbon tetrachloride, inhalation is the most important or the only route of uptake. The exception is hydrogen cyanide. Fumes and solid particles floating in the air are also taken up by inhalation. Evaporation depends on the vapour pressure of the chemical and on its temperature. The higher the temperature the higher is the rate of evaporation. Surface is a very important factor and that is why spilled solvents evaporate so rapidly.

From the physical condition of the chemical one cannot predict the volatility. Thus methylmercury chloride which is a solid, is 5.7 times more volatile than the liquid metallic mercury. Compared with an undiluted chemical, evaporation is less from its solution, though in an enclosed space the final air concentration depends only on the vapour pressure at a given temperature. Fumes in the laboratory mainly derive from acids and their formation is also temperature dependent (external heat or reaction heat). The grinding of solid chemicals, and all the solid chemicals in the form of powder, can form solid aerosols. Ventilation of the premises decreases inhalation exposure by the replacement of contaminated air by fresh air, though too strong a draught can increase evaporation or the formation of solid aerosols (dusting of the chemical).

Working processes as the source of exposure

A high proportion of exposure to toxic chemicals is of accidental origin. Fire promotes the evaporation of volatile chemicals, favours toxic fume formation (*eg* nitrous oxide) and toxic gases (carbon monoxide, phosgene). Breakage of glass containers and spillage of volatile liquids also results in increased evaporation.

Pipetting by mouth can result in swallowing of the toxic chemical, but if the chemical is volatile, vapours dissolve in the saliva or can be inhaled. Grinding of solids or measuring light powders in a draught increase the danger of the formation of solid aerosols. Inhalation exposure is a likely consequence when chemicals are heated outside a fume cupboard or in a fume cupboard with inadequate ventilation.

In analytical work chemicals are used in small quantities and thus the potential hazard is reduced, though for many purposes solvents, acids and alkalies are used in larger quantities, without dilution or in a concentrated form. If one considers only the health hazard, the preparation of a stock solution is always more dangerous than the final analytical reagent.

Distillation or other purification procedures and synthesis, are the working processes where larger quantities of toxic chemicals might be handled. All the known laboratory methylmercury poisonings affected chemists who synthesised the compound.

The fate and effects of toxic compounds after absorption

The process of distribution within the body

Without absorption by the organism, the toxic chemical cannot exert any toxic effect. Even chemical burns and irritation of the mucous membrane or skin is a result of penetration from the outside into the surface layer of the body but most chemicals act inside the body.

Once a chemical is absorbed, blood plays an important part in its distribution. In the blood the chemical is either transported in the red blood cells or in the plasma. Thus methylmercury or atomic mercury is mainly carried by the cells and inorganic mercury (Hg^{2+}) by the plasma. Interestingly, the brain uptake for methylmercury and atomic mercury is relatively higher than for inorganic mercury and it seems that the ability of a chemical to pass through the membrane of the red cells is related to its ability to cross other biological membranes.

The passage of chemicals from the blood into organs is a complex process. An uncharged molecule can diffuse easily through membranes, but other molecules usually react with physiological carriers. In some cases the reaction product is small enough to pass through membranes, in other cases, as the bile excretion of some compounds indicates, protein binds the chemical and delivers it to the liver. Diffusion and metabolically dependent transport are the two forms of transport. These two processes seem to contribute to the accumulation of mercury in the kidneys.

There are two special barriers in the body which provide protection against some toxic chemicals. Thus the so called blood–brain barrier is able to protect the brain against the uptake of compounds which are not lipid-soluble and are ionised. The difference in the brain uptake of metallic mercury and mercuric mercury is explained by this selective behaviour of the blood–brain barrier. The placental barrier located between the maternal and foetal circulation as far as the mercury compounds are concerned behaves similarly.

Exposure and damage

After a single or repeated administration of a toxic chemical, its concentration in various organs usually shows differences. Up to a certain concentration the function of an organ is not affected by the toxic chemical. However, as the concentrations in the organs increase, one organ will be the first to attain a concentration which affects its function. This organ is called the critical organ for the toxic compound. The concept of 'critical' implies that the critical effect initiated by the critical concentration in the critical organ will influence and dominate the clinical course of intoxication. For one toxic chemical there might be more than one critical organ. Acute exposure to cadmium fumes damages the lungs, long-term exposure to the same metal results in kidney damage. The critical organ is not necessarily the organ which will contain the highest amount or the highest concentration of the chemical. The main storage organ for lead is the bone, but the critical organ is the haemopoetic system which forms the blood and, in infants, the central nervous system. Irrespective of the chemical type of mercury, the kidney is always the organ which contains the highest concentration of mercury, but the kidney is the critical organ only for mercuric mercury. For metallic mercury or methylmercury the critical organ is the nervous system.

The identification of the critical organ for a toxic chemical is based on the dose effect relationship. Of the possible effects, the earliest one is selected which is crucial for the progress of intoxication. The prevention of this effect should result in the protection of the whole organism against the toxic effect. Threshold limits should aim to prevent the occurrence of this critical effect which means that the exposure should be low enough to prevent the accumulation of the toxic chemical to the critical level in the critical organ. Thus, the exposure to lead is not harmful if the haem synthesis is not affected.

Body burden and excretion

If exposure is repeated, the body burden increases until uptake and excretion are equal, that is, the body burden will be in a steady state condition. The time needed to reach the steady state depends on the extent of daily excretion of the toxic chemical in relation to body burden. If the toxic chemical is excreted completely between two working days it will not accumulate in the body from one working day to the next. Carbon monoxide is an example for this type of toxic compounds. Methylmercury has a half time in man of 70 days which assures that the steady state at constant level of daily exposure is reached approximately after one year. It might be that in different organs the half time is different. Thus the biological half time of lead in bone is about 10–20 years, but in blood and soft tissues it is only 20–30 days.

After a single dose or after the end of any length of exposure, the body burden declines. The decrease in the body burden is the result of the excretion of the chemical. The three main routes of excretion are exhalation, urinary and faecal excretion.

Exhalation is the route of excretion for those gases, such as carbon monoxide, which are not metabolised in the body. As a general rule every chemical which can be inhaled as vapour or gas can also be exhaled, and the concentration in the exhaled air depends on the exposure and on the rate of metabolism to non-exhalable compounds. However not only inhaled chemicals can be exhaled. Exhalation is very noticeable if a solvent is given to an experimental animal either orally or by injection.

For the majority of toxic chemicals excretion in the faeces or urine is the most important route and the non-volatile compounds are excreted only by these routes, though small quantities might appear in tears, sweat or milk.

Some part of the oral dose is excreted with the faeces without being absorbed. Only a small proportion of the toxic inorganic metal compounds are actually absorbed, though with an organic radical their absorption (like that of methylmercury) might be nearly complete. The chemical which is absorbed into the blood stream might occur in the gastrointestinal lumen through the normal shedding of the inner surface of the gut, but the most important source of faecal excretion is the bile. The biliary excretion after oral exposure is helped by the fact that all the blood which collects absorbed chemicals in the intestines first passes through the liver before it reaches the general circulation. Biliary excretion is influenced by the molecular weight of the compound and by its polarity. The optimum molecular weight is between 500 and 1000; between 300 and 500 mainly polar compounds are excreted. There are mechanisms in the liver which facilitate the biliary excretion of compounds by increasing their polarity and their molecular weight. A compound in the bile might be excreted with the faeces or might be reabsorbed again. The reabsorption of biliary methylmercury contributes to its long biological half time. If biliary excretion is the major route of elimination of the chemical from the body, diseases of the liver which affect the function of liver cells or obstruct the bile flow will increase the half time of the chemical and increase the hazard to health.

Extraction of chemicals by the kidneys is facilitated by the fact that one fourth of the blood pumped per minute into the circulation by the heart passes through the kidneys and about half of this is submitted to a filtration mechanism which is the first step in urine formation. Not all the chemicals which appear in the filtrate (tubular urine) are excreted, some diffuse into the tubular cells which line the tubules in which the pre-formed urine flows in the direction of the ureter. These special kidney cells take up chemicals from the tubular urine by diffusion and from the surrounding capillaries by diffusion and by active metabolic processes. Urinary excretion depends on the ability of the kidney to extract the chemicals from the blood, and on the ability of tubular cells to accumulate or release the toxic compound. The availability of small molecules which can complex the toxic chemical can increase the urinary excretion because they compete for the chemical with protein binding sites in the cells. Thus the excretion of heavy metals can be increased very significantly by the administration of complexing agents, such as the excretion of inorganic mercury by BAL or D-penicillamine, or the excretion of lead by EDTA.

Biotransformation

The absorbed chemicals are in contact with highly reactive biological compounds and catalytic systems. Metallic mercury is mainly oxidised in the blood, and it depends on the rate of oxidation as to how much highly diffusable metallic mercury reaches the brain. However, the main organ for the metabolic conversion of chemicals is the liver which can oxidise, hydrolyse, reduce or conjugate many toxic compounds. The result of these metabolic processes is usually a compound which can be more easily excreted in the bile or urine. However, this metabolic process can also change the toxic character of the compound.

The biotransformation might produce a more toxic derivative. Aniline is converted to *p*-aminophenol which is a potent methaemoglobin forming agent. Carbon disulphide reacts with amino compounds to give copper complexing dithiocarbamates. From carbon tetrachloride highly reactive free radicals are formed, which are at least partly responsible for its hepatotoxic effect. The toxicity of methanol resulting in the damage of the optic nerve is based on the formation of formaldehyde. Parathion is a weak acetylcholinesterase inhibitor but it is oxidised to a powerful inhibitor: paraoxon.

The metabolic processes which increase the toxicity of one compound, can decrease the toxicity of another. Butanol is oxidised through aldehyde to CO_2. Cyanide reacts with sulphur donors to give the innocuous thiocyanate. Conjugation makes phenols suitable for urinary excretion. The split of the carbon to mercury bond in phenylmercury or in alkoxyalkyl mercury gives mercuric mercury which is less dangerous to health than a stable organomercury compound, like methylmercury.

The problem of complex exposure

Exposure to a single chemical is not a rule but an exception. Exposure to more than one compound might not change the risk, but there are many examples which indicate some type of interaction. Exposure to two solvents might result in a more advanced narcotic effect or a more severe liver damage than exposure to one of them, as the effect of one is added to the effect of the other. This additive effect is very frequent. Thus direct methaemoglobin forming agents exert an additive effect on the formation of methaemoglobin, an inactive form of the oxygen carrying pigment in the blood. Many of the organophosphorus compounds have additive effects when they are given in combination. The critical organ or in a broader sense the critical biochemical mechanism is the same for the compounds with additive effects.

There is another form of interaction which is based on the effect of one compound on the metabolism of the other. In the body many chemicals are metabolised and the resulting compounds can be more or less toxic than the parent compound. If these metabolic processes are stimulated, the same exposure to a chemical can result in a higher or lower risk.

Phenobarbitone is a powerful compound which potentiates many metabolic processes on foreign chemicals. After phenobarbitone treatment carbon tetrachloride, carbon disulphide, chloroform or nitrosamine become more but aflatoxin less toxic. Chemicals which inhibit the enzymes responsible for the formation of a more toxic compound, have the opposite effect and they decrease the risk. Thus CS_2 or diethyldithiocarbamate decreases the toxicity of carbon tetrachloride.

There are other forms of interactions which do not fit into the normal biotransformation processes. Methaemoglobin forming agents protect against the toxic effect of cyanide. Interaction can change even the toxicity of metals, thus cadmium pre-treatment has a protective effect against the renotoxic doses of mercuric chloride, and selenium protects against cadmium and mercury and vice versa.

As interaction can change the risk imposed by a certain level of exposure, from the theoretical point of view it would be justified to adopt TLVs to fit the shift in the dose-response curve. However, our knowledge on the mechanisms of toxic effects, on the conditions of interactions and even the methods of measuring exposure are far from perfect to put into practice such corrections.

The consequences of toxic exposure

The symptoms and signs of intoxication are diverse and are neither always specific for, nor characteristic of a toxic chemical. However, as the exact meaning of toxicity includes the definition of a harmful effect in a biological system, like the human body, the following discussion might help to illustrate how the body responds to toxic chemicals. It also aims to help understanding of the meaning of toxic effects in relation to the listed chemicals in Chapter 8.

The respiratory tract and the lungs

The upper respiratory tract through the bronchial system, which, like the branches of a tree connect the trunk with the leaves, is connected with an intricate network of honeycomb-like airfilled spaces, the alveoli. Between the alveoli and the blood capillaries there is only a very thin membrane. This is the site of gas exchange between the inhaled air and the blood for the two physiological gases: oxygen and carbon dioxide, but this is also the place for the uptake and release of toxic gases.

Though the bronchial system takes no part in the physiological gas exchange, every water soluble compound can be dissolved in the watery layer of their mucous membrane. However, high water solubility offers a protection as the dissolved compound is more easily cleared from the lungs. Though the immediate irritative effect of the water soluble sulphur dioxide might be more pronounced than that of the poorly water soluble nitrogen dioxide, the latter is more likely to produce pulmonary oedema (a retention of water in the lungs). Lung oedema and respiratory irritation have a pronounced detrimental effect on gas exchange. If the respiratory irritation is too strong, the lumen of the bronchial tree constricts and increases the resistance to the respiratory airflow. Irritative chemicals such as sulphur dioxide, sulphuric acid mist, ozone, formaldehyde, phosgene, bromine, methyl bromide and nitrogen dioxide are all able to increase the respiratory resistance and many are able to cause oedematous changes in the lungs.

The bronchial tree has a mechanism which removes particles from the lungs. The damage to this promotes sensitivity to infections or dusts. These secondary effects with the primary irritant effect can lead to irreversible structural changes in the lungs manifested by a decrease in the surface suitable for gas exchange.

Skin and eyes
In laboratory circumstances the contact of the skin with chemicals is a frequent though not a necessary consequence of work. Compounds, which irritate the skin or mucous membranes in the vapour form, can cause more severe damage if splashed on the skin. Irritation is a general term which covers a multiplicity of processes: for example the removal of the lipid layer by solvents or alkaline detergents; dehydration of the skin by acids, anhydrides and alkalis; precipitation of proteins by heavy metals; oxidation by acids, peroxide or chlorine; the dissolution of keratin by soap, alkalis or sulphides. If the contact lasts longer or is repeated frequently the injury called contact dermatitis becomes more and more severe.

There are so called sensitisers which have the same effect as the primary irritants but the first signs appear only after 5–7 days exposure. Chromate, formalin, phenylenediamine and turpentine belong to this group.

Halogenated compounds act on the skin where there is discontinuity: chloracne is composed of plugged sebaceous glands and suppurative inflammation of the hair folliculi. There are substances such as coal tar, creosote, Rhodamine N and bergamot oil which sensitise the skin to sunlight.

Skin can be affected by a chemical taken orally or another way. Thus thallium results in the loss of hair and arsenic in the proliferation of the keratin layer.

Vapours which irritate the skin are more likely to irritate the conjunctiva or other mucous membranes. The irritative effect is more pronounced when the irritant is rubbed into the eye with a dirty hand, or acid, alkalis, lime, solvents or detergents are splashed into the eye. Chemical burns can affect the conjunctiva and the cornea. In the latter case scar or vascularisation as a part of the healing process might affect the visual function.

Chemicals can affect not only the surface of the eye: chronic exposure to dinitrophenol is able to cause opacity in the lens, thallium or methanol can damage the optic nerve.

Gastrointestinal tract and liver

The gastrointestinal tract can be the route of uptake of a toxic chemical, but can also be the site of action if corrosive chemicals are swallowed. Every chemical which can cause chemical burns on the skin, can injure the oesophagus and the stomach. Bleeding, perforation and deformation are the outcome of such a contact effect.

Other chemicals like arsenic, barium, chloromethane, fluorides or sublimates lead to gastroenteritis and increased motility of the intestines.

Liver is the first organ exposed to a chemical absorbed from the gastrointestinal tract. However, its vulnerability is based more on the extraordinary metabolic activity of the liver cells which can structurally modify many foreign chemicals. If the chemical is detoxified, the liver is the last organ to be exposed to the more toxic parent compounds; if the change is in the opposite direction, the liver is the first organ to be exposed to the more toxic forms.

The uptake of non-polar chlorinated hydrocarbons by the liver is facilitated by their lipid solubility, but their toxicity mainly depends on the metabolic transformation with the formation of highly reactive metabolic products.

Chemicals can injure the liver in many ways: they can increase its water content, fat content and they can destroy the cells. The liver has an ability to regenerate, that is, to replace the dead cells, but repeated toxic effects result in a liver full of scarring and unable to carry out the normal physiological functions. These functions contribute to the general metabolic processes of the whole organism (lipid, protein and sugar metabolism), the formation of the bile and the extraction of chemicals from the plasma for bile secretion. Hepatotoxic agents, besides causing morphological changes, can depress the metabolic activity of the liver, the bile flow and the excretion of bilirubin and other compounds. Bilirubin is a metabolic product of haem, the active part of haemoglobin and different haem enzymes. Approximately 300 mg bilirubin is formed per day (more in the case of the destruction of red blood cells) and a failure of its biliary excretion due to 1,1,2,2-tetrachloroethane, carbon tetrachloride or arsenic can cause severe liver damage resulting in jaundice. Nearly all the solvents have some potential for causing some degree of liver injury which might be noticed only if enzymes, leaked from the liver cells into the blood, are estimated or some functional tests are carried out. If the liver once suffers chemical damage or infective disease, its threshold to the effect of hepatotoxic agents is decreased.

Blood

The effect of a chemical on blood can be direct or indirect. In the former case the chemical acts on the blood itself and in the latter case on the formation of the formed elements or the plasma. The formed elements, red blood cells, white blood cells and platelets, are suspended in a volume ratio of 1: 1 in the plasma which is an aqueous solution of proteins, amino acids, salts and other small molecules. The main function of 4.5–5 million red blood cells per μl is the transport of oxygen from the lungs to the tissues. This function is carried out by a special protein called haemoglobin which makes up 97 per cent of the dry weight of these cells. The number of white cells is very much smaller, only 7000 per μl, their function is the defense against infection. The smaller platelets number 300 000 per μl contribute to another defensive mechanism: blood coagulation.

One of the most common effects of toxic chemicals on blood manifests itself in a change of the number of formed elements. There are so called haemolytic poisons, like arsine, phenylhydrazine or a large group of aromatic amino and nitro compounds which are able to destroy the red blood cells (haemolysis). In the case of arsine the effect can be so dramatic that the large quantity of haemoglobin released from the cells aggravates the effect of arsine on the liver and kidneys.

The aromatic amino and nitro compounds are able to oxidise the ferrous iron of the haem part of haemoglobin. This results biochemically in the formation of an inactive haemoglobin called methaemoglobin and clinically in a deteriorated supply of oxygen to the tissues. Sometimes methaemoglobin formation is useful, as methaemoglobin competes for cyanide with essential ferric enzymes in the oxygen utilisation system.

One of the most common forms of inactive haemoglobin is carboxyhaemoglobin, the reaction product of CO and haemoglobin. The inactivation is based on the affinity of CO to haemoglobin which is approximately 240 times higher than that of oxygen. Carboxyhaemoglobin is fully dissociable and when the CO concentration decreases in the inhaled air, CO is replaced on the haemoglobin by oxygen. The reactivation of methaemoglobin needs the participation of reductive enzymes present in the red blood cells.

In the case of lead intoxication the cause of anaemia is outside the blood. Lead interferes with the synthesis of haem, that is the non-protein part of haemoglobin. The first block in the synthesis of haem results in the increased urinary excretion of delta-aminolaevulinic acid, the substrate of one of the inhibited enzymes, the other step of interference results in the excretion of a degradation product (coproporphyrin). The estimation of these two compounds is widely used in the clinical diagnosis of lead intoxication or in the measurement of exposure to lead.

The best known toxic chemical affecting the white blood cells and platelets is benzene. Benzene can shift their numbers in two directions. At first the numbers of white blood cells and platelets decrease, for example, after ionising radiation. The most common manifestation at this stage is leakage of blood from the capillaries even in the case of minor trauma. If the exposure is high and long lasting, the effect of benzene can move in the opposite direction as a result of a malignant proliferation of the tissues which produce the blood cells.

Cell death caused in any part of the body, mostly in the liver, can leak enzyme proteins into the plasma, and thus their activity in blood can be used as a measure of tissue damage. Chronic liver damage caused by alcohol or other solvents might result in a disorder in the synthesis of plasma proteins with, as the first sign, a decrease in the albumin concentration. A general decrease in the concentration of plasma proteins might be the consequence of loss of proteins through the kidneys.

Nervous system

The nervous system is divided into two parts: the central and the peripheral. The central part is the brain and the spinal cord; the peripheral part consists of the motor and sensory nerves.

The uptake of chemicals by the central nervous system depends on the vascularisation of the brain and the characteristics of the chemicals. Non-polar, lipid soluble compounds, such as halogenated solvents easily penetrate the barrier between the blood and the brain tissue. Their effect, like that of chloroform, is anaesthetic–narcotic, that is, they depress the function of the central nervous system. This in severe cases leads to respiratory and/or cardiac failure. Though the biochemical mechanisms is quite different, the general effect of anoxaemic poisons is very similar to the narcotic ones. Chemical anoxaemia is caused either by the failure of the oxygen transport in the blood by the inactivation of haemoglobin (CO, aromatic amino or nitro compounds) or by the utilisation of oxygen in the tissues (cyanide, azide).

Though it is a solvent, carbon disulphide has a more selective effect on the central nervous system than other solvents. CS_2 interferes with the metabolism of catecholamines, which belong to those compounds which participate in the transmission of an impulse from one nerve cell to the next. This effect is more pronounced in a certain part of the brain, the same that is mainly affected by manganese. Cerebral oedema caused by triethyltin, encephalopathy caused by lead, ataxy, blindness, deafness, *etc* caused by methylmercury and behavioural disorders and tremor caused by mercury vapour are based on a more selective interaction between the toxic chemical and biochemical processes in the brain.

Many of the chemicals like CS_2 which affect the central nervous system also affect the peripheral nerves. Central nervous symptoms like drowsiness and headaches are the consequence of mild intoxication with anticholinesterase organophosphorus compounds, but some of them are able to cause delayed peripheral neuropathy in the form of paralysis. Triorthocresylphosphate though it is not a potent anticholinesterase inhibitor caused a mass epidemic of paralysis in Morocco when engine-oil treated with triorthocresylphosphate was sold and used as cooking oil. Acrylamide, a different type of compound, injures not only the motor but also the sensory nerves. Contact of the skin with phenol is able to cause a local loss of sensation which cancels the alarming effect of pain in the case of further contamination.

Kidneys
One of the main functions of the kidneys is the excretion of metabolic products (mainly from protein metabolism) and the regulation of the salt and water balance. Both processes are linked to the formation of urine which is also an important route for the clearance of toxic compounds and their metabolites. During the process of urinary excretion the chemicals filter, diffuse or are transported through cells in the kidneys, but some chemicals remain bound to cell components at least for a time.

Heavy metals, such as mercury, or chlorinated hydrocarbons like carbon tetrachloride, are the most prominent groups of compounds which can damage the kidney cells. The consequence is either a change in the composition or in the volume of urine or both. For example, in severe cases of mercuric chloride intoxication there might be no urine at all, in others, the occurrence of serum proteins in the urine, indicates a defect in the filter-system of the kidneys, which normally retains proteins. However, every cellular injury results in the excretion of cell fragments which increases the protein concentration of urine.

Heavy metals like cadmium, mercury, uranium or lead are also able to increase the excretion of amino acids, glucose, or phosphate which is the consequence of an impairment in the function of tubular cells to reabsorb these small molecules from the preformed urine. As the half time of cadmium is 10–40 years in the kidneys, a short term exposure to cadmium will affect the kidney content of cadmium for a lifetime. Even without occupational exposure, that is when cadmium is taken up only from food, the cadmium content of the kidneys increases up to 50–60 years of age. At this age probably the ageing process results in loss of kidney cells with their cadmium.

Excretion of toxic compounds with the urine is a powerful protective mechanism. In some cases the concentration of toxic chemicals in the urine exposes the lower part of the urinary tract to higher concentrations than from the blood. 2-Naphthylamine or benzidine proved to be so potent in causing bladder tumours by this way, that their use in the UK is now prohibited.

Cardiovascular system

Heart or respiratory failure is the cause of death in many acute intoxications. Chemicals such as carbon monoxide, phosgene, nitrogen dioxide, hydrogen sulphide, chlorinated compounds or solvents, which can injure cells in other organs can also injure the heart. In the not too severe cases of anoxaemia caused by CO, cyanide or methaemoglobin forming agents, the toxic effect is shown only by a decrease in the blood pressure or an increase in the frequency of heart beats.

Ageing of the cardiovascular system is one of the cardinal processes of the ageing in general. Lead, carbon monoxide or carbon disulphide are able to speed up atherosclerosis and they can aggravate the outcome if a partial blockage develops in the blood supply of the heart muscle.

Bone and muscles

The whole body is supported by the skeletal system. Loss of calcium by exposure to cadmium, when aggravated by nutritional deficiency, can result in the softening and deformation of the bones. The compression of nerves by deformed bones can be painful. In a district in Japan, river water used for irrigating paddy fields became contaminated with cadmium which caused an epidemic of the disease called Ouch Ouch. In contrast to cadmium, fluoride and phosphorus can increase the fragility of bones. Interestingly lead, which is mainly stored in the bones, has no similar effects.

In severe carbon monoxide intoxication the loss of muscle power may trap the victim at the site of exposure, as he will be unable to move.

Chapter 7

Medical Services and First Aid

Medical services

Throughout this book much emphasis is given to the importance of safety at work. To many people this conception conjures up images of hard helmets and other devices to protect against physical hazards, while the more insidious and sometimes greatly more dangerous effects upon health of exposure to toxic chemicals are often overlooked.

In the past there has been a conspicuous neglect of these dangers in chemical laboratories where a great variety of reagents are in daily use, some toxic and others innocent. Furthermore, whenever such hazards were recognised little guidance has been available to managers either on how to obtain medical advice in this field or on what precautions are necessary to protect the health of laboratory staff against the risks of acquiring an occupationally induced disease.

The purpose of introducing a section on medical services into the new edition of this book is to draw attention to some of the commoner health hazards which may threaten in a chemical laboratory and, more importantly perhaps, to give advice on where management may turn for help in protecting the well-being of their staff.

In 1973 the Employment Medical Advisory Service was established and is now incorporated as the medical arm of the Health and Safety Executive. The introduction of this service reflects the greatly increased concern demanded by society for the welfare of all persons at work and a laboratory manager can no longer afford to neglect his responsibility for ensuring a healthy working environment.

The doctors who staff the Employment Medical Advisory Service are all specialists in occupational health and are freely available for consultation on any aspects of medical care in the work place. A manager who has any doubts or queries either on the safety of his laboratory or on the first aid cover which he needs to provide should invite the nearest Employment Medical Adviser to visit the laboratory for an opinion on what measures should be taken. A list of the Regional Offices of this service and their addresses through which the nearest Employment Medical Adviser can be approached is given at the end of the chapter.

In the same way the local district inspector of factories can also give valuable advice, to ensure a safe working environment. Under the recent Health and Safety at Work Act both these officers have powers to enter a laboratory to examine the premises without notice and in the final analysis the district inspector has wide powers of action against any place of employment if he is not satisfied with the conditions which he finds. It is worth noting that both these officers may also be approached by any staff association or trades union or indeed by any single individual who is dissatisfied with the safety of his working conditions.

In conjunction with the Employment Medical Adviser the laboratory manager should consider whether there is a case for instituting regular medical checks on some or all of the laboratory staff. For example *benzene* which is a common bench reagent can induce a serious blood disease if regularly inhaled at low concentrations over long periods (*see* also Chapter 6 p 86). Apart from ensuring that it is only used under conditions of efficient extraction ventilation there may be a case for periodic blood examinations of all staff regularly handling benzene. Whether this is necessary and the frequency with which it needs to be done can be decided from the recommendations of the Employment Medical Adviser. Advice may also be needed on other *carcinogenic chemicals* in use, some of which may be covered by the Carcinogenic Substances Regulations of 1967 with which the laboratory manager should be familiar. A number of these are potent bladder carcinogens and the matter of obtaining screening tests of urine for cells suggestive of cancer should be considered (*see* also Chapter 6 p 76 and p 87). Medical advice may also be needed for laboratory staff handling *lead* compounds. Periodic estimations of blood lead or urinary excretion levels may be desirable to ensure there is no undue lead absorption. Assistance in obtaining and interpreting urinary and blood lead analyses will be given by the Employment Medical Adviser. In passing it should be noted that the most hazardous route of entry into the body for lead is by inhalation of fumes or mist; absorption through the skin can be ignored (*see* also Chapter 6 p 75).

Inorganic mercury poses a constant and insidious hazard in most laboratories. Mercury vaporises at room temperature and its inhalation over long periods will produce progressive and irreversible disease of the nervous system. Its physical properties are such that droplets of mercury can become widely and inaccessibly dispersed throughout a laboratory liberating pockets of surprisingly high vapour concentrations. These present a very real danger to the health of laboratory workers. Special benches should be constructed wherever mercury is handled to allow recovery of all spillages however small and all work with mercury should be performed in fume cupboards or under local extraction ventilation to prevent the escape of its vapour into the room. Bottles of mercury must always be stoppered when not in use and if any quantity is stored it should be in a room with good extraction ventilation.

The laboratory manager should discuss with the Employment Medical Adviser the question of routine urine monitoring for mercury workers as a measure of their mercury absorption. He will be given advice on whether this precaution is necessary, on which of the staff should be monitored and help in obtaining facilities to get the tests done and in the interpretation of the results.

These are just a few examples of the medical precautions which may be needed to preserve the health of staff at work in a laboratory. It is by no means an exhaustive list and the prudent manager should make full use of the advisory services which are available without charge to help him. He should also be aware of the hazards presented by the chemicals which characterise his laboratory.

Dermatitis

In all fields of employment dermatitis presents a problem of such magnitude and is the subject of so many misconceptions that it justifies an attempt to explain its causation in simple terms before outlining the principles of its prevention.

The size of the problem is illustrated by the following figures supplied by the Department of Health and Social Security for the year 1973–74. During this period 10 000 spells of certified sickness totalling 480 000 days of absence from work were reported as due to non-infected dermatitis of external origin.* To this massive total may be added a large but unrecorded number of cases not severe enough to warrant time away from work.

In striking contrast is the figure of 180 000 working days lost during the

* Prescribed disease No. 42, *ie* occupational dermatitis attracting the payment of industrial injury benefit.

same period by absences due to all the other 46 prescribed industrial diseases combined.

Laboratory technicians, exposed during their work to a great variety of chemicals each one potentially harmful to the skin, are especially vulnerable to occupational dermatitis which is the most frequent cause of medical disability arising out of their employment.

In the minds of many people 'dermatitis' is synonymous with occupational skin disease. Inside medical circles however the term means no more than 'inflammation of the skin'. It is not a disease entity in itself and its entry on medical certificates as a diagnosis without an indication of its cause should be avoided whenever possible. Regrettably such restraint is not always observed and leads to much confusion and disappointment when sick certificates carrying 'dermatitis' as a diagnosis are presented for industrial injury benefit.

Many causes of dermatitis have no connection with occupation and it is worth observing that skin disease which appears to be connected with work may have its origin in activities followed out of working hours or be a manifestation of some other more general medical condition.

In general terms, medical classification divides occupational skin disease into two main groups

1. Contact dermatitis

2. Sensitisation dermatitis.

Both are caused by contact with substances which for one reason or another interfere with normal skin physiology; the number which can be incriminated is almost infinite. Some are sufficiently powerful to damage the skin of all persons working with them if contact occurs. Others, usually regarded as innocuous, will give rise to skin disease if handled by an individual with an abnormal sensitivity.

Informed judgements may be offered on the likelihood of an unfamiliar substance inducing skin disease, based upon past experiences and a knowledge of its chemistry but in the final analysis only contact between worker and substance will supply the answer.

1. Contact dermatitis

This diagnosis is applied to an inflammation of the skin caused by exposure to a substance that attacks its surface. Depending upon the degree of damage suffered the inflammation may be acute or chronic and will not heal until contact with the offending agent is stopped. The distribution of the skin affected will be largely determined by the physical nature of the dermatitic agent. Liquids and solids tend to affect the skin of the hands and forearms while fumes or dusts produce a much more diffuse pattern usually involving the face and neck. After an attack of contact dermatitis the behaviour of the skin is unpredictable. Whether it will respond in the same manner to a repetition of contact will depend upon the concentration of the substance to which it is exposed, the duration of that exposure and its resistance to further attack. Small exposures may be tolerated without further trouble, and a return to work with a chemical known to have induced a previous attack of contact dermatitis is justified provided it is done with caution and excessive or prolonged exposure is avoided. After two recurrences all further exposure should be banned.

There are many well recognised skin irritants amongst the chemicals to be found in a laboratory though their action upon the skin of different individuals is not constant. Resistance of the skin to attack is the main factor in determining whether a contact dermatitis will or will not follow exposure. Because of their diversity and number it is impossible to give a comprehensive list of chemicals which are classed as irritants. They include all corrosive chemicals and most solvents because of their defatting and desiccating action on the skin. For this reason the practice of removing stains from the hands by cleaning with solvents is always to be deprecated. Other conditions which render the skin liable to attack are constant immersion in water, constant friction or pressure and the 'soggy' skin which develops when natural perspiration is unable to escape during prolonged use of rubber gloves.

As a general rule people with fair skins or red hair are more vulnerable to contact dermatitis than their darker colleagues.

Persons who suffer from any form of chronic non-occupational skin disease such as eczema should avoid taking up employment in a laboratory where they will be exposed to chemicals whose action will be potentiated by pre-existing disease.

2. Sensitisation dermatitis

Unlike contact dermatitis sensitisation dermatitis develops following an activation of the natural immunity of the skin. Superficially the events that take place may be compared with the response made by the body to vaccination against smallpox. In the latter there is the deliberate introduction of minute quantities of an infectious virus so that the body will marshal and prepare its defences to meet a much more serious threat if and when exposure to the disease itself occurs. Similarly in sensitisation dermatitis the offending compound must penetrate the skin where it becomes attached to one of its protein elements. This combination stimulates the immune defences present in the skin so that if further exposure to the same substance follows a reaction, manifested by damage and inflammation of the skin, takes place between the foreign chemical and the reinforced defences. It is this sequence of events that is implied by the word 'sensitisation'. A colourful analogy is the destruction inflicted upon a battlefield in the course of a battle.

Typically, a symptomless period varying from a few weeks to many months, during which time the immunological defences of the skin are being primed, follows first exposure to the sensitising compound. Once this priming has reached a certain stage of advancement, sensitisation dermatitis will break out if contact continues. It is not uncommon for a sensitising agent also to be a skin irritant producing a contact dermatitis by direct damage to the skin surface. When the two conditions coexist the diagnosis is difficult and treatment complicated.

Healing of the skin after sensitisation dermatitis is often prolonged and irregular and will only follow complete removal of the sufferer from any further exposure. Unlike contact dermatitis the area of skin affected is usually extensive and on occasions may involve the whole body surface. If for any reason contact with the sensitising agent continues, the skin may become so responsive that it reacts violently to its presence at almost molecular concentrations, and in extreme cases the subject may not be able to enter a room in which the sensitising agent is present without developing a recurrence.

From this brief and oversimplified description it follows that a diagnosis of sensitisation dermatitis is more serious in its implications than is one of contact dermatitis. Once the former diagnosis has been made, no further exposure should be permitted to the sensitising agent when a return to work is allowed and, if no other alternative is available, the only solution may be a change of occupation.

Many substances are recognised to be skin sensitisers but because of individual susceptibilities it is not possible to predict who is liable to become sensitised or who will remain unaffected. Another difficulty which characterises sensitisation dermatitis is its appearance in an individual who may have worked harmlessly for many years with a substance to which he suddenly becomes intolerant. One authority quotes the five most common causes of sensitisation dermatitis in order of frequency as:

1. Hair dyes (*p*-phenylenediamine)

2. Nickel compounds

3. Rubber compounds

4. Epoxy- and phenol-formaldehyde plastics

5. Antihistame skin medicaments derived from 1,2-diaminoethane

(*see* also Chapter 6, p 83)

Local preventative measures

It does not require much perspicacity to appreciate that the most effective protection against dermatitis is to avoid all contact with potential irritant or sensitising chemicals. This ideal cannot always be achieved but the objective of preventative measures is to approach it as closely as possible. Clean working conditions, properly planned bench operations and careful attention to the familiar and simple principles of skin hygiene all contribute to reducing the incidence of skin disease. Cuts and abrasions should be treated without delay and kept covered with a clean dressing until fully healed; in all this the help of an efficient first-aider is invaluable.

For work with corrosives or with other known irritants, impermeable gloves should be worn and care must be taken to ensure that they are intact. It is not however good practice to wear such gloves for too long since the skin of the hands becomes soggy and vulnerable to damage. Nor is it wise to allow laboratory gloves to remain in use for long without replacement or to permit their indiscriminate use by all members of the laboratory staff. To do so will merely encourage contamination inside the glove and help to spread skin infections from one user to the next.

Barrier creams

There are many varieties of barrier cream on the market for which enthusiastic claims are made by the manufacturers. They are harmless in themselves and may help to keep the skin in good condition, but there is little convincing evidence that they provide effective protection against irritant dusts or vapours. Against sensitisation reactions they offer no protection. Possibly their most useful function is to keep the skin clean and help in removing grease and oil stains.

Certainly they cannot replace more conventional measures such as good personal hygiene and careful working methods.

Treatment

Any persistent skin disease should be seen by a doctor and, if its cause is occupational, no resumption of contact with the causal agent should be allowed until healing is complete. If the condition proves to be a sensitisation dermatitis there should be a complete ban on any further work with the sensitiser unless or until medical approval to resume contact is obtained.

If an excessive incidence of dermatitis occurs in a laboratory and the cause is not apparent or easily eliminated the manager may always turn to the local Employment Medical Adviser for advice and assistance in detecting the cause and controlling exposure to it.

Occasionally the agent responsible for an attack of dermatitis cannot be identified and this ignorance may result in mystifying recurrences. In such cases patch testing of the skin under medical supervision with a selection of substances normally encountered at work is advisable but the physician conducting the tests needs special training in their technique and interpretation.

First aid

Before setting out to describe the arrangements for first aid which should be provided in any chemical laboratory it is as well to make a few introductory remarks which all persons responsible for organising first aid facilities and their application should consider.

In a working environment, wherever it may be or whatever the tasks performed inside it, there are broadly speaking two categories of incident for which first aid may be required:

1. Incidents arising from natural causes, and

2. Incidents arising as a direct result of the work performed.

Those in the first of these two categories are by and large unavoidable and are as likely to occur inside the work-place as outside it. Familiar examples range from a simple faint through the scale of common but acute illnesses up to the serious emergency presented by a heart attack. For all of them there are the appropriate first aid measures which in extreme cases may be life-saving. However, apart from ensuring that trained first aid assistance is readily available there is little else that can be done in the way of precautionary measures to prevent their occurrence. Their treatment is not the concern of this chapter and to learn about them the first aider must turn to *The first aid manual*, published jointly by the St John and St Andrews Ambulance Association and the Red Cross.

By contrast those incidents in the second category are avoidable and must be viewed as the direct result of a failure to observe the principles of safe working conditions. Since common things happen most commonly such incidents are usually of such a trivial nature that no treatment is called for and they are dismissed as of no significance; only rarely will they be of such a serious nature that urgent first aid is necessary. Nevertheless whatever their outcome such accidents always indicate a need to examine and perhaps correct the method of work from which they arose. Against incidents of this kind first aid is only a second line of defence and the provision of a first aid service however comprehensive or well prepared must never be allowed to supplant constant supervision and improvement of working methods by all members of a laboratory staff. The theme of working safely has already been the subject of preceding chapters of this book but its re-emphasis in the context of first aid can do no harm.

In the UK only laboratories operating as part of a factory complex are included in legislation prescribing the provision of first aid facilities (Factories Act 1961). No regulations have yet been made under the Health & Safety at Work Act imposing a legal obligation to provide trained first aiders in those laboratories which do not form part of a factory complex but there is little doubt that any Factory Inspector will regard the availability of adequate first aid facilities as an integral part of the Act's overall requirement to provide a safe working environment. This view adds further emphasis to the need for organising an efficient first aid service in every laboratory.

Because of the tremendous variation in the size of laboratories and the multifarious nature of the work on which they are engaged it is not possible to offer in a single short chapter advice which can be universally applied without consideration of individual circumstances. For this reason any laboratory manager who is unsure of the scale of first aid cover appropriate to the operations for which he is responsible would be wise to invite the nearest Employment Medical Adviser to visit the laboratory and act upon the recommendations which he makes.

The number of trained first aiders available in any single laboratory must obviously depend on the numbers of staff employed and upon the dangers presented by the work on which it is engaged, but as a basic rule at least one fully trained first aider, usually recruited from the staff, should be on duty at all times during working hours. It is equally important to ensure that everybody employed in the laboratory should know who is the nominated first aider and where is his or her usual place of work. This information can be displayed on a board easily visible in every room where bench work is taking place.

In this context it should not be forgotten that there will be times when the nominated first aider will be away from work. For such absences a deputy needs to be appointed who should undergo the same training as the colleague for whom he substitutes. If a particularly hazardous operation such as work with hydrogen cyanide or hydrofluoric acid is to be undertaken the laboratory manager should consider whether to make it a rule that no work commences unless a first aider is on duty.

Finally, it is worth pointing out that even in silent hours there may be cleaners at work who are not familiar with the chemicals by which they are surrounded and who are at special risk because of their isolation. Their safety is the responsibility of the laboratory manager.

Training of the first aider
Like every other skill the practice of first aid can only be learned by training and practice. None of the many wall-charts and simple diagrams available for display illustrating the treatments recommended for first aid emergencies can in any way approach the level of knowledge and confidence acquired through a properly conducted course of training. For many years the St John Ambulance Association, St Andrews Ambulance Association and the Red Cross have offered such courses and there will certainly be an opportunity for any prospective first aider to attend one without having to travel far from his home or place of work. Every first aider should also be in possession of *The first aid manual*. Information on these courses and where they are held is obtainable from local branches of the three bodies.

In the same way the latter offer refresher courses for already certificated first aiders in order to keep their knowledge up to date. Any laboratory manager is well advised to take advantage of these facilities in order to reassure both himself and his staff that when needed first aid will be given by a qualified person.

The objective of this training is not to produce instant doctors but rather to turn out a person able to assess the seriousness of an emergency rapidly and apply with confidence a few basic manoeuvres designed to prevent a deterioration in the casualty's condition until more professional medical attention can be obtained. The essence of its success is speed and simplicity of action combined with confidence and common sense. The measures to be learned are few but the first aider must be familiar with them and be prepared to use them in difficult and sometimes cramped physical surroundings. The well ordered conditions in which training sessions take place are rarely repeated on those occasions when the first aider is called upon to exercise his skills in earnest.

Equipment

All industrial laboratories which form part of a factory complex are required by law to maintain a prescribed number of first aid boxes or cupboards which must be clearly labelled and readily accessible. Regulations require their contents to be as follows:

1. A copy of the leaflet giving advice on first aid treatment (form F1008) issued by the Health and Safety Executive and available from HMSO.

2. A sufficient number (not less than 12) of small sterilised unmedicated dressings for injured fingers.

3. A sufficient number (not less than 6) of medium-sized sterilised unmedicated dressings for injured hands or feet.

4. A sufficient number (not less than 6) of large sterilised unmedicated dressings for other injured parts.

5. A sufficient number (not less than 24) of adhesive wound dressings of an approved type and of assorted sizes.

6. A sufficient number (not less than 4) of triangular bandages of unbleached calico, the longest size of which measures not less than 51 inches (128 cm) and each of the other sides not less than 36 inches (90 cm).

7. A sufficient supply of adhesive plaster.

8. A sufficient supply of absorbent sterilised cotton in half-ounce packets.

9. A sufficient supply of approved eye-ointment in a container of an approved type and size.

10. A sufficient number (not less than 4) of sterilised eye-pads in separate sealed packets.

11. A rubber bandage or pressure bandage.

12. Safety pins.

As yet there is no legal requirement for laboratories which are not covered by the provisions of the Factories Act to install first aid boxes or cupboards but the Health and Safety at Work Act allows regulations to be made on any matter relating to the health of persons at work, and it is likely that in the future their provision will become mandatory in all places of work. Even without this likely development it is right that laboratories should provide one or more suitably stocked first aid boxes. They should be placed in the care of the nominated first aider who should ensure that the contents are always kept up to standard. Often the latter may disappear because of unauthorised borrowings but the remedy is not to keep the boxes locked or hidden in a safe place to preserve their inviolability.

The only item which can be confidently omitted from a non-statutory box is the rubber bandage or pressure bandage, the correct use of which calls for advanced medical knowledge if its application to control heavy bleeding is not to make the casualty's condition worse.

In addition to first aid boxes there should always be one or more eyewash-bottles strategically placed about the laboratory and filled with at least 450 cm³ (16 ozs) of clean water. Many suitable varieties of bottles are available commercially or one can simply be constructed from standard laboratory equipment (*see Fig. 7.1*). However, prevention is better than cure and eye protection in the form of goggles should always be readily available for use to protect against splashes entering the eyes of staff handling liquid reagents, Indeed the laboratory manager would be wise to make a cardinal rule that no member of the staff should handle liquid chemicals whatever their nature without wearing some form of eye protection.

Like the boxes, each bottle should be in the care of the nominated first aider who will be responsible for ensuring that the water is changed at least once a month. A number of sophisticated buffer solutions have been recommended for use in eyewash-bottles but they should be avoided and have no advantage over clean water. It is important that not only the first aider but all members of the laboratory staff are familiar with the proper use of eyewash-bottles since speed of action to wash out the eye is of overriding importance in the treatment of eye injuries occasioned by splashes.

First aid boxes and eyewash-bottles are the only essential pieces of first aid equipment that should always be provided. There is on the market a large range of more sophisticated equipment such as oxygen-giving sets and bellows designed to administer artificial respiration. Whether these are added to the first aid armoury must depend largely on the size of the laboratory, the discretion of the laboratory manager and the opinion of his advisers, but the prudent director should be wary of purchasing expensive equipment from which there will be little or no practical benefit.

Last, but by no means least, it is incumbent upon every manager to ensure that all reagents in the laboratory are clearly labelled.

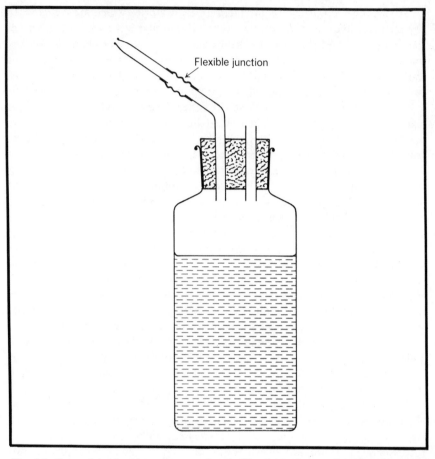

Flexible junction

Fig. 7.1. Eyewash-bottle. Note the use of a flexible dispensing nozzle. Over-enthusiastic irrigation in the heat of the moment with an unyielding nozzle can inflict added damage to the eye.

Treatments

It is not the purpose of this chapter to replace the St John Ambulance first aid training manual and therefore the section in the previous edition of this book devoted to illustrating methods of artificial respiration has been omitted deliberately. These techniques are an essential part of the first aider's knowledge, but along with other basic practical measures can only be learned from an experienced first aid instructor. It cannot be emphasised too strongly that every first aider should attend a suitable training course before he can be expected to exercise his skills with any degree of competence.

Despite the importance of nominating and training individual members of the staff in the skills of first aid it should be impressed upon all members of the laboratory staff that each of them should be prepared to provide first aid assistance if called upon to do so, and it is therefore wise for everyone to be familiar with a number of simple and effective treatments which can be applied without specialised training. Such measures are described below under the heading of 'Standard Treatments'; each is designed to prevent a casualty's condition from worsening until more skilled aid is to hand, and should be seen as a 'team' responsibility in which every member of the laboratory staff has a part to play.

Most accidents with chemicals which are likely to occur in a laboratory will fall into one of the following categories:

1. Splashes of the skin (including chemical burns).

2. Splashes of the eyes.

3. Inhalation of gases or dusts.

4. Ingestion of chemicals.

5. Burns.

Since there are few specific antidotes against the great majority of chemicals, the standard treatments about to be described are designed for the type of accident irrespective of the chemical involved. Thus all splashes of the skin should be treated in the same way. Similarly each of the other four groups of accident calls for its own particular treatment. In Chapter 8 there are brief monographs on the hazardous properties and effects upon the body of many of the chemicals which are commonly used in laboratories. Under each one reference is made to first aid treatment. When only the entry 'Standard Treatment' is made the reader should turn to the page indicated against it where a summary of the correct first aid treatment will be found. It is essential that these simple standard treatments are applied with the minimum of delay.

Standard treatments

1. Splashes of the skin (For summary *see* p 108)

In general dangerous splashes are either corrosive (acids and alkalis) inflicting chemical burns, or are toxic by absorption through the skin (phenol, aniline). In all such emergencies continuous drenching with water and removal of all contaminated clothing is the essential treatment. All chemicals must be regarded as hazardous unless proved otherwise and if any doubt exists the drenching must be started immediately before further information is sought. In such circumstances it is better to err on the side of caution at the expense of the victims's discomfort rather than allow a potentially hazardous chemical time to exert its action. Similarly, modesty in removing contaminated clothing must not be allowed to interfere with the treatment.

Drenching should continue for up to five minutes or until the first aider has satisfied himself that no further danger exists. This treatment must take priority over all other actions. Water is safe, normally available in large quantities and is simple to apply. In an emergency these features make it the treatment of choice over all other remedies which may be advocated. However, if the substance is known to be insoluble in water gentle cleaning of the surface with soap while the drenching continues will help to remove the splash. If it is necessary to refer the casualty to hospital details of the treatment and the chemical responsible should be provided.

2. Splashes of the eye (For summary *see* p 109)
Here again, the same principles apply as recommended under splashes of the skin. Flooding with large quantities of water is invariably the treatment of choice and should be continued for at least 10 minutes or longer if considered necessary. Strong alkalis are particularly damaging to the eye and must be treated with great care. A special feature of eye injuries is the spasm which may develop in the lids to keep them firmly shut. To overcome this the first aider must gently prise the lids apart with his fingers to ensure the water bathes the eyeball and continue to hold them apart until the treatment is completed. Flooding with water is best done by holding the casualty's head under a gently running tap or by using one of the specially provided eyewash-bottles. Care must be taken not to use a powerful jet of water which may cause added damage to the eye. Regard to the casualty's clothing must not interfere with the thoroughness of the treatment. All injuries to the eye require medical attention and if removal to hospital is arranged information on the chemical responsible with brief details of the treatment given should accompany the casualty. As a precaution against this type of accident protective eye-shields or glasses should be used whenever liquid chemicals are in use.

3. Inhalation hazards (For summary *see* p 109)
Of the routes by which toxic chemicals can enter the body, inhalation is the most dangerous and rapidly acting. Fortunately most toxic gases are either acutely irritating to the tissues of the respiratory tract or possess a warning odour which is detectable at concentrations well below the danger level. This last property however is an unreliable indicator of their presence since the nose quickly becomes insensitive to smell and a serious casualty may result if the initial warning is ignored. In all cases of gassing the essential treatment is to remove the subject into fresh air. Provided breathing has not stopped and the casualty is conscious there is little else that needs to be done except to maintain careful observation on his condition. If breathing has stopped artificial respiration should be administered until it resumes naturally or medical attention becomes available. In cases where hydrogen cyanide is the

gas responsible, mouth-to-mouth artificial respiration must *NOT* be employed because of danger to the rescuer. In its place the Silvester method must be used (for details see *The first aid manual*). Oxygen may be given to all cases of gassing if a giving set is available and the first aider trained in its use. If the victim is unconscious he must be placed in a face-down position in order to maintain a clear airway to his lungs and prevent the inhalation of vomit or foreign bodies such as false teeth into the breathing spaces. In all cases of serious gassing or if the first aider is not happy about a casualty's condition removal to hospital must be arranged. Information on the gas responsible and brief details of the treatment given should accompany the casualty.

4. Ingestion (For summary *see* p 109)
This route of entry presents the least hazard since it is unlikely that any significant quantity of a harmful liquid or solid will be swallowed without deliberate intent. However, improbable accidents do occur though normally the offending chemical will be spat out and not swallowed. If it is corrosive, burns of the mouth may result and for such injuries repeated mouth-washes with water is the correct treatment. In this context it must be emphasised that pipetting by mouth is a practice that should be absolutely forbidden in any properly supervised laboratory.

If a poisonous substance has been swallowed it should be diluted by giving quantities of water or milk to drink and the casualty moved to hospital as rapidly as possible. It is important in such cases to supply information on the chemical involved with brief details of the treatment given and an estimate of the quantity consumed.

The question of inducing vomiting is a difficult one. In general, it should not be attempted since it may result in further damage to the delicate tissues of the upper food passages if the poison swallowed is corrosive. Tickling the back of the throat or drinking strong solutions of common salt are the traditional methods recommended but these are not without their own risks and more often than not are unsuccessful.

5. Burns
Chemical burns from corrosive reagents should be treated as advised under the standard treatment for splashes of the skin.

Dry burns from bench fires and scalds from boiling water or hot chemicals are a familiar hazard even in the most elementary of laboratories and their treatment is described below.

Depending upon the gravity of the burn the damage inflicted will vary from superficial reddening of the skin to extensive surface blistering and death of underlying tissues. However serious, the correct first aid treatment is no more than covering the burnt surface with loosely applied dry sterile dressings.

These form part of the standard contents of all first aid boxes. If blisters are present every effort should be made not to burst them. To reduce the dangers of infection, handling of the burnt area must be reduced to a minimum and any temptation to clean its surface must be resisted. All burns or scalds of more than a trivial nature should be referred to hospital.

If the victim's clothes are on fire the flames must be extinguished either by drenching with water or by throwing the victim to the ground and smothering them with any garment that comes to hand such as the rescuer's lab coat. It is hardly necessary to emphasise that these moves must be resolute and swift.

Special treatments

There are a number of chemicals whose properties are so dangerous that specific antidotes have been developed against them. Brief details of the treatments recommended are to be found under their respective monographs in Chapter 8. One or two of them however are in sufficiently common use to warrant an amplified description of the preparations needed and the treatment to be given if a casualty occurs. Where this is the case a reference to the appropriate page number of the expanded account is made in the summarised version to be found in Chapter 8.

In all laboratories where dangerous chemicals of this order are handled it is a wise precaution to set up a liaison with the nearest hospital to agree on a plan of action for the immediate admission and treatment of a casualty. The first aider should also make himself familiar beforehand with the measures he must take on the spot if the need arises.

Hydrogen cyanide
The immediate urgency is to remove the casualty well away from the neighbourhood of the gas. To do this the rescuer must ensure that he also does not fall victim to its action. If life is to be saved it is unlikely that there will be time to don a breathing apparatus but whenever possible a cord or similar device should be attached to his waist so that he can be dragged clear if he succumbs. Obviously circumstances will dictate the manner of the rescuer's actions but it must be emphasised that inadequate preparations and unsuccessful heroics will worsen the emergency. If after reaching fresh air the casualty is still conscious and breathing he must be made to lie down quietly and await removal to hospital. No other treatment is required apart from keeping a close observation upon his general condition. Only if the breathing stops should the following measures be adopted:

1. Loss of consciousness
Place in the prone position with the mouth down. Ensure a clear breathing passage.

2. Breathing arrested

Apply artificial respiration by the Silvester method. Do not use the mouth-to-mouth method because of the danger of inhaling hydrogen cyanide gas from the victim's lungs. Cardiac massage may be necessary if the casualty's heart has stopped beating.

3. Amyl nitrite*

Break two amyl nitrite capsules in quick succession beneath the victim's nose so that the vapour is inhaled. If the breathing has stopped this must be done by an assistant during the inhalation cycle of the artificial respiration.

4. Kelo-Cyanor resuscitation kits

These packs are specifically designed against cyanide poisoning and are available from retail chemists. They include a specific antidote for intravenous injection, and one or more packs should be provided by the laboratory either for on-the-spot use by trained personnel or to accompany the casualty to hospital. The question of who is entitled to administer an intravenous injection in an emergency of this gravity is a contentious one. It may be possible for a nursing sister or an experienced first aider to receive proper training in this technique from a local doctor or the accident department of the nearest hospital. Whether it is necessary to prepare in this way will depend upon the degree of hazard present in a laboratory and the discretion of the laboratory manager, but any arrangements to do so must always be made in conjunction with a doctor who, for an emergency of this gravity, is willing to accept responsibility in writing for the competence and judgement of the person selected to give the intravenous injection.

Soluble cyanides

Cyanide salts and solutions of cyanides present considerably less risk than does hydrogen cyanide gas since only by ingestion are they dangerous and ingestion is unlikely to occur in significant quantities except by deliberate intention; poisoning by inhalation of these preparations is not a hazard. This however does not mean they should be treated without due circumspection. Solutions of over 1 per cent strength should always be kept under strict control; those under 1 per cent constitute, in practice, little danger but always demand careful handling.

If ingestion of cyanide solution does take place, amyl nitrite should be administered as for the gas and the casualty rapidly moved to hospital.

In the past it has been the practice to prepare as cyanide antidotes

---

* Note: Amyl nitrite capsules are available through retail chemists. They have a limited stock life and note must be taken of the expiry date for renewal. They should be stored at temperatures not exceeding 15 °C (60 °F).

solutions A (ferrous sulphate)* and B (sodium carbonate)* to be taken by mouth against cyanide ingestion. This remedy has now been superseded by the Kelo-cyanor treatment but there can be no objection to their combined use if the laboratory manager prefers after consultation with medical advice.

There should be an effort to estimate the quantity drunk and a note made of its concentration to obtain some idea of the amount of cyanide consumed. Only if the patient has collapsed and is in extremis should the question of an intravenous injection of 'Kelo-cyanor' be considered. Although cyanide can be absorbed through the skin splashes of cyanide solutions do not constitute a major emergency; nevertheless they should be washed off immediately with large volumes of water.

The Health and Safety Executive have prepared a special wall-poster against cyanide poisoning (No. SHW 385 revised). Copies are available through HMSO offices and give clear instructions on treatment. The CIA has also prepared a label, No. 15, which gives details of first aid care and advice on medical treatment. It is intended for accompanying the victim to hospital. Wherever there is a significant risk of cyanide poisoning in a laboratory either by inhalation or ingestion these notices should be available and the former prominently displayed.

Nitriles (To identify dangerous nitriles refer to entry in Chapter 8)

Hydrogen fluoride and hydrofluric acid

Hydrogen fluoride and its concentrated or dilute solutions in water, which we know as hydrofluoric acid, inflict destructive and extremely painful burns on any tissue with which they come into contact. With concentrated preparations these effects are rapidly apparent but in dilute solutions they are often delayed and may not be noticed for a number of hours. All laboratory staff who are likely to handle hydrogen fluoride or hydrofluoric acid should be warned of this insidious property and instructed to take immediate action if they are splashed.

Because of its serious nature any laboratory handling hydrogen fluoride or the acid regularly should set up a liaison with the local hospital to arrange for

---

* A. 158 grams ferrous sulphate crystals ($FeSO_4.7H_2O$) and 3 grams BP citric acid crystals in a litre of cold distilled water (the solution must be inspected regularly and be replaced if any deterioration occurs).
  B. 60 grams anhydrous sodium carbonate ($Na_2 CO_3$) dissolved in a litre of distilled water.
  50 ml of solution A is placed in a 170 ml (6 oz) wide-necked bottle closed by a polythene-covered cork and labelled clearly 'CYANIDE ANTIDOTE A'. 50 ml of solution B is similarly bottled and labelled 'CYANIDE ANTIDOTE B'. Both bottles should bear the legend 'Mix the whole contents of bottles "A" and "B" and swallow the mixture'.
  The merit of the basic ferrous hydroxide suspension that is swallowed is that it is likely to induce vomiting while at the same time forming insoluble non-toxic iron complexes with the cyanide.

the treatment and admission of casualties rather on the same lines as recommended for cyanide poisoning. The symptoms which may start immediately or may be delayed commence as a dull throbbing which builds up to become a particularly severe and persistent pain due to death of the underlying tissues. If not treated it can result in extensive and permanent damage which may involve the underlying bone. Accompanying the pain there may be a visible reddening of the damaged skin.

The first aid treatment consists of removing all contaminated clothing and flooding the skin with large volumes of running water. This should be continued for at least one minute. Thereafter 2 per cent calcium gluconate gel* should be applied liberally to the affected areas and continuously massaged into the skin for at least 15 minutes or until medical aid becomes available. Should the symptoms recur a further application of the gel should be made in the same way and continued until the pain subsides. If the acid has penetrated below the nails the ointment should be liberally applied over and around the nail and the area continuously massaged for at least 15 minutes.

All hydrogen fluoride or hydrofluoric acid splashes of any size must be referred to hospital after washing the skin and starting treatment with the ointment. The CIA has issued a label No. 14 newly revised in 1975 which explains clearly and precisely how to treat hydrofluoric burns and all laboratories which handle the acid should have ready a special fluoride pack containing the ointment and a copy of the revised CIA label. A pack should always accompany the casualty to hospital. The Health and Safety Executive has produced a wall-poster (F2250) on the treatment of hydrofluoric acid burns. It is obtainable through any HMSO office and should be displayed prominently in the vicinity of any work with hydrogen fluoride.

Splashes of hydrogen fluoride or hydrofluoric acid in the eyes are particularly painful and dangerous; they must be treated immediately by flooding for at least 15 minutes with large volumes of running water from a tap. (For details of treatment *see* Splashes of the eye p 102.)

The urgent removal of the patient to hospital is essential since failure to treat correctly may all too easily result in the loss of an eye.

Although there are no special treatments available against them there are two compounds commonly encountered in chemical laboratories which deserve special mention because of their lethal properties. They are:

Phenol

Treatment for phenol splashes is the same as the standard treatment for Splashes of the skin given on p 10. Phenol is caustic and rapidly absorbed

---

* 2 per cent gluconate gel may be obtained from Industrial Pharmaceutical Service Ltd, Hampden Road, Sale, Cheshire, who are the sole suppliers.

through the skin. If a large surface is splashed there is a real danger of death from collapse and kidney damage. It is essential to remove all contaminated clothing and flood with water for at least 15 minutes. The casualty must be removed to hospital promptly and information given on the details of the accident.

Hydrogen sulphide

Hydrogen sulphide gas is commonly encountered in many laboratories. In concentrations above a few hundred ppm it can be rapidly fatal. There is no specific antidote against it and first-aid treatment is the same as that recommended for Inhalation accidents (*see* p 102). Strict safety precautions must be in operation whenever it is generated.

There are, of course, a number of other chemicals with equally dangerous properties but since they are not part of the normal stock-in-trade of chemical laboratories they are not listed in this chapter. Appropriate warnings against them are found under the monographs of individual chemicals given in Chapter 8.

Summary of standard treatments

Splashes of the skin

If the chemical responsible is hydrogen fluoride, hydrofluoric acid or a related compound turn to p 106 for instructions.

1. Flood the splashed surface thoroughly with large quantities of running water and continue for at least 10 minutes, or until satisfied that no chemical remains in contact with the skin. Removal of splashes with solvents, solutions and chemicals known to be insoluble in water will be facilitated by the use of soap.
2. Remove all contaminated clothing, taking care not to contaminate yourself in the process.
3. If the situation warrants it, arrange for transport to hospital or refer for medical advice to the nearest doctor. Provide information to accompany the casualty on the chemical responsible and brief details of the first aid treatment given.

Splashes of the eye

If the chemical responsible is hydrogen fluoride, hydrofluoric acid or a related compound turn to p 106 for instructions.

1. Flood the eye thoroughly with large quantities of gently running water either from a tap or from one of the eyewash-bottles provided and continue for at least 10 minutes.
2. Ensure the water bathes the eyeball by gently prising open the eyelids and keeping them apart until the treatment is completed.
3. All eye injuries from chemicals require medical advice. Arrange transport to hospital and supply information to accompany the casualty on the chemical responsible and brief details of the treatment already given.

Inhalation of gases

If the gas responsible is hydrogen cyanide turn to p 104 for instructions.

1. Remove the casualty out of the danger area after first ensuring your own safety.
2. Loosen clothing. Administer oxygen if available.
3. If the casualty is unconscious place in a face-down position and watch to see if breathing stops.
4. If breathing has stopped apply artificial respiration by the mouth-to-mouth method. If the gas responsible is HCN *only use* the Silvester method.
5. If the emergency warrants it remove the patient to hospital and provide information on the gas responsible with brief details of the first aid treatment given.

Ingestion of poisonous chemicals

1. If the chemical has been confined to the mouth give large quantities of water as a mouth wash. Ensure the mouth wash is not swallowed.
2. If the chemical has been swallowed give copious drinks of water or milk to dilute it in the stomach.
3. Do not induce vomiting.
4. Arrange for transport to hospital. Provide information to accompany the casualty on the chemical swallowed with brief details of the treatment given and if possible an estimate of the quantity and concentration of the chemical consumed.

Regional offices of the employment medical advisory service

Region	*Address*
Northern	Government Buildings, Kenton Bar, Newcastle upon Tyne NE1 2YX Telephone 0632 863411
North Eastern	87 Manningham Lane, Bradford BD1 3TB Telephone 0274 32903
Eastern and SE Midlands	4 Dunstable Road, Luton LU1 1DX Telephone 0582 415722
London and South Eastern	Atlantic House, Farringdon Street, London EC4A 4Ba Telephone 01 583 5020 (Ext 303)
South Western	Beacon Tower, Fishponds Road, Fishponds, Bristol BS16 3HA Telephone 0272 659573
Wales	St Davids House, Wood Street, Cardiff CF1 1PB Telephone 0222 43984
West Midlands	Auchinleck House (5th Floor), Broad Street, Birmingham B15 1DL Telephone 021 643 8441/4
North Western	Quay House, Quay Street, Manchester M3 3JE Telephone 061 832 7137
Scotland	Portcullis House, India Street, Glasgow G2 4NR Telephone 041 221 3828

Hazardous Chemicals

Chapter 8

Hazardous Chemicals

This chapter includes monographs describing briefly the hazardous properties and effects upon the human body of over 460 flammable, explosive, corrosive and/or toxic substances or groups of substances commonly used in chemical laboratories; it also recommends first aid and fire-fighting procedures in the case of accidents and suggests methods of dealing with spillages.

In contrast to the same chapter in the first edition, it now incorporates references to dangerous reactions or preparations of many of these substances, and a large number of other materials that can be classed loosely as 'research chemicals', with other reagents or under specified physical or experimental circumstances. The publication of *Handbook of reactive chemical hazards* by L. Bretherick (Butterworth, 1975) has made this possible, and the inclusion in this book of his chapter 'Reactive Chemical Hazards' has added further to its scope. With Mr Bretherick's approval, this chapter also acts as an index to much of the information about actual incidents outlined in his important work. The letter (B) followed by a page number, after the name of the substance, indicates a reference to the *Handbook* page, and also to the page number of useful generalisations about the hazards of groups of compounds (*eg* acetylenic peroxides, halogen azides, tetrazoles) in its first section.

Monographs have been confined to chemicals in fairly general use in laboratories. Sophisticated chemicals important only to specific industries (*eg* agriculture or plastics manufacture) are generally well catered for in the literature automatically supplied with these products by the manufacturers. The fact that other chemicals commonly used in laboratories are not included does not mean that they are harmless. To devote space to the large number of substances with low hazard properties would greatly increase the size of the chapter, thereby reducing the speed with which information can be found in it.

The nomenclature used is based upon that currently recommended by the International Union of Pure and Applied Chemistry (IUPAC). Cross-references are used extensively to cover the survival of many older but well-established names, and synonyms are also given after the main monograph title. Since all the chemicals are arranged in alphabetical order, their names do not appear in the Subject Index.

Properties

In a book of this kind, compromises must be made in the interests of clarity, conciseness and speed of reference. These characteristics are vital in an emergency. The temptation to list more of the physical properties of a toxic material than will assist rapid identification by the person confronted with a casualty has been resisted. Boiling points of liquids and melting points of low-melting solids have been included along with short qualitative descriptions of appearance, volatility and odour when these apply. Water plays a vital role in first aid measures and disposal procedures, and reference is therefore made to the solubility, miscibility or reactivity of the material with water.

As a guide to those planning to use volatile or dusty hazardous chemicals regularly in some laboratory operation, the threshold limit values (TLV) accepted by the American Conference of Governmental Hygienists (1976) have been included where these have been determined. These values, expressed as parts of gas or vapour per million parts (ppm) of contaminated air by volume at 25 °C and 760 mmHg pressure, and also as milligrams per cubic metre (mg m^{-3}) of air, and as mg m^{-3} for dusts, 'represent conditions under which it is believed that nearly all workers may be repeatedly exposed day after day, without adverse effect'. An eight-hour working day is implied. The word 'skin' in brackets after the TLV figure indicates that absorption by direct contact or absorption of airborne vapour or dust by skin, mucous membranes and eyes can occur, and that protection against poisoning through these routes has been assumed to have been provided in arriving at the figure. It should be noted that TLV figures are reviewed annually and are therefore liable to be changed from time to time.

Warnings

Capital letters are used to draw attention immediately to the main risks associated with the material being described. In the first edition, these warnings followed closely the wording adopted in *Marking containers of hazardous chemicals* (Chemical Industries Association, 4th Edition 1970) but United Kingdom entry into the European Economic Community has required the adoption of the EEC 'Dangerous Substances Directive' based on the multilingual Yellow Book *Dangerous chemical substances and proposals concerning their labelling* published by the Council of Europe. Where chemicals with monographs are included in the list of chemicals covered by the Directive, they now carry the 'user' risk phrases required by this; the warnings for other chemicals have been based on the same phrases, although their use is not enforced by the Directive.

The suppliers of laboratory chemicals within the EEC are required by this Directive, in respect of the substances specified therein, to put on each container a label bearing the name of the material, its origin, the danger symbol(s) (*see Fig. 8.1*), and the risk phrases. Safety advice phrases, comparable with the injunctions used in this book, are also prescribed in the Directive, but their inclusion on labels of small containers is still optional. If a chemical is covered by the UK Pharmacy and Poisons Act 1933, and later legislation, it must also carry the warning POISON.

Difficulties with the warning words used for flammable substances in the UK have not yet been resolved. Although it is widely agreed that 'flammable' is a better word than 'inflammable' to use for materials that burn readily in air, 'inflammable' remains the word used in the older legislation, *eg* The Petroleum (Consolidation) Act 1928. The EEC Directive wordings for degrees of flammability have been adopted in the monographs, as have the ranges of flash point defining those degrees. Thus, FLAMMABLE substances are those with a flash point between 21 °C and 55 °C, HIGHLY FLAMMABLE substances have flash points between 0 °C and 21 °C, and EXTREMELY FLAMMABLE substances have flash points below 0 °C and a boiling point below 35 °C. According to the Highly Flammable Liquids and Liquefied Gases Regulations 1972, liquids with a flash point below 32 °C are designated as HIGHLY FLAMMABLE in the UK. This unfortunate difference will have to be accepted until the issue is settled by international agreement.

The injunctions that follow the risk phrases do not always use EEC safety advice phraseology as it is thought useful to keep the distinction between the

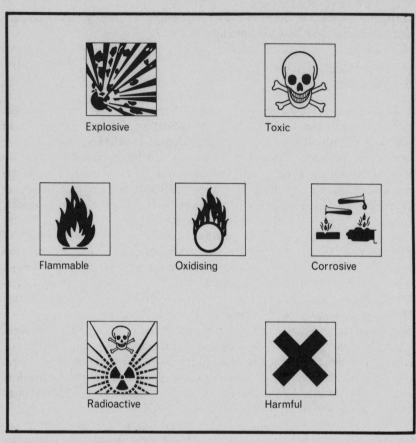

Fig. 8.1. EEC directive warning symbols.

words 'avoid' and 'prevent', the latter being used to indicate that inhalation, eye or hand protection is important.

Toxic effects

The new chapter 'Chemical hazards and toxicology' has provided a contemporary interpretation of the words 'toxic' and 'toxicity', and suggested a sensible attitude towards the risks of poisoning that may arise in a laboratory.

The use of less familiar medical terms has been avoided as far as possible in describing the toxic effects of harmful materials. Where the information is available and relevant to laboratory operations, chronic effects (those produced by prolonged exposure to small amounts of toxic materials) are included as well as the acute (short-term) effects. The effects are usually listed in each monograph in the order inhalation, eye contact, skin contact, mouth contact and ingestion.

Whereas the toxicologist uses six grades of toxicity (see Chapter 6, p 70) the simple warning requirements of labelling call for a two-degree classification of the chemicals into 'toxic' or 'harmful' substances or alternatively into 'corrosive' or 'irritant' substances. According to the Council of Europe *Yellow book* toxic substances are those offering a serious risk of acute or chronic poisoning by any route while 'harmful' calls attention to a risk which, although minor is nevertheless real. Corrosive substances are those which destroy living tissues; irritant substances are those which can cause inflammation.

The literature drawn upon in trying to present a reasonable account of toxic effect does not always cover all the possible accidents that might occur. This is particularly true of mouth contact and ingestion and where evidence of such incidents is lacking certain effects are assumed to be probable. The rash use of an ordinary pipette to measure out a volume of a particular corrosive acyl chloride may not have occurred in the past, but one day it will—and it seems wise to anticipate probable effects and prescribe commonsense first aid treatment for a case of mouth contact and swallowing. A simple aspirator bulb must be used to avoid this particular risk in pipetting corrosive or poisonous liquids.

The use of protective clothing, gloves, goggles, breathing apparatus and so on is discussed in the chapter 'Planning for safety' (p 24). The level at which the quantity of a dangerous material that is being handled necessitates the wearing of protective clothing and equipment will always be a matter of debate. Although eye protection may be insisted upon, how many university laboratories would require a student to wear rubber gloves, let alone a respirator, when he is pouring a few cm^3 of aniline into a test-tube? Teaching laboratories are the schools for gaining first-hand experience in handling dangerous materials at a low level of risk.

Because the level of risk varies so widely in different laboratories, it is not possible to prescribe protective wear under each individual chemical; injunctions are qualified by using the word 'prevent' instead of 'avoid' when the danger from, say, skin absorption is of a high order, and the wearing of gloves becomes more important. Familiarity with the chapter 'Planning for safety', together with the warnings and injunctions for the material to be handled, should tell the laboratory worker or his supervisor when and what protection is appropriate. However, in a few of the monographs reference is made to special protection that is important.

Hazardous reactions

Normally, a brief indication of how an incident occurred is followed by a page reference to Bretherick's *Handbook of reactive chemical hazards*—for example (B625)—which takes the reader back to the original report of the incident.

First aid

The general first aid principles that have been adopted, and the practice recommended in each monograph, have been described at some length in the preceding chapter which should be read and understood by those likely to use this book in an emergency. The emphasis that Chapter 7 rightly places on sound general first aid training for representatives of laboratory staff has led us to drop the repetitive presentation of first aid procedures in the monographs that were characteristic of the first edition. The time has come for the routine procedures for dealing with inhalation, eye, skin and ingestion casualties to be part of trained first aid knowledge, to be imparted to those in the laboratories for whom they act. A limited number of special procedures remain and are recorded.

Fire hazards

Chapter 4 has dealt fully with the nature of the risks involved in using flammable materials, precautions that can be taken to minimise these, and the methods of dealing with a fire once it has broken out. The short note **Fire hazard** in the monograph on each flammable substance provides three figures (if available):

(a) flash point (closed cup unless otherwise stated),

(b) explosive limits in terms of the range of percentage flammable component in a mixture with air that presents an explosion risk, and

(c) ignition temperature, which is the minimum temperature required to initiate or maintain self-sustained combustion independent of the source of heat.

The booklet *Laboratory waste disposal manual* published by the Manufacturing Chemists' Association has been the main source for the figures presented for these properties.

The note concludes with recommendations for the types of extinguishant that can be used in fighting a fire involving the substance concerned. Here we have drawn extensively from Fire Service Circular No 6/1970 issued by the Home Office.

Spillage disposal

Recommendations for dealing with spillages of hazardous materials have been based largely on the procedures suggested in the chart *How to deal with spillages of hazardous chemicals* published by BDH Chemicals Ltd and, to a lesser extent, on the Manufacturing Chemists' Association's booklet referred to above.

In interpreting these recommendations, the following points should be noted:

1. A suitable non-flammable dispersing agent, and hand pump sprays for dealing with small spillages of water-insoluble liquids are commercially available. It is generally desirable to use one volume of the dispersing agent for every two volumes of flammable, water-insoluble liquid that has been spilt, together with 10 volumes of water. When this has been worked to an emulsion with a brush and run to waste, diluting greatly with running water, there is no risk of a flammable vapour mixture developing in a drainage system. Less dispersant is needed with non-flammable water-insoluble liquid spillages. This procedure lends itself to laboratory spillages of up to $2\frac{1}{2}$ litres (Winchester quart) proportions.

2. Generally speaking, local authorities will accept small quantities of various chemicals into their sewage systems, provided these are adequately diluted so as not to interfere with the purification system. It is important to know the acceptable dilutions in the case of poisonous materials such as arsenic, lead and mercury compounds. In no case should the amount of water-insoluble flammable liquid thus disposed of be sufficient to create an explosion hazard.

3. It is usually most convenient to allow volatile liquids of low or moderate toxicity to evaporate to atmosphere in an area where no nuisance can be caused by lachrymatory or foul-smelling materials. Lachrymatory materials should be either destroyed (alcoholic sodium or potassium hydroxide will often do this) or buried deeply. The great majority of harmful materials are rendered innocuous on burial by hydrolysis or slow dispersion in the soil by the action of rain water; these processes are assisted if the spillage has been absorbed on sand. However, it is important that the burial place selected should not be close to an area draining into a water supply system if large amounts of spillage are involved.

Abietates metal

Hazardous reactions   Finely divided abietates of Al, Zn, Ca, Na, Co, Pb and Mn are subject to spontaneous heating and ignition (B76).

Acetal – *see* 1,1-Diethoxyethane

Acetaldehyde (ethanal)

Colourless liquid with pungent fruity odour; miscible with water; bp 21 °C.

EXTREMELY FLAMMABLE
MAY FORM EXPLOSIVE PEROXIDES

Avoid breathing vapour. Avoid contact with eyes. TLV 100 ppm (180 mg m^{-3}).

Toxic effects   The vapour causes irritation of eyes, headache and drowsiness. The liquid, if swallowed, causes severe irritation of the digestive organs. **Chronic effects** Repeated inhalation of vapour can cause delirium, hallucinations, loss of intelligence, *etc,* as in chronic alcoholism.

Hazardous reactions   Extremely reactive with acid anhydrides, alcohols, halogens, ketones, phenols, amines, ammonia, hydrogen cyanide and hydrogen sulphide; exothermic polymerisation with acetic acid; fires and explosions with air, oxygen and hydrogen peroxide under various circumstances (B307–308).

First aid   *Vapour inhaled:* standard treatment (p 109).
Affected eyes: standard treatment (p 109).
If swallowed: standard treatment (p 109).

Fire hazard   Flash point −38 °C; explosive limits 4–57%; Ignition temp. 185 °C. *Extinguish fire with* water spray, dry powder, carbon dioxide or vaporising liquids.

Spillage disposal   Shut off all possible sources of ignition. Instruct others to keep at a safe distance. Wear breathing apparatus and gloves. Mop up with plenty of water and run to waste diluting greatly with running water. Ventilate area well to evaporate remaining liquid and dispel vapour.

Acetic acid

Colourless liquid with pungent acrid odour; bp 118 °C; glacial acetic acid freezes to a crystalline solid in cool weather, miscible with water.

FLAMMABLE
CAUSES BURNS

Avoid breathing vapour. Avoid contact with eyes and skin. TLV 10 ppm (25 mg m^{-3}).

Toxic effects   The vapour irritates the respiratory system. The vapour irritates and the liquid burns the eyes severely. The liquid is very irritating to the skin and can cause burns and ulcers. If taken by mouth causes internal irritation and damage.

Hazardous reactions   Causes exothermic polymerisation of acetaldehyde; violent or explosive reactions with oxidants BrF_5, CrO_3, $KMnO_4$ or Na_2O_2 (B310).

First aid   *Vapour inhaled:* standard treatment (p 109).
Affected eyes: standard treatment (p 109).
Skin contact: standard treatment (p 108).
If swallowed: standard treatment (p 109).

Fire hazard   Flash point 43 °C; explosive limits 4–16%; ignition temp. 426 °C. *Extinguish fire with* water spray, dry powder, carbon dioxide or vaporising liquids.

Spillage disposal   Shut off all possible sources of ignition. Wear face shield and gloves. Mop up with plenty of water and run to waste diluting greatly with running water. Ventilate area well to evaporate remaining liquid and dispel vapour.

Acetic anhydride

Colourless liquid with strong acrid odour; bp 140 °C; reacts slowly with water to form acetic acid.

FLAMMABLE
CAUSES BURNS

Avoid breathing vapour. Prevent contact with eyes and skin. TLV 5 ppm (20 mg m^{-3}).

Toxic effects   The vapour irritates the respiratory system. The vapour irritates and the liquid burns the eyes severely, with delayed damage. The liquid irritates and may burn the skin severely, with blistering and peeling. If swallowed, causes immediate irritation, pain and vomiting.

Hazardous reactions   Vigorously oxidised by CrO_3, metal nitrates, nitric acid, $KMnO_4$; violent reactions with glycerol/$POCl_3$, perchloric acid/water, and with water when added slowly to mixture with acetic acid (B398–400).

First aid   *Vapour inhaled:* standard treatment (p 109).
Affected eyes: standard treatment (p 109).
Skin contact: standard treatment (p 108).
If swallowed: standard treatment (p 109).

Fire hazard   Flash point 54 °C; explosive limits 3–10%; ignition temp. 380 °C.
Extinguish fire with water spray, dry powder, carbon dioxide or vaporising liquids.

Spillage disposal   Shut off all possible sources of ignition. Instruct others to keep at a safe distance. Wear breathing apparatus and gloves. Absorb on sand, shovel into bucket(s), transport to safe place and tip into large volume of water; leave to decompose before decanting water to waste, diluting greatly with running water. Site of spillage should be ventilated after washing thoroughly with water and soap or detergent.

Acetone

Colourless, mobile liquid with characteristic odour; bp 56 °C; miscible with water.

HIGHLY FLAMMABLE
IRRITATING TO EYES

Avoid breathing vapour. Prevent contact with eyes. TLV 1000 ppm (2400 mg m^{-3}).

Toxic effects Inhalation of vapour may cause dizziness, narcosis and coma. The liquid irritates the eyes and may cause severe damage. If swallowed may cause gastric irritation, narcosis and coma.

Hazardous reactions Vigorously oxidised by air in the presence of active carbon, nitric/sulphuric acid mixtures, BrF_3, Br_2, nitrosyl chloride, nitrosyl perchlorate, nitryl perchlorate, chromyl chloride, CrO_3, F_2O_2, nitric acid, hydrogen peroxide, peroxomonosulphuric acid; violent reactions with bromoform or chloroform and base, SCl_2 (B362–363).

First aid *Vapour inhaled:* standard treatment (p 109).
Affected eyes: standard treatment (p 109).
If swallowed: standard treatment (p 109).

Fire hazard Flash point −18 °C; explosive limits 3–13%; ignition temp. 538 °C. *Extinguish fire with* water spray, dry powder, carbon dioxide or vaporising liquids.

Spillage disposal Shut off possible sources of ignition. Wear face shield and gloves. Mop up with plenty of water and run to waste diluting greatly with running water. Ventilate area well to evaporate remaining liquid and dispel vapour.

Acetone cyanohydrin
– *see* 2-Hydroxy-2-methyl propiononitrile

Acetonitrile (methyl cyanide)

Colourless, volatile liquid with similar odour to acetamide; bp 80 °C; miscible with water.

HIGHLY FLAMMABLE
SERIOUS RISK OF POISONING
BY INHALATION OR SWALLOWING
GIVES OFF POISONOUS VAPOUR

Do not breathe vapour. Avoid contact with skin and eyes. TLV 40 ppm (70 mg m^{-3}).

Toxic effects Inhalation of vapour may cause fatigue, nausea, diarrhoea and abdominal pain; in severe cases there may be delirium, convulsions, paralysis and coma.
Evidence is lacking on the effects of skin absorption and ingestion, but these may be similar to those resulting from inhalation.

Hazardous reactions Violent or explosive reactions with dinitrogen tetraoxide (in presence of indium), *N*-fluoro compounds, nitric acid, sulphuric acid (B297).

First aid	*Vapour inhaled:* standard treatment (p 109). *Affected eyes:* standard treatment (p 109). *Skin contact:* standard treatment (p 108). *If swallowed:* standard treatment (p 109).
Fire hazard	Flash point 6 °C; explosive limits 4–16%; ignition temp. 524 °C. *Extinguish fire with* foam, dry powder, carbon dioxide or vaporising liquids.
Spillage disposal	Shut off all possible sources of ignition. Instruct others to keep at safe distance. Wear breathing apparatus and gloves. Mop up with plenty of water and run to waste diluting greatly with running water. Ventilate area well to evaporate remaining liquid and dispel vapour.

Acetylacetone – *see* Pentane-2,4-dione

Acetyl azide

Hazardous reactions	Treacherously explosive (B299).

Acetyl bromide

Colourless to yellow liquid; bp 77 °C; decomposed by water with formation of hydrobromic acid and acetic acid.

CAUSES BURNS
IRRITATING TO SKIN, EYES AND RESPIRATORY SYSTEM

Avoid breathing vapour. Prevent contact with skin and eyes.

Toxic effects	The vapour irritates all parts of the respiratory system. The vapour irritates the eyes severely. The liquid burns the skin and eyes. If taken by mouth, there is immediate and severe internal irritation and damage.
First aid	*Vapour inhaled:* standard treatment (p 109). *Affected eyes:* standard treatment (p 109). *Skin contact:* standard treatment (p 108). *If swallowed:* standard treatment (p 109).
Spillage disposal	Instruct others to keep at a safe distance. Wear breathing apparatus and gloves. Spread soda ash liberally over the spillage and mop up cautiously with water — run to waste, diluting greatly with running water.

Acetyl chloride

Colourless, fuming, volatile liquid with a pungent odour; bp 51 °C; rapidly decomposed by water with formation of hydrochloric acid and acetic acid.

HIGHLY FLAMMABLE
CAUSES BURNS
IRRITATING TO SKIN, EYES AND RESPIRATORY SYSTEM

Avoid breathing vapour. Prevent contact with eyes and skin.

Toxic effects The vapour irritates all parts of the respiratory system. The vapour irritates the eyes severely. The liquid burns the skin and the eyes. If taken by mouth, there is severe internal irritation and damage.

Hazardous reactions Violent decomposition in preparation from PCl_3 and acetic acid; violent reactions with water, dimethyl sulphoxide (B294).

First aid *Vapour inhaled:* standard treatment (p 109).
Affected eyes: standard treatment (p 109).
Skin contact: standard treatment (p 108).
If swallowed: standard treatment (p 109).

Fire hazard Flash point 4 °C; ignition temp. 390 °C. *Extinguish fire with* water spray, foam, dry powder, carbon dioxide or vaporising liquids.

Spillage disposal Shut off all possible sources of ignition. Instruct others to keep at a safe distance. Wear breathing apparatus and gloves. Spread soda ash liberally over the spillage and mop up cautiously with water — run to waste, diluting greatly with running water.

Acetylene

Colourless gas; bp −83 °C; commercial gas has garlic-like odour due to impurities; slightly soluble in water.

FORMS VERY SENSITIVE EXPLOSIVE METALLIC COMPOUNDS
HIGHLY FLAMMABLE

Avoid breathing gas.

Toxic effects Inhalation of gas may cause dizziness, headache, nausea. Mixed with oxygen it can have narcotic properties but it is primarily an asphyxiant.

Hazardous reactions The gas explodes, alone or mixed with air, under various circumstances; contact with copper results in formation of explosive copper acetylide; reacts violently or explosively with Br_2 and Cl_2; explodes on contact with potassium (B285–287).

First aid *Vapour inhaled:* standard treatment (p 109).

Fire hazard Explosive limits 3–82%; ignition temp. 335 °C. Since the gas is supplied in a cylinder, turning off the valve will reduce any fire involving it; if possible cylinders should be removed quickly from an area in which a fire has developed.

Acetylene dichloride
– *see* **1,2-Dichloroethylene**

Acetylene tetrabromide
– *see* **1,1,2,2-Tetrabromoethane**

Acetylene tetrachloride
– *see* 1,1,2,2-Tetrachloroethane

Acetylenic compounds Review of class (B3).

Acetylenic peroxides Review of group (B4).

Acetyl nitrate

Hazardous reactions Thermally unstable and may decompose violently or explode under a variety of circumstances (B299).

Acetyl nitrite

Hazardous reactions Unstable and liable to explode (B298).

Acraldehyde (acrolein, acryl aldehyde, propenal)

Colourless to yellow, volatile liquid with pungent, choking odour; bp 53 °C; somewhat soluble in water.

EXTREMELY FLAMMABLE
GIVES OFF POISONOUS VAPOUR
IRRITATING TO SKIN, EYES AND RESPIRATORY SYSTEM

Avoid breathing vapour. Prevent contact with eyes and skin. TLV 0.1 ppm (0.25 mg m^{-3}).

Toxic effects The vapour irritates all parts of the respiratory system and may cause unconsciousness. Short exposure may cause pain to the nose and eyes in addition to intense irritation. Assumed to have extremely poisonous and irritant action if taken by mouth.

Hazardous reactions Liable to polymerise violently, especially in contact with strong acid or basic catalysts (B353).

First aid *Vapour inhaled :* standard treatment (p 109).
Affected eyes : standard treatment (p 109).
Skin contact : standard treatment (p 108).
If swallowed : standard treatment (p 109).

Fire hazard Flash point −26 °C; explosive limits 3–31%; ignition temp. 278 °C. *Extinguish fire with* water spray, dry powder, carbon dioxide or vaporising liquids.

Spillage disposal Shut off all possible sources of ignition. Instruct others to keep at a safe distance. Wear breathing apparatus and gloves. Mop up with plenty of water and run to waste, diluting greatly with running water. Ventilate area well to evaporate remaining liquid and dispel vapour.

Acrylamide

White crystalline solid; soluble in water; may polymerise with violence on melting (85 °C).

HARMFUL IN CONTACT WITH SKIN
IRRITATING TO SKIN, EYES AND RESPIRATORY SYSTEM

Avoid breathing dust. Avoid contact with skin and eyes. TLV (skin) (0.3 mg m^{-3}).

Toxic effects Irritates the skin and eyes. Expected, from animal experiments, to affect the central nervous system as a result of skin absorption. (Ch 6 p 87)

First aid *Affected eyes:* standard treatment (p 109).
Skin contact: standard treatment (p 108).
If swallowed: standard treatment (p 109).

Spillage disposal Wear face-shield or goggles, and gloves. Mop up with plenty of water and run to waste diluting greatly with running water.

Acrylic acid

Colourless solid and liquid with acrid odour; mp 14 °C; bp 141 °C; miscible with water.

FLAMMABLE
CAUSES BURNS
IRRITATING TO SKIN, EYES AND RESPIRATORY SYSTEM

Avoid breathing vapour. Prevent contact with skin and eyes.

Toxic effects Severely irritates the skin, eyes and respiratory system in concentrated solutions or as a liquid; assumed to be extremely irritant if taken by mouth.

Hazardous reaction Liable to polymerise violently.

First aid *Affected eyes:* standard treatment (p 109).
Vapour inhaled: standard treatment (p 109).
Skin contact: standard treatment (p 108).
If swallowed: standard treatment (p 109).

Fire hazard Flash point 52 °C; ignition temp. 429 °C. *Extinguish fire with* water spray, dry powder, carbon dioxide or vaporising liquids.

Spillage disposal Shut off all possible sources of ignition. Wear face-shield and gloves. Mop up with plenty of water and run to waste, diluting greatly with running water. Ventilate area well to evaporate remaining liquid and dispel vapour.

Acrylonitrile (vinyl cyanide)

Colourless, volatile liquid with ethereal odour; bp 77 °C; 1 part dissolves in about 15 parts water at 25 °C.

HIGHLY FLAMMABLE
SERIOUS RISK OF POISONING BY INHALATION OR SWALLOWING
GIVES OFF POISONOUS VAPOUR

Prevent inhalation of vapour. Prevent contact with skin and eyes. Do not expose to flame or acids. TLV (skin) 20 ppm (45 mg m^{-3}).

Toxic effects Vapour may cause dizziness, nausea and unconsciousness. The liquid may cause dermatitis and acute effects by absorption. Poisonous if taken by mouth. Action similar to cyanides. **Chronic effects** Low concentrations of vapour inhaled over a long period may cause flushing of face, nausea, giddiness and jaundice.

Hazardous reactions Violent or explosive polymerisation promoted by a variety of reagents including strong acids and bases, Br_2 and silver nitrate (B348–350).

First aid Special treatment for hydrogen cyanide (p 104).

Fire hazard Flash point 0 °C (open cup); explosive limits 3–17%; ignition temp. 481 °C. *Extinguish fire with* water spray, dry powder, carbon dioxide or vaporising liquids.

Spillage disposal Shut off all possible sources of ignition. Instruct others to keep at a safe distance. Wear breathing apparatus and gloves. Mop up with plenty of water and run to waste diluting greatly with running water. Ventilate area well to evaporate remaining liquid and dispel vapour.

Acyl azides Review of group (B4).

Acyl nitrates Review of group (B5).

Acyl nitrites Review of group (B6).

Alkali metal derivatives of hydrocarbons
Review of group (B6).

Alkali metals Handling and uses (B6).

Alkenes Reactions with oxides of nitrogen (B7).

Alkylaluminium derivatives Review of class (B8).

Alkylboranes Review of group (B10).

Alkyl hydroperoxides Review of group (B10).

Alkylmetal halides Review of class (B11).

Alkylmetals Review of group (B11).

Alkyl nitrates Review of group (B12).

Alkylnon-metal halides Review of class (B13).

Alkylnon-metal hydrides Review of group (B13).

Alkylnon-metals Review of group (B14).

Alkyl perchlorates Review of group (B15).

Allene – *see* Propadiene

Allyl alcohol (prop-2-en-1-ol, vinyl carbinol)

Colourless liquid with pungent odour; bp 97° C; miscible with water.

HIGHLY FLAMMABLE
GIVES OFF VERY POISONOUS VAPOUR
IRRITATING TO SKIN, EYES AND RESPIRATORY SYSTEM

Prevent breathing of vapour. Prevent contact with skin and eyes. TLV (skin) 2 ppm (5 mg m^{-3}).

Toxic effects	The vapour irritates eyes and respiratory tract severely. The liquid irritates the eyes and skin. Causes severe irritation of digestive organs and kidneys if swallowed.
First aid	*Vapour inhaled:* standard treatment (p 109). *Affected eyes:* standard treatment (p 109). *Skin contact:* standard treatment (p 108). *If swallowed:* standard treatment (p 109).
Fire hazard	Flash point 21 °C; explosive limits 3–18%; ignition temperature 378 °C. *Extinguish fire with* water spray, dry powder, carbon dioxide or vaporising liquids.
Spillage disposal	Shut off all possible sources of ignition. Instruct others to keep at a safe distance. Wear breathing apparatus and gloves. Mop up with plenty of water and run to waste diluting greatly with running water. Ventilate area well to evaporate remaining liquid and dispel vapour.

Allyl benzenesulphonate

Hazardous reaction	Explosion of vacuum distillation residue (B569).

Allyl bromide (3-bromopropene)

Colourless liquid with an unpleasant smell; bp 71 °C; almost insoluble in water.

> EXTREMELY FLAMMABLE
> GIVES OFF VERY POISONOUS VAPOUR
> IRRITATING TO SKIN, EYES AND RESPIRATORY SYSTEM

Prevent breathing of vapour. Prevent contact with eyes and skin.

Toxic effects The vapour irritates all parts of the respiratory system and may cause dizziness and headache. The vapour and liquid irritate the eyes. Assumed to be very irritant and poisonous if taken by mouth.

First aid *Vapour inhaled:* standard treatment (p 109).
Affected eyes: standard treatment (p 109).
Skin contact: standard treatment (p 108).
If swallowed: standard treatment (p 109).

Fire hazard Flash point −1 °C; explosive limits 4–7%; ignition temp. 295 °C. *Extinguish fire with* water spray, foam, dry powder, carbon dioxide or vaporising liquids.

Spillage disposal Shut off all possible sources of ignition. Instruct others to keep at a safe distance. Wear breathing apparatus and gloves. Apply non-flammable dispersing agent if available and work to an emulsion with brush and water — run this to waste, diluting greatly with running water. If dispersant not available, absorb on sand, shovel into bucket(s) and transport to safe open place for atmospheric evaporation or burial. Site of spillage should be washed thoroughly with water and soap or detergent.

Allyl chloride (3-chloropropene)

Colourless, volatile liquid with unpleasant, pungent odour; bp 45 °C; immiscible with water.

> EXTREMELY FLAMMABLE
> GIVES OFF POISONOUS VAPOUR

Prevent breathing of vapour. Prevent contact with skin and eyes. TLV 1 ppm (3 mg m^{-3}).

Toxic effects The vapour irritates all parts of the respiratory system and inhalation may cause headache, dizziness and, in high concentrations, unconsciousness. The vapour and liquid irritate the eyes. Muscular pains may follow skin absorption. Assumed to be very irritant and poisonous if taken by mouth.

First aid *Vapour inhaled:* standard treatment (p 109).
Affected eyes: standard treatment (p 109).
Skin contact: standard treatment (p 108).
If swallowed: standard treatment (p 109).

Fire hazard Flash point −32 °C; explosive limits 3–11%; ignition temp. 392 °C. *Extinguish fire with* water spray, foam, dry powder, carbon dioxide or vaporising liquids.

Spillage disposal Shut off all possible sources of ignition. Instruct others to keep at a safe distance. Wear breathing apparatus and gloves. Apply non-flammable dispersing agent if available and work to an emulsion with brush and water — run this to waste diluting greatly with running water. If dispersant not available, absorb on sand, shovel into bucket(s) and transport to safe open place for atmospheric evaporation or burial. Site of spillage should be washed thoroughly with water and soap or detergent.

Allyl compounds Review of group (B16).

Allyl ethyl ether

Hazardous reaction   Explosion of peroxide at end of distillation (B450).

Allyl glycidyl ether

Colourless liquid; bp 154 °C; soluble in water.

HARMFUL VAPOUR
CAUSES IRRITATION OF SKIN AND EYES

Avoid breathing vapour. Avoid contact with skin, eyes and clothing. TLV 5 ppm (22 mg m^{-3}).

Toxic effects   The vapour irritates the respiratory system and has caused pulmonary oedema. The vapour irritates the eyes and skin. The liquid irritates the skin and alimentary system and may cause depression of the central nervous system.

First aid   *Vapour inhaled:* standard treatment (p 109).
Affected eyes: standard treatment (p 109).
Skin contact: standard treatment (p 108).
If swallowed: standard treatment (p 109).

Spillage disposal   Shut off all possible sources of ignition. Wear face-shield or goggles, and gloves. Mop up with plenty of water and run to waste diluting greatly with running water.

Allyl iodide

Yellowish to brown liquid with unpleasant pungent odour; bp 102 °C; immiscible with water.

HIGHLY FLAMMABLE
HARMFUL VAPOUR
IRRITATING TO SKIN, EYES AND RESPIRATORY SYSTEM

Avoid breathing vapour. Prevent contact with skin and eyes.

Toxic effects   The vapour irritates the eyes and the respiratory system. The liquid irritates the eyes and skin. Assumed to be very irritant and poisonous if taken by mouth.

First aid   *Vapour inhaled:* standard treatment (p 109).
Affected eyes: standard treatment (p 109).
Skin contact: standard treatment (p 108).
If swallowed: standard treatment (p 109).

Fire hazard   Flash point below 21 °C. *Extinguish fire with* water spray, foam, dry powder carbon dioxide or vaporising liquids.

Spillage disposal   Shut off all possible sources of ignition. Instruct others to keep at a safe distance. Wear breathing apparatus and gloves. Apply non-flammable dispersing agent if available and work to an emulsion with brush and water — run this to waste, diluting greatly with running water. If dispersant not available, absorb on sand, shovel into bucket(s) and transport to safe open place for burial. Site of spillage should be washed thoroughly with water and soap or detergent.

Allyl isothiocyanate

Hazardous preparation   Explosion at end of reaction between allyl chloride and sodium thiocyanate in autoclave (B388).

Aluminium (metal)

Hazardous reactions   Violent or explosive reactions with numerous oxidants including ammonium nitrate, ammonium peroxodisulphate, Na_2O_2, CuO, bromates; under certain conditions, the metal reacts violently with butanol, many halocarbons, halogens, HCl, iron/water, mercury salts, PCl_5, propan-2-ol, silver chloride, sulphur (B159–165).

Aluminium bromide anhydrous

Colourless to yellow or brown solid; violently decomposed by water with evolution of hydrogen bromide.

HARMFUL BY INHALATION
CAUSES BURNS

Avoid breathing dust. Prevent contact with skin and eyes.

Toxic effects   Inhalation of the dust produces irritation or burns of the respiratory system. The dust will cause painful eye burns. Heat is produced on contact with moist skin resulting in thermal and/or acid burns. Severe internal burns and damage will result if taken by mouth.

First aid   *Dust inhaled:* standard treatment (p 109).
Affected eyes: standard treatment (p 109).
Skin contact: standard treatment (p 108).
If swallowed: standard treatment (p 109).

Spillage disposal   Wear goggles and gloves. Mix with sand, shovel into dry bucket, transport to safe open area and add, a little at a time, to a large quantity of water; after reaction complete, run to waste diluting greatly with running water.

Aluminium chlorate

Hazardous reaction   Explosion when aqueous solution was evaporated (B169).

Aluminium chloride anhydrous

Yellow or off-white pieces, granules or powder; violently decomposed by water, with the formation of hydrogen chloride.

CAUSES BURNS
IRRITATING TO SKIN, EYES AND RESPIRATORY SYSTEM

Avoid breathing dust. Prevent contact with eyes and skin.

Toxic effects   Inhalation of dust produces irritation or burns of the mucous membranes. The material will cause painful eye burns. When moisture is present on the skin, heat is produced on contact, resulting in thermal and acid burns. If taken by mouth, the immediate local reaction causes severe burns.

Hazardous reactions   Violent reactions with water, ethylene oxide, nitrobenzene/phenol (B168–169).

First aid   *Dust inhaled :* standard treatment (p 109).
Affected eyes : standard treatment (p 109).
Skin contact : standard treatment (p 108).
If swallowed : standard treatment (p 109).

Spillage disposal   Wear goggles and gloves. Mix with sand, shovel into dry bucket, transport to a safe open area and add, a little at a time, to a large quantity of water; after reaction complete, run to waste diluting greatly with running water.

Aluminium hydride

Hazardous reactions   May explode spontaneously at ambient temperatures; violent decomposition in certain methyl ethers in presence of carbon dioxide (B170).

Aluminium lithium hydride (lithium tetrahydroaluminate)

White microcrystalline powder and lumps; decomposes above 125 °C; reacts rapidly with water with evolution of hydrogen.

CONTACT WITH WATER LIBERATES HIGHLY FLAMMABLE GASES
IRRITATING TO SKIN AND EYES

Avoid contact with skin, eyes and clothing. Keep container tightly closed.

Toxic effects   Reaction with moisture forms corrosive lithium hydroxide which irritates the skin and eyes.

131

Hazardous reactions	May ignite when ground in mortar; use in dehydrating bis(2-methoxyethyl) ether has resulted in explosions; other experiments involving boron trifluoride diethyl etherate, dibenzoyl peroxide, 1,2-dimethoxyethane, ethyl acetate, fluoroamides also resulted in explosions; vigorous reactions with pyridine and tetrahydrofuran also reported (B172–173).
First aid	*Affected eyes:* standard treatment (p 109). *Skin contact:* standard treatment (p 108). *If swallowed:* standard treatment (p 109).
Fire hazard	This arises usually by contact with small quantities of water. Such a fire is best extinguished by smothering with sand and disposing in the manner given below. Do not use extinguishers.
Spillage disposal	Instruct others to keep at a safe distance. Wear face-shield or goggles, and gloves. Cover with dry soda ash, shovel into dry bucket, transport to safe open area and add, a little at a time, to a large excess of dry propan-2-ol. Leave to stand for 24 hours and run to waste diluting greatly with running water.

Aluminium sodium hydride (sodium tetrahydroaluminate)

White or grey crystalline powder; begins to melt at 183 °C and decomposes completely at 230–240 °C; decomposed by water with evolution of hydrogen.

CONTACT WITH WATER LIBERATES HIGHLY FLAMMABLE GASES
CAUSES BURNS

Protect eyes. Avoid contact with skin or clothing. Keep container tightly closed.

Toxic effects	Reaction with moisture forms corrosive sodium hydroxide which irritates or burns the skin and eyes.
Hazardous preparation	Violent explosion when being synthesised from its elements in tetrahydrofuran (B173).
First aid	*Affected eyes:* standard treatment (p 109). *Skin contact:* standard treatment (p 108). *If swallowed:* standard treatment (p 109).
Fire hazard	This arises usually from contact with small amount of water. Fire is best extinguished by smothering with dry sand and disposing in the manner given below. Do not use extinguishers.
Spillage disposal	Instruct others to keep at a safe distance. Wear face-shield or goggles and gloves. Cover with dry sand, shovel into dry bucket and add, a little at a time to an excess of dry propan-2-ol. Leave to stand for 24 hours and run liquid to waste diluting greatly with running water.

Aluminium tetrahydroborate (aluminium borohydride)

Hazardous reaction	Vapour is spontaneously flammable in air and explodes in oxygen (B167).

Aluminium triazide

Hazardous reaction May detonate by shock (B174).

Aluminium triformate

Hazardous reaction Explosion when aqueous solution was being evaporated (B346).

Amidosulphuric acid (sulphamic acid; sulphamidic acid)

White crystals melting at about 205 °C with decomposition; one part dissolves in about six parts water at 0 °C, in about two parts at 80 °C.

IRRITATING TO SKIN AND EYES

Avoid contact with skin and eyes.

Toxic effects The dust or solution irritates the eyes. Prolonged contact with the skin may cause irritation. It is not especially toxic when taken by mouth.

First aid *Affected eyes:* standard treatment (p 109).
Skin contact: standard treatment (p 108).
If swallowed: standard treatment (p 109).

Spillage disposal Wear face-shield or goggles, and gloves. Clear up with dust pan and brush. May be disposed of after mixing with sand as normal refuse or flushed away to waste with water.

Aminium perchlorates Review of group (B20).

4-Amino-*N*,*N*-diethylaniline and salts
– see *N*,*N*-Diethyl-*p*-phenylenediamine and salts

2-Aminoethanol (ethanolamine, 2-hydroxyethylamine)

Colourless, viscous liquid with ammoniacal smell; bp 170 °C; miscible with water.

HARMFUL VAPOUR
IRRITATING TO SKIN AND EYES

Avoid breathing vapour. Avoid contact with skin, eyes and clothing.
TLV 3 ppm (6 mg m^{-3}).

Toxic effects As the vapour pressure is low, it is unlikely to cause irritation of respiratory system except when the liquid is hot. The liquid irritates the eyes and may irritate the skin — and alimentary system if taken by mouth.

First aid *Vapour inhaled:* standard treatment (p 109).
Affected eyes: standard treatment (p 109).
Skin contact: standard treatment (p 108).
If swallowed: standard treatment (p 109).

Spillage disposal Wear face-shield or goggles, and gloves. Mop up with plenty of water and run to waste diluting greatly with running water.

Aminoguanidinium nitrate

Hazardous reaction Violent explosion when aqueous solution was being evaporated on a steam bath (B255).

2-Aminophenol

White to brown, light-sensitive crystals; sparingly soluble in water.

HARMFUL SUBSTANCE IF TAKEN INTERNALLY
OR IF IN CONTACT WITH SKIN

Avoid contact with skin and eyes.

Toxic effects May cause dermatitis and cyanosis by skin absorption. Assumed to be poisonous if taken by mouth.

First aid *Affected eyes:* standard treatment (p 109).
Skin contact: standard treatment (p 108).
If swallowed: standard treatment (p 109).

Spillage disposal Small amounts may be swept up and dispersed in a large volume of water which is then run to waste, diluting greatly with running water.

2-Aminopropiononitrile

Hazardous reaction Stored material exploded after polymerisation to yellow solid (B361).

2-Aminothiazole

Hazardous reactions Material ignited in drying oven; violent explosion when nitrated with nitric/sulphuric acids (B352).

Amminemetal oxosalts Review of class (B17).

Ammonia (gas)

Colourless gas with characteristic pungent odour. It is supplied to laboratories in cylinders of various sizes in liquid (bp $-33\,^{\circ}$C) form.

> TOXIC BY INHALATION
> IRRITATING TO SKIN, EYES AND RESPIRATORY SYSTEM

Avoid breathing gas. TLV 25 ppm (18 mg m^{-3}).

Toxic effects	The gas irritates all parts of the respiratory system. The gas irritates the eyes severely.
Hazardous reactions	Mixtures with air have exploded; violent reactions or explosive products with halogens or interhalogens; violent reactions with boron halides; explosive reaction with chlorine azide; causes explosive polymerisation of ethylene oxide; forms explosive compounds with $AuCl_3$, Hg, NCl_3, silver compounds, stibine, $TeCl_4$, tetramethylammonium amide (B801–802).
First aid	*Vapour inhaled:* standard treatment (p 109). *Affected eyes:* standard treatment (p 109).
Fire hazard	Explosive limits 16–25%; ignition temperature 651 $^{\circ}$C. Since the gas is supplied in a cylinder, turning off the valve will reduce any fire involving it; if possible cylinders should be removed quickly from an area in which a fire has developed.
Disposal	Surplus gas or leaking cylinder can be vented slowly into water-fed scrubbing tower or column, or into a fume cupboard served by such a tower.

Ammonia (solutions)

Ammonia solution is commonly supplied to laboratories as a 35% solution in water (0.88 specific gravity). In warm weather this strong solution develops pressure in its bottle and the cap must be released with care.

> CAUSES BURNS
> IRRITATING TO SKIN, EYES AND RESPIRATORY SYSTEM

Avoid breathing vapour. Prevent contact with eyes and skin. TLV 25 ppm (18 mg m^{-3}).

Toxic effects	The vapour irritates all parts of the respiratory system. The solution causes severe eye burns. The solution burns the skin. If swallowed, the solution causes severe internal damage.
First aid	*Vapour inhaled:* standard treatment (p 109). *Affected eyes:* standard treatment (p 109). *Skin contact:* standard treatment (p 108). *If swallowed:* standard treatment (p 109).
Spillage disposal	Wear goggles and gloves (and rubber boots or overshoes if spillage is large). Mop up with plenty of water and run to waste diluting greatly with running water.

Ammonium amidosulphate (ammonium sulphamate)

Hazardous reaction   Vigorous exothermic hydrolysis of 60% solution with acid (B818).

Ammonium azide

Hazardous reaction   Explodes on rapid heating (B814).

Ammonium bromate

Hazardous reaction   An explosive salt (B205).

Ammonium chlorate

Hazardous reactions   Occasionally explodes spontaneously, always above 100 °C; cold saturated solution may decompose explosively (B649).

Ammonium dichromate – *see* Chromates and dichromates

Ammonium fluoride – *see* Fluorides (water-soluble)

Ammonium fluorosilicate – *see* Hexafluorosilicic acid and salts

Ammonium hydrogen difluoride – *see* Fluorides (water-soluble)

Ammonium iodate

Hazardous reaction   Decomposed violently on touching with scoop (B808).

Ammonium nitrate

Colourless crystals: mp 169 °C, decomposing at about 210 °C.

CONTACT WITH COMBUSTIBLE MATERIAL MAY CAUSE FIRE

Keep out of contact with all combustible material.

Hazardous reactions	Reviews of fire and explosion hazards of the salt; reactions with acetic acid, alkali metals, powdered Al, Sb, Bi, Cd, Cr, Co, Cu, Fe, Pb, Mg, Mn, Ni, Sn, Zn, brass and stainless steel, charcoal, organic fuels, potassium nitrite, potassium permanganate, sulphur and urea (B812–814).
Fire hazard	Mixtures of ammonium nitrate and combustible materials are readily ignited; mixtures with finely divided combustible materials can react explosively. *Extinguish fire with* water spray.
Spillage disposal	Mop up with plenty of water and run to waste diluting greatly with running water. Ensure that site of spillage is thoroughly washed down to eliminate future fire risks.

Ammonium oxalate – *see* Oxalates

Ammonium perchlorate – *see* Perchlorates

Ammonium periodate

Hazardous reaction	Exploded while being transferred by scoop (B808).

Ammonium permanganate

Hazardous reactions	Friction sensitive when dry and explodes at 60 °C (B809).

Ammonium peroxodisulphate (ammonium persulphate)

White crystals, mp 120 °C with decomposition; soluble in water.

CONTACT WITH COMBUSTIBLE MATERIAL MAY CAUSE FIRE

Keep out of contact with all combustible materials. Avoid contact with skin, eyes and clothing.

Hazardous reactions	Mixture with water and powdered Al may explode; in slightly acid concentrated solution, iron dissolved vigorously; mivture with sodium peroxide explodes on grinding in mortar (B820).
Fire hazard	Mixtures of ammonium persulphate with combustible materials are readily ignited. *Extinguish fire with* water spray.
Spillage disposal	Mop up with plenty of water and run to waste diluting greatly with running water.

Ammonium persulphate
– see Ammonium peroxodisulphate

Ammonium picrate

Hazardous reactions Explodes on heating or impact (B488).

Ammonium sulphamate
– see Ammonium amidosulphate

Ammonium sulphide solution

Yellow liquid with offensive odour; contact with acid liberates poisonous hydrogen sulphide.

CONTACT WITH ACID LIBERATES A TOXIC GAS
CAUSES BURNS

Avoid breathing vapour. Prevent contact with eyes and skin.

Toxic effects Vapour inhaled in high concentration may cause unconsciousness. Lower concentrations cause headache, giddiness and loss of energy some time after exposure. The liquid severely irritates and may burn the eyes. The liquid irritates and may burn the skin. The liquid causes severe internal damage if taken by mouth.

First aid *Vapour inhaled:* standard treatment (p 109).
Affected eyes: standard treatment (p 109).
Skin contact: standard treatment (p 108).
If swallowed: standard treatment (p 109).

Spillage disposal Wear goggles and gloves. Mop up with plenty of water and run to waste diluting greatly with running water.

Ammonium vanadate
– see Vanadium compounds

Amyl acetates *– see* Pentyl acetate

Amyl alcohol (mixed isomers)

Colourless liquid. Ordinary amyl alcohol is mainly primary iso-amyl alcohol, bp 132 °C.

FLAMMABLE
IRRITATING TO SKIN, EYES AND RESPIRATORY SYSTEM
HARMFUL IF TAKEN INTERNALLY

Avoid contact with skin and eyes.

Toxic effects Vapour may irritate the eyes and respiratory system. Liquid irritates eyes severely and may irritate skin. The liquids, if swallowed, may cause headache, vertigo, nausea, vomiting, excitement and delirium followed by coma.

First aid *Vapour inhaled:* standard treatment (p 109).
Affected eyes: standard treatment (p 109).
Skin contact: standard treatment (p 108).
If swallowed: standard treatment (p 109).

Fire hazard Flash points 41 °C; explosive limits 1–9%. Ignition temp. about 350 °C. *Extinguish fire with* water spray, dry powder, carbon dioxide or vaporising liquids.

Spillage disposal Shut off all possible sources of ignition. Wear face-shield and gloves. Apply non-flammable dispersing agent if available and work to an emulsion with brush and water — run this to waste diluting greatly with running water. If dispersant not available absorb on sand, shovel into bucket(s) and transport to safe open area for atmospheric evaporation or burial. Ventilate area well to evaporate remaining liquid and dispel vapour.

t-Amyl alcohol – *see* 2-Methylbutan-2-ol

Amyl alcohols (n- and s-) – *see* Pentanols

Amyl nitrite (isopentyl nitrite)

Colourless or pale yellow, highly volatile liquid with pungent fruity odour; bp 99 °C: advisable to store in refrigerator; immiscible with water.

HIGHLY FLAMMABLE
HARMFUL BY INHALATION

Avoid breathing vapour. Avoid contact with skin and eyes.

Toxic effects If inhaled may cause headache, flushing of face, weakness and collapse. If swallowed, similar effects may be expected.

First aid *Vapour inhaled:* standard treatment (p 109).
Affected eyes: standard treatment (p 109).
If swallowed: standard treatment (p 109).

Fire hazard   Flash point 10 °C; ignition temp. 209 °C. *Extinguish fire with* dry powder, carbon dioxide or vaporising liquids.

Spillage disposal   Shut off all possible sources of ignition. Instruct others to keep at a safe distance. Wear breathing apparatus and gloves. Apply non-flammable dispersing agent if available and work to an emulsion with brush and water — run this to waste diluting greatly with running water. If dispersant not available, absorb on sand, shovel into bucket(s) and transport to safe open area for atmospheric evaporation or burial. Site of spillage should be washed thoroughly with water and soap or detergent.

Aniline

Colourless to brown liquid; bp 185 °C; immiscible with water.

TOXIC IN CONTACT WITH SKIN
GIVES OFF POISONOUS VAPOUR

Avoid breathing vapour. Avoid contact with eyes and skin. TLV (skin) 5 ppm (19 mg m^{-3}).

Toxic effects   Inhalation of the vapour or absorption through the skin causes headache, drowsiness, cyanosis, mental confusion and, in severe cases, convulsions. The liquid is dangerous to the eyes and the above effects are also experienced if it is swallowed. **Chronic effects** Prolonged exposure to the vapour, or slight skin exposure over a period, affects the nervous system and the blood, causing fatigue, loss of appetite, headache and dizziness. (*see* note in Ch 6, p 85.)

Hazardous reactions   Vigorously oxidised by a number of oxidants including perchloric acid, fuming nitric acid, sodium peroxide and ozone. Violent reaction with BCl_3 (B490).

First aid   *Vapour inhaled:* standard treatment (p 109).
Affected eyes: standard treatment (p 109).
Skin contact: standard treatment (p 108).
If swallowed: standard treatment (p 109).

Spillage disposal   Wear breathing apparatus (or face-shield if amount is small) and gloves. Mix with sand and shovel mixture into a suitable vessel (glass, polythene or enamel) for dispersion in an excess of dilute hydrochloric acid (1 volume concentrated acid diluted with 2 volumes of water). Allow to stand, with occasional stirring, for 24 hours and then run acid extract to waste, diluting greatly with running water and washing the sand. Sand can be treated as normal waste.

Anilinium salts

The commoner anilinium salts — the hydrochloride and sulphate — are soluble in water. They are colourless to greyish-brown in colour.

POISONOUS DUSTS

Avoid contact with eyes and skin.

Toxic effects   The acute and chronic effects of aniline poisoning are described under **Aniline**. Skin absorption does not occur so readily with the salts or their solutions as with aniline itself, but they are dangerous to the eyes, partly because of the acidity, and cause intense irritation. If taken by mouth, the effects of aniline poisoning (headache, drowsiness, cyanosis) will be apparent.

First aid *Affected eyes:* standard treatment (p 109).
Skin contact: standard treatment (p 108).
If swallowed: standard treatment (p 109).

Spillage disposal Small amounts may be swept up and dispersed in a large volume of water which is then run to waste, diluting greatly with running water.

Anisidines (aminoanisoles)

o-Anisidine is a pale yellow to orange liquid; bp 224 °C; immiscible with water. *p*-Anisidine is a pale yellow to brown solid; mp 59 °C; insoluble in water.

SERIOUS RISK OF POISONING
BY INHALATION, SWALLOWING OR SKIN CONTACT
IRRITATING TO SKIN AND EYES

Avoid breathing vapour. Avoid contact with skin and eyes. TLV (skin) (0.5 mg m^{-3}).

Toxic effects These are not recorded but are assumed to be similar to those of aniline poisoning, *ie* headache, drowsiness and cyanosis.

First aid *Vapour inhaled:* standard treatment (p 109).
Affected eyes: standard treatment (p 109).
Skin contact: standard treatment (p 108).
If swallowed: standard treatment (p 109).

Spillage disposal Wear breathing apparatus (or face-shield if amount is small) and gloves. Mix with sand and shovel mixture into suitable glass, polythene or enamel vessel for dispersion in an excess of dilute hydrochloric acid (1 volume concentrated acid diluted with 2 volumes water). Allow to stand, with occasional stirring, for 24 hours and then run acid extract to waste, diluting greatly with running water and washing the sand. Sand can be treated as normal waste.

Antimony compounds (water-soluble)

Most soluble antimony compounds are colourless crystals or powder; the pentachloride is a reddish, fuming liquid with an offensive smell.

POISON
IRRITATING TO SKIN, EYES AND RESPIRATORY SYSTEM

TLV (as Sb) (0.5 mg m^{-3}).

Toxic effects All soluble antimony compounds must be considered to be poisonous when taken by mouth. Some compounds cause skin irritation and dermatitis. If taken by mouth, soluble antimony compounds may cause burning of the mouth and throat, choking, nausea and vomiting. Stibine, which may be formed by the action of acidic reducing agents on antimony-containing materials, is an extremely poisonous gas, causing blood destruction and damage to liver and kidneys. Insoluble antimony compounds, such as the oxide and sulphide, are not toxic.

First aid
Skin contact : standard treatment (p 108).
If swallowed : standard treatment (p 109).

Spillage disposal
The disposal of these in any quantity must be considered carefully in the light of local conditions and regulations. Burial in an isolated area can be considered as can gradual disposal at very high dilution into a sewage system permitting this. Both the chlorides $SbCl_3$ and $SbCl_5$ are readily hydrolysed by water.

Antimony(III) nitride

Hazardous reactions
Impure material explodes mildly on heating in air, or on contact with water or dilute acids (B873).

Aqua regia

Hazardous storage
Pressure develops in screw-capped bottles (B21).

Arsenic compounds

Most arsenic compounds are colourless powders or crystals — they include arsenites and arsenates of many metals; syrupy arsenic acid and arsenic trichloride are liquids. All must be considered to be extremely poisonous. The metal itself 'has not been recognised as a noteworthy hazard': (*see* Ch 6)

POISON
SERIOUS RISK OF POISONING BY INHALATION OR SWALLOWING
DANGER OF CUMULATIVE EFFECTS
IRRITATING TO SKIN, EYES AND RESPIRATORY SYSTEM

Do not inhale dust or fume. Prevent contact with skin and eyes. TLV (as As) (0.5 mg m^{-3}).

Toxic effects
The inhalation of dust or fume irritates the mucous membranes and leads to arsenical poisoning. Certain compounds, especially the trichloride and arsenic acid, irritate the eyes and skin, and absorption causes poisoning. If swallowed, arsenic compounds irritate the stomach severely and affect the heart, liver and kidneys; nervousness, thirst, vomiting, diarrhoea, cyanosis and collapse may be symptoms. Arsine — *see* below. **Chronic effects** The inhalation of small concentrations of dust or fume over a long period will cause poisoning; skin contact over a long period may cause ulceration.

First aid
Dust or fume inhaled : standard treatment (p 109).
Affected eyes : standard treatment (p 109).
Skin contact : standard treatment (p 108).
If swallowed : standard treatment (p 109).

Spillage disposal
The disposal of these in any quantity must be considered carefully in the light of local conditions and regulations. Deep burial mixed with sand in an isolated area or consignment to deep sea water in a heavy container can be considered as can gradual disposal at very high dilution into a sewage system permitting this.

Arsine (arsenic trihydride; hydrogen arsenide)

Colourless gas with garlic odour; slightly soluble in water; formed whenever nascent hydrogen is in contact with an aqueous solution of arsenic.

> VERY TOXIC BY INHALATION
> FLAMMABLE

Prevent inhalation of gas. TLV 0.05 ppm (0.2 mg m^{-3}).

Toxic effects	A few inhalations may be fatal, death resulting from anoxia or pulmonary oedema. Symptoms of poisoning include headache, weakness, vertigo and nausea. Damage is caused to kidneys and liver. (*see* note in Ch 6, p 84.)
Hazardous reactions	Ignites in chlorine; explodes with fuming nitric acid (B178).
First aid **Disposal**	*Gas inhaled:* standard treatment (p 109); prompt medical attention vital. Leaking laboratory cylinders should be removed to open space, wearing self-contained breathing apparatus, and allowed to discharge as slowly as possible. Consult suppliers.

Auramine

Yellow flakes or powder. In the United Kingdom the manufacture of this substance is controlled by The Carcinogenic Substances Regulations 1968.

> SERIOUS RISK OF POISONING BY INHALATION OR SWALLOWING
> POSSIBLE RISK OF VERY SERIOUS IRREVERSIBLE EFFECTS

Prevent contact with skin, eyes and clothing.

Toxic effects	Absorption through the skin may result in dermatitis and burns, nausea and vomiting.
First aid	*Skin contact:* standard treatment (p 108). *If swallowed:* standard treatment (p 109).
Spillage disposal	Wear face-shield or goggles, gloves and breathing apparatus. Mix with sand and transport to safe, open area for burial. Site of spillage should be washed thoroughly with water and soap or detergent.

Azides Review of group (B24).

Azidoacetaldehyde

Hazardous reaction	Decomposed vigorously below 80 °C at 5 mbar (5×10^2 Pa) (B300).

Azidoacetone and oxime

Hazardous reactions   Azidoacetone has exploded after 6 months' storage in the dark (B358); distillation residue of its oxime exploded violently (B361).

2-Azidocarbonyl compounds Review of group (B24).

N-Azidodimethylamine

Hazardous reaction   Rather explosive (B321).

Azidodimethylborane

Hazardous reaction   Explodes on warming (B318).

Azidosilane

Hazardous preparation (B805).

5-Azidotetrazole

Hazardous reactions   The compound and its salts are explosive (B230).

Aziridine (ethyleneimine)

Hazardous reactions   Erroneous preparative procedure—liable to polymerise explosively; reaction with hypochlorite gives explosive chloroaziridine; forms explosive silver derivatives (B314).

α-Azoisobutyronitrile

Hazardous reactions   Decomposes when heated; explosive decomposition when technical material being recrystallised from acetone (B554).

Azo compounds Review of group (B24).

Azo-*N*-nitroformamidine

Hazardous reaction   Decomposes explosively at 165 °C (B307).

Barium bromate

Hazardous reaction   Decomposes almost explosively at 300 °C (B193).

Barium compounds

Practically all barium compounds are colourless crystals or powders. All barium compounds, except the sulphate, must be considered to be poisonous when taken by mouth.

POISON
HARMFUL IF TAKEN INTERNALLY

TLV (soluble compounds) (0.5 mg m^{-3}).

Toxic effects   If ingested, soluble barium compounds cause nausea, vomiting, stomach pains and diarrhoea. (*see* note in Ch 6, p 84.)

Hazardous reactions   Dangerous reactions of the metal, hydride, hydroxide, nitrate, oxide, sulphate, sulphide are indicated (B193–196).

First aid   *If swallowed:* standard treatment (p 109).
Spillage disposal   The sulphate may be brushed up and treated as normal refuse. Soluble barium salts should be mopped up with water and the solution run to waste, diluting greatly with running water.

Barium diazide

Hazardous reaction   Impact sensitive when dry (B195).

Barium peroxide

Hazardous reactions   May ignite H_2S, hydroxylamine and organic materials especially in presence of water (B195–196).

Benzal chloride – *see* Benzylidene chloride

Benzaldehyde

Hazardous reaction   Violent oxidation by 90% performic acid (B527).

Benzene (benzol; coal naphtha)

Colourless, volatile liquid with characteristic odour; bp 80 °C; immiscible with water.

EXTREMELY FLAMMABLE
GIVES OFF VERY POISONOUS VAPOUR
DANGER OF CUMULATIVE EFFECTS

Avoid inhalation of vapour. Prevent contact with skin and eyes. TLV (skin) 10 ppm (32 mg m^{-3}).

Toxic effects	Inhalation of the vapour causes dizziness, headache and excitement; high concentrations may cause unconsciousness. The vapour irritates the eyes and mucous membranes. The liquid is absorbed through the skin and poisoning may result from this. Assumed to be extremely poisonous if taken by mouth. **Chronic effects** Repeated inhalation of low concentrations over a considerable period may cause severe, even fatal, blood disease. (*see* references to toxicity of benzene in Ch 6 and 7.)
Hazardous reactions	Complex with silver perchlorate exploded on crushing in mortar; certain mixtures with 84% of nitric acid are highly sensitive to detonation; mixture with liquid oxygen is explosive; benzene solution of rubber exploded when ozonised; reacts vigorously or explosively with other oxidants, interhalogens, uranium hexafluoride (B483–484).
First aid	*Vapour inhaled:* standard treatment (p 109). *Affected eyes:* standard treatment (p 109). *Skin contact:* standard treatment (p 108). *If swallowed:* standard treatment (p 109).
Fire hazard	Flash point −11 °C; explosive limits 1.4–8%; ignition temp 562 °C. *Extinguish fire with* foam, dry powder or vaporising liquids.
Spillage disposal	Shut off all possible sources of ignition. Instruct others to keep at a safe distance. Wear breathing apparatus and gloves. Apply non-flammable dispersing agent if available and work to emulsion with brush and water — run this to waste diluting greatly with running water. If dispersant not available, absorb on sand, shovel into bucket(s) and transport to safe open area for atmospheric evaporation. Site of spillage should be washed thoroughly with water and soap or detergent.

Benzenediazonium-2-carboxylate

Hazardous reactions	Internal salt explosive and reacts explosively with aniline and violently with aryl isocyanides (B522).

Benzenediazonium chloride

Hazardous reactions	Dry salt explosive (B476) as is zinc chloride complex (B590).

Benzenediazonium nitrate

Hazardous reactions	Highly sensitive to friction and impact and explodes at 90 °C (B483).

Benzenediazonium-4-sulphonate

Hazardous reaction   Dry internal salt exploded violently on touching (B473).

Benzene-disulphonic and -sulphonic acids – *see* Sulphonic acids

Benzenesulphinyl chloride

Hazardous reaction   Bottle, undisturbed for months, exploded (B477).

Benzenesulphonyl azide

Hazardous reaction   Pure azide decomposed smoothly at 105 °C but crude exploded on heating (B482).

Benzenesulphonyl chloride (benzenesulphonchloride)

Colourless to brown liquid; bp 251 °C with decomposition; reacts with water to form benzene-sulphonic acid and hydrochloric acid.

IRRITATING TO SKIN, EYES AND RESPIRATORY SYSTEM
CAUSES BURNS

Prevent contact with skin and eyes.

Toxic effects   Irritates the eyes severely and causes skin burns. Causes severe internal irritation if taken by mouth.

Hazardous reaction   Violent reaction with dimethyl sulphoxide (B477).

First aid   *Affected eyes:* standard treatment (p 109).
Skin contact: standard treatment (p 108).
If swallowed: standard treatment (p 109).

Spillage disposal   Wear face-shield or goggles, and gloves. Spread soda ash liberally over the spillage and mop up cautiously with water—run to waste diluting greatly with running water.

Benzenethiol (thiophenol)

Hazardous preparation   Violent explosion when being made from benzenediazonium chloride (B489).

Benzidine and salts

The use of these compounds in the United Kingdom is now prohibited under The Carcinogenic Substances Regulations 1968. Inhalation or absorption through the skin of the dust has been recognized as a cause of bladder tumours. It is not therefore considered appropriate to deal with their hazards more fully in this book.

Benzonitrile

Colourless liquid, bp 191 °C; sparingly soluble in water.

HARMFUL BY INHALATION
HARMFUL IF TAKEN INTERNALLY
OR IF IN CONTACT WITH SKIN OR EYES

Avoid inhalation of vapour. Avoid contact with skin or eyes.

Toxic effects	No record has been found of instances of poisoning by this material, but its constitution is such that it must be assumed to be toxic by inhalation, skin and eye contact, and ingestion.
First aid	*Vapour inhaled:* standard treatment (p 109). *Affected eyes:* standard treatment (p 109). *Skin contact:* standard treatment (p 108). *If swallowed:* standard treatment (p 109).
Spillage disposal	Instruct others to keep at a safe distance. Wear breathing apparatus and gloves. Apply dispersing agent if available and work to an emulsion with brush and water — run this to waste diluting greatly with running water. If dispersant not available absorb on sand, shovel into bucket(s) and transport to safe open area for burial. Site of spillage should be washed thoroughly with water and soap or detergent.

p-Benzoquinone (quinone)

Yellow crystals, with characteristic, irritating odour. Slightly soluble in water.

VERY IRRITANT DUST

Avoid breathing dust. Avoid contact with skin and eyes. TLV 0.1 ppm (0.4 mg m^{-3}).

Toxic effects	The dust irritates the respiratory system severely. Skin or eye contact is very irritating and can cause severe local damage. Must be considered highly irritant and dangerous if taken by mouth.
First aid	*Affected eyes:* standard treatment (p 109). *Skin contact:* standard treatment (p 108). *If swallowed:* standard treatment (p 109).
Spillage disposal	Wear face-shield or goggles, and gloves. Mop up with plenty of water and run to waste diluting greatly with running water.

Benzotriazole

Hazardous reaction	Large batch exothermally decomposed and then detonated during distillation at 160 °C/2.5 mbar (2.5 × 10² Pa) (B482).

Benzotrichloride – *see* Benzylidyne chloride

Benzotrifluoride – *see* Benzylidyne fluoride

Benzoyl azide

Hazardous reaction	Crude material exploded violently between 120 °C and 165 °C (B525).

Benzoyl chloride

Colourless, fuming liquid with pungent smell; bp 197 °C; reacts with water forming benzoic acid and hydrochloric acid.

CAUSES BURNS
IRRITATING TO SKIN, EYES AND RESPIRATORY SYSTEM

Avoid breathing vapour. Prevent contact with eyes and skin.

Toxic effects	The vapour irritates the respiratory system. The vapour irritates and the liquid burns the eyes severely. The liquid is very irritating to the skin and can cause burns. If taken by mouth there is immediate irritation and damage.
Hazardous reaction	Violent reaction with dimethyl sulphoxide (B524).
First aid	*Vapour inhaled:* standard treatment (p 109). *Affected eyes:* standard treatment (p 109). *Skin contact:* standard treatment (p 108). *If swallowed:* standard treatment (p 109).
Spillage disposal	Wear goggles and gloves. Spread soda ash liberally over the spillage and mop up cautiously with plenty of water — run to waste diluting greatly with running water.

Benzoyl nitrate

Hazardous reactions	Unstable liquid which explodes on rapid heating and may also explode on exposure to light (B524).

Benzoyl peroxide – *see* Dibenzoyl peroxide

1,1-Benzoylphenyldiazomethane
– see **2-Diazo-2-phenylacetophenone**

Benzvalene (tricyclo[3.1.0.0²,⁶]hex-3-ene)

Hazardous reaction   Exploded violently when scratched (B484).

Benzylamine

Colourless liquid; bp 185 °C; miscible with water.

> CAUSES BURNS
> IRRITATING TO SKIN, EYES AND RESPIRATORY SYSTEM

Avoid contact with skin and eyes.

Toxic effects   These have not been recorded to any extent, but it has been found that benzylamine causes skin burns; by inference it must be assumed to damage the eyes and cause internal irritation and damage if taken by mouth.

First aid   *Affected eyes:* standard treatment (p 109).
Skin contact: standard treatment (p 108).
If swallowed: standard treatment (p 109).

Spillage disposal   Instruct others to keep at a safe distance. Wear breathing apparatus and gloves. Mop up with plenty of water and run to waste diluting greatly with running water. Ventilate area well to evaporate remaining liquid and dispel vapour.

Benzyl bromide (α-bromotoluene)

Colourless to pale yellow liquid; bp 198 °C; immiscible with water.

> CAUSES BURNS
> IRRITATING TO SKIN, EYES AND RESPIRATORY SYSTEM

Avoid breathing vapour. Prevent contact with skin and eyes.

Toxic effects   The vapour irritates the respiratory system. Low vapour concentrations cause lachrymation and severe irritation to the eyes. The vapour irritates the skin and the liquid causes burns. Can be assumed to cause severe internal irritation and damage if taken by mouth.

First aid   *Vapour inhaled:* standard treatment (p 109).
Affected eyes: standard treatment (p 109).
Skin contact: standard treatment (p 108).
If swallowed: standard treatment (p 109).

Spillage disposal Instruct others to keep at a safe distance. Wear breathing apparatus and gloves. Apply dispersing agent if available and work to an emulsion with brush and water — run this to waste diluting greatly with running water. If dispersant not available, absorb on sand, shovel into bucket(s) and transport to safe open area for burial. Site of spillage should be washed thoroughly with water and soap or detergent.

Benzyl chloride (α-chlorotoluene)

Colourless to brown-yellow liquid with acrid smell; bp 179 °C; immiscible with water.

CAUSES BURNS
IRRITATING TO SKIN, EYES AND RESPIRATORY SYSTEM

Avoid breathing vapour. Prevent contact with skin and eyes. TLV 1 ppm (5 mg m^{-3}).

Toxic effects The vapour irritates the respiratory system. Low vapour concentrations cause lachrymation and severe irritation to the eyes. The vapour irritates the skin and the liquid causes burns. Can be assumed to cause severe internal irritation and damage if taken by mouth.

First aid *Vapour inhaled:* standard treatment (p 109).
Affected eyes: standard treatment (p 109).
Skin contact: standard treatment (p 108).
If swallowed: standard treatment (p 109).

Spillage disposal Instruct others to keep at a safe distance. Wear breathing apparatus and gloves. Apply non-flammable dispersing agent and work to an emulsion with brush and water—run this to waste diluting greatly with running water. If dispersant not available, absorb on sand, shovel into bucket(s) and transport to safe open area for burial. Site of spillage should be washed thoroughly with water and soap or detergent.

Benzyl chloroformate

Colourless or yellow, fuming, oily liquid; decomposes slowly at room temperature and needs refrigerated storage; bp 103 °C at 22 mmHg; insoluble in water.

HARMFUL VAPOUR
CAUSES BURNS

Prevent contact with skin, eyes and clothing.

Toxic effects Vapour causes severe irritation of eyes and respiratory system. Liquid blisters the skin and severely damages the eyes.

First aid *Vapour inhaled:* standard treatment (p 109).
Affected eyes: standard treatment (p 109).
Skin contact: standard treatment washing with soap and water (108).
If swallowed: standard treatment (p 109).

Spillage disposal Instruct others to keep at a safe distance. Wear breathing apparatus and gloves. Apply non-flammable dispersing agent and work to an emulsion with brush and water — run this to waste diluting greatly. If dispersant not available, absorb on sand and transport to safe open area for burial.

Benzyl cyanide – *see* Phenylacetonitrile

Benzylidene chloride (benzal chloride, ααα-trichlorotoluene)

Colourless liquid; bp 205 °C: immiscible with water.

IRRITATING TO SKIN, EYES AND RESPIRATORY SYSTEM

Avoid breathing vapour. Avoid contact with eyes and skin.

Toxic effects	The vapour irritates the respiratory system. The vapour and liquid irritate the eyes and may cause conjunctivitis. The liquid may irritate the skin. If taken by mouth, internal irritation and damage must be assumed.
First aid	*Vapour inhaled:* standard treatment (p 109). *Affected eyes:* standard treatment (p 109). *Skin contact:* standard treatment (p 108). *If swallowed:* standard treatment (p 109).
Spillage disposal	Instruct others to keep at a safe distance. Wear breathing apparatus and gloves. Apply dispersing agent if available and work to an emulsion with brush and water — run this to waste diluting greatly with running water. If dispersant not available, absorb on sand, shovel into bucket(s) and transport to safe open area for burial. Site of spillage should be washed thoroughly with water and soap or detergent.

Benzylidyne chloride (benzotrichloride)

Colourless to yellow, fuming liquid; bp 214 °C; immiscible with water.

HARMFUL BY INHALATION, IF TAKEN INTERNALLY OR IF IN CONTACT WITH SKIN

Avoid breathing vapour. Avoid contact with eyes and skin.

Toxic effects	The vapour irritates all parts of the respiratory system. The vapour and liquid irritate the eyes and skin. If taken by mouth, internal irritation and damage must be assumed.
First aid	*Vapour inhaled:* standard treatment (p 109). *Affected eyes:* standard treatment (p 109). *Skin contact:* standard treatment (p 108). *If swallowed:* standard treatment (p 109).
Spillage disposal	Instruct others to keep at a safe distance. Wear breathing apparatus and gloves. Apply dispersing agent if available and work to an emulsion with brush and water — run this to waste diluting greatly with running water. If emulsifier not available, absorb on sand, shovel into bucket(s) and transport to safe open area for burial. Site of spillage should be washed thoroughly with water and soap or detergent.

Benzylidyne fluoride (benzotrifluoride)

Colourless liquid with an aromatic odour; bp 101 °C; immiscible with water.

HIGHLY FLAMMABLE
HARMFUL BY INHALATION AND IN CONTACT WITH SKIN

Avoid breathing vapour. Avoid contact with skin and eyes.

Toxic effects	Animal experiments indicate the risk of central nervous system depression through inhalation, absorption and ingestion.
First aid	*Vapour inhaled:* standard treatment (p 109). *Affected eyes:* standard treatment (p 109). *Skin contact:* standard treatment (p 108). *If swallowed:* standard treatment (p 109).
Fire hazard	Flash point 12 °C. *Extinguish fire with* water spray, foam, dry powder, carbon dioxide or vaporising liquids.
Spillage disposal	Shut off all possible sources of ignition. Instruct others to keep at a safe distance. Wear breathing apparatus and gloves. Apply non-flammable dispersing agent and work to an emulsion with brush and water — run this to waste diluting greatly with running water. If dispersant not available, absorb on sand, shovel into bucket(s) and transport to safe open place for atmospheric evaporation or burial. Site of spillage should be washed thoroughly with water and soap or detergent.

Benzyloxyacetylene

Hazardous reaction	Explosion if heated above 60 °C during vacuum distillation (B567).

Beryllium (metal)

Hazardous reactions	Heavy impact flashes mixtures of powdered Be with CCl_4 or C_2HCl_3; incandescent reaction when heated with phosphorus (B197).

Beryllium compounds

White powder or crystals.

POISONOUS DUST

Prevent inhalation of dust. Prevent contact with eyes and skin. TLV (as Be) (0.002 mg m^{-3}).

Toxic effects	Particles penetrating the skin through wounds and abrasions may cause local damage difficult to heal. Symptoms of poisoning, indicated by respiratory troubles or cyanosis, may develop within a week or after a latent period of even several years.

First aid   *Dust inhaled:* standard treatment (p 109).
                       Affected eyes: standard treatment (p 109).
                       Skin contact: standard treatment (p 108).
                       If swallowed: standard treatment (p 109).

Spillage   The disposal of these in any quantity must be considered carefully in the
disposal   light of local conditions and regulations. Deep burial mixed with sand in an
                       isolated area or consignment to deep sea water in a heavy container can be
                       considered as can gradual disposal at very high dilution into a sewage
                       system permitting this.

Bis(4-aminophenyl)methane

(di-(4-aminophenyl)methane)

Light brown solid; sparingly soluble in water.

> HARMFUL BY INHALATION
> HARMFUL IF TAKEN INTERNALLY
> OR IF IN CONTACT WITH SKIN

Avoid contact with skin and eyes.

Toxic effects   The toxicity of this base is not well documented but, because of its chemi-
                       cal relationship with, for example, *p*-toluidine (*qv*) similar effects may be
                       assumed.

First aid   *Affected eyes:* standard treatment (p 109).
                       Skin contact: standard treatment (p 108).
                       If swallowed: standard treatment (p 109).

Spillage   Wear face-shield or goggles, and gloves. Mix with sand and shovel mix-
disposal   ture into a glass, enamel or polythene vessel for dispersion in dilute
                       hydrochloric acid (1 volume concentrated acid diluted with 2 volumes of
                       water). Allow to stand, with occasional stirring for 24 hours and then run
                       extract to waste, diluting greatly with running water and washing the sand.
                       The residual sand can be dealt with as normal refuse.

Bis-*o*-azidobenzoyl peroxide

Hazardous   Exploded on touching with metal spatula (B599).
reaction

Bis(2-chloroethyl)amine

Hazardous   Violent explosion when ethereal solution was being evaporated under
reaction   vacuum (B313).

Bis(2-chloroethyl) ether (di(2-chloroethyl) ether)

Colourless liquid with pungent odour; bp 178 °C; practically insoluble in water; liable to form explosive peroxides, on exposure to air and light, which must be decomposed before the ether is distilled to small volume.

> FLAMMABLE
> SERIOUS RISK OF POISONING BY INHALATION,
> SWALLOWING OR SKIN CONTACT

Avoid breathing vapour. Avoid contact with skin and eyes. TLV (skin) 10 ppm (60 mg m^{-3}).

Toxic effects The vapour irritates the respiratory system and high concentrations may result in lung damage after a latent period of some hours. The vapour and liquid irritate the eyes and may cause conjunctivitis. Assumed to be poisonous if taken by mouth.

First aid *Vapour inhaled:* standard treatment (p 109).
Affected eyes: standard treatment (p 109).
Skin contact: standard treatment (p 108).
If swallowed: standard treatment (p 109).

Fire hazard Flash point 55 °C; ignition temp. 369 °C. *Extinguish fire with* dry powder, carbon dioxide or vaporising liquid.

Spillage disposal Shut off all possible sources of ignition. Instruct others to keep at a safe distance. Wear breathing apparatus and gloves. Apply non-flammable dispersing agent if available and work to an emulsion with brush and water — run this to waste diluting greatly with running water. If dispersant not available, absorb on sand, shovel into bucket(s) and transport to safe open area for burial. Site of spillage should be washed thoroughly with water and soap or detergent.

Bis(2-cyanoethyl)amine
– see **3,3′-iminodipropiononitrile**

Bis(4-isocyanatophenyl)methane
(di(4-isocyanatophenyl)methane; 4,4′-methylene bisphenyl isocyanate)

Yellow crystals or fused solid with irritating smell; mp 37 °C; hydrolysed by water — by storing at 5 °C, a tendency to form polymeric solids is reduced to a minimum.

> HARMFUL VAPOUR AND DUST

Avoid breathing vapour. Avoid contact with skin, eyes and clothing. TLV 0.02 ppm (0.2 mg m^{-3}).

Toxic effects The vapour is irritating to the eyes and respiratory system, as is the dust. The solid or molten material irritates the eyes and skin and must be considered poisonous if taken by mouth.

First aid *Vapour inhaled:* standard treatment (p 109).
Affected eyes: standard treatment (p 109).
Skin contact: standard treatment (p 108).
If swallowed: standard treatment (p 109).

Spillage disposal Wear face-shield or goggles, and gloves. Shovel into dry bucket(s), transport to safe open area and add, a little at a time, to a large quantity of water; after reaction is complete, run to waste, diluting greatly with running water.

Bismuth (metal)

Hazardous reactions The finely divided metal reacts violently with BrF_5, and fuming nitric acid; violent or explosive reaction with fused ammonium nitrate; its reaction with perchloric acid may be explosive; that with nitrosyl fluoride or iodine pentafluoride may be accompanied by incandescence **(B198)**.

Bismuth nitride

Hazardous reactions Exploded on shaking, heating, or on contact with water or dilute acids **(B199)**.

Bismuth pentafluoride

Hazardous reaction Reacts vigorously with water and may ignite **(B198)**.

Bistrichloroacetyl peroxide

Hazardous reaction Explodes on standing at room temperature **(B381)**.

Bis(triethyltin)acetylene

Hazardous reaction Compound with stannic chloride is highly explosive **(B605)**.

Bistrifluoroacetyl peroxide

Hazardous reaction Explodes on standing at room temperature **(B381)**.

Bleaching powder
– *see* Calcium hypochlorite

Borazine

Hazardous reaction Sealed ampoules exploded in daylight (B191).

Boron

Hazardous reactions Ignites in Cl_2 or F_2 at ambient temperature. Reacts explosively when ground with silver fluoride at ambient temperature. Violence of interaction with fused metal nitrates increased by presence of nitrites. Violent reactions with other oxidants, *eg* nitrosyl fluoride, Na_2O_2, PbO (B181).

Boron tribromide

Colourless, fuming liquid with a pungent odour; bp 90 °C; reacts violently with water.

> TOXIC BY INHALATION
> REACTS VIOLENTLY WITH WATER
> CAUSES BURNS
> IRRITATING TO SKIN, EYES AND RESPIRATORY SYSTEM

Avoid breathing vapour. Prevent contact with skin and eyes. Do not put water into container. TLV 1 ppm (10 mg m^{-3}).

Toxic effects The vapour irritates all parts of the respiratory system. The vapour irritates the eyes. The liquid burns the skin and eyes. If taken by mouth, there would be severe internal burning.

Hazardous reactions Reacts violently when poured into an excess of water and explosively when water is poured into it. Mixture with sodium metal explodes on impact (B182).

First aid *Vapour inhaled:* standard treatment (p 109).
Affected eyes: standard treatment (p 109).
Skin contact: standard treatment (p 108).
If swallowed: standard treatment (p 109).

Spillage disposal Instruct others to keep at a safe distance. Wear breathing apparatus and gloves. Spread soda ash liberally over the spillage and mop up cautiously with plenty of water — run to waste diluting greatly with running water.

Boron trichloride

Colourless fuming liquid or gas with a pungent odour; bp 12.5 °C; reacts rapidly with water forming boric and hydrochloric acids.

> TOXIC BY INHALATION
> CAUSES BURNS
> IRRITATING TO SKIN, EYES AND RESPIRATORY SYSTEM

Avoid breathing gas. Prevent contact with skin and eyes. Do not put water into container.

Toxic effects The gas irritates the eyes, skin and respiratory system. The liquid irritates or burns the skin and burns the eyes. If taken by mouth there would be severe internal burning.

Hazardous reaction Reacts violently with aniline (B183).

First aid *Gas inhaled:* standard treatment (p 109).
Affected eyes: standard treatment (p 109).
Skin contact: standard treatment (p 108).
If swallowed: standard treatment (p 109).

Spillage disposal In warm weather, it will exist as a gas, in which case instruct others to keep out of the affected area; wear breathing apparatus and organise adequate ventilation. In cool weather, a spillage of the liquid can be covered with excess of soda ash (wearing breathing apparatus and gloves) and then mopped up cautiously with water and run to waste.

Boron trifluoride

Colourless, fuming gas with pungent, suffocating odour; bp −100 °C. Decomposes in water forming fluoroboric and boric acids.

TOXIC BY INHALATION
IRRITATING TO SKIN, EYES AND RESPIRATORY SYSTEM

Prevent inhalation of gas. Prevent contact with skin and eyes. TLV 1 ppm (3 mg m^{-3}).

Toxic effects The gas irritates the skin, eyes and respiratory system; at high concentrations it may burn the skin.

Hazardous reaction Reacts with hot alkali or alkaline earth (not Mg) metals with incandescence (B183).

First aid *Gas inhaled:* standard treatment (p 109).
Affected eyes: standard treatment (p 109).
Skin contact: standard treatment (p 109).

Disposal Surplus gas or leaking cylinder can be vented slowly into water-fed scrubbing tower or column, or into a fume cupboard served by such a tower.

Boron trifluoride complexes

The liquid complexes formed between boron trifluoride and acetic acid, diethyl ether, methanol and propan-1-ol all display hazards and toxic effects associated with their constituents. All are readily hydrolysed by water, corrosive and, to some degree, flammable.

Boron triiodide

Hazardous reactions Strongly exothermal reaction with ammonia. Incandescent reaction with warm red or white phosphorus. Violent reaction with limited amounts of water (B186).

Bromine

Dark reddish-brown fuming liquid; bp 59 °C; slightly soluble in water.

> GIVES OFF VERY POISONOUS VAPOUR
> CAUSES SEVERE BURNS

Prevent breathing of vapour. Prevent contact with eyes and skin. TLV 0.1 ppm (0.7 mg m⁻³).

Toxic effects	The vapour irritates all parts of the respiratory system. The vapour severely irritates the eyes and mucous membranes. The liquid burns the skin and eyes. If taken by mouth, severe local burns and internal damage would result.
Hazardous reactions	Bromine reacts with varying degrees of violence with a large number of compounds and elements including acetone, acrylonitrile, ammonia, BrF₃, copper(I) hydride, diethyl ether, *N,N*-dimethylformamide, ethanol/phosphorus, fluorine, germane, hydrogen, metal acetylides and carbides, metal azides, Li, Na, K, Rb, Al, Hg, Ti, methanol, ozone, F₂O, P, tetracarbonylnickel, trialkylboranes, F₂O₃ (B206–210).
First aid	*Vapour inhaled:* standard treatment (p 109). *Affected eyes:* standard treatment (p 109). *Skin contact:* standard treatment (p 108). *If swallowed:* standard treatment (p 109).
Spillage disposal	Instruct others to keep at a safe distance. Wear breathing apparatus and gloves. Spread soda ash liberally over the spillage and mop up cautiously with plenty of water — run to waste diluting greatly with running water.

Bromine azide

Hazardous reactions	Very shock-sensitive in solid, liquid and vapour forms. Liquid explodes on contact with As, Na, Ag foil, P (B205).

Bromine pentafluoride

Pale yellow fuming liquid with pungent odour; bp 40 °C. It reacts vigorously and possibly explosively with water.
TLV 0.1 ppm (0.7 mg m⁻³).

and

Bromine trifluoride

Colourless to grey-yellow, fuming liquid with pungent choking smell; bp 127 °C; mp 8.8 °C; reacts vigorously with water; extremely reactive, etching glass and setting fire to paper, wood and other organic material.

> REACTS VIOLENTLY WITH WATER
> CAUSES SEVERE BURNS

Prevent inhalation of vapour. Prevent contact with skin, eyes, clothing and all combustible material.

Toxic effects   The vapours severely irritate and may burn the eyes, skin and respiratory system. The liquids burn all human tissue and cause severe damage.

Hazardous reactions   Review of hazards and necessary precautions in use of BrF_3 indicated; violent or explosive reactions occur with ammonium halides, antimony chloride oxide, CO, acetone, diethyl ether, toluene, uranium, water, silicone grease, CCl_4, benzene, other organic materials; incandescent reactions with Br_2, I_2, As, Sb — also with powdered Mo, Nb, Ta, Ti, V, B, C, P, S; BrF_5 reacts violently with strong nitric and sulphuric acids, Cl_2, I_2, ammonium chloride, KI, Sb, As, B, Se, Te, Al, Ba, Bi, Co, Cr, Ir, Fe, Li, Mn, C, P, S, As_2O_5, B_2O_3, CaO, CO, Cr_2O_3, I_2O_5, MgO, Mo_2O_3, P_2O_5, SO_2, W_2O_3; fire or explosion in contact with acetic acid, ammonia, benzene, ethanol, hydrogen, hydrogen sulphide, methane, cork, grease, paper, wax, *etc*; violent reaction or explosion with water (B201–3).

First aid   *Vapour inhaled :* standard treatment (p 109).
Affected eyes : standard treatment (p 109).
Skin contact : standard treatment (p 108).
Mouth contact : standard treatment (p 109).

Spillage disposal   Instruct others to keep at a safe distance. Wear breathing apparatus and gloves. Absorb on sand, shovel into bucket(s), transport to safe open area and tip slowly into large volume of water when decomposed, decant to waste diluting greatly with running water. Site of spillage should be ventilated after washing thoroughly with water.

Bromine trioxide

Hazardous reactions   Unstable — violently explosive in presence of trace impurities (B206).

N-Bromoacetamide

Hazardous reaction   Decomposes rapidly when hot in presence of moisture and light (B302).

Bromoacetic acid

Colourless to pale brown solid; mp 50 °C; soluble in water.

CAUSES SEVERE BURNS

Prevent contact with skin and eyes.

Toxic effects   Contact of the solid or solution with the eyes causes severe burns. The effect on the skin is not immediate and blisters may not appear for 12 hours or more after contact. Can be assumed to cause severe internal irritation and damage if taken by mouth.

First aid   *Affected eyes :* standard treatment (p 109).
Skin contact : standard treatment (p 108).
If swallowed : standard treatment (p 109).

Spillage
disposal

Wear goggles and gloves. Spread soda ash liberally over the spillage and mop up cautiously with plenty of water — run to waste diluting greatly with running water.

Bromoacetone oxime

Hazardous
reaction

Explodes during distillation (B360).

α-Bromoacetophenone – *see* Phenacyl bromide

Bromoacetylene

Hazardous
reaction

Unstable — may burn or explode on contact with air (B280).

1-Bromoaziridine

Hazardous
reaction

Unstable — explodes during or shortly after distillation (B302).

Bromobenzene

A colourless liquid with an aromatic smell; bp 156 °C: immiscible with water.

FLAMMABLE
HARMFUL BY INHALATION
HARMFUL IN CONTACT WITH SKIN

Avoid breathing vapour. Avoid contact with skin and eyes.

Toxic effects

Little is known about the toxic properties, but its relationship with benzene suggests caution in handling. The vapour may be narcotic in high concentrations. It should be assumed to be poisonous through skin absorption and if taken by mouth.

First aid

Vapour inhaled: standard treatment (p 109).
Affected eyes: standard treatment (p 109).
Skin contact: standard treatment (p 108).
If swallowed: standard treatment (p 109).

Fire hazard

Flash point 51 °C; ignition temp. 566 °C. *Extinguish fire with* dry powder, carbon dioxide or vaporising liquids.

Spillage disposal Shut off all possible sources of ignition. Instruct others to keep at a safe distance. Wear breathing apparatus and gloves. Apply non-flammable dispersing agent and work into an emulsion with brush and water — run this to waste diluting greatly with running water. If dispersant not available, absorb on sand, shovel into bucket(s) and transport to safe open place for burial. Site of spillage should be washed thoroughly with water and soap or detergent.

p-Bromobenzoyl azide

Hazardous reaction Explodes violently above mp 46 °C (B520).

Bromochloromethane (methylene chlorobromide)

Colourless liquid with sweetish odour; bp 69 °C; insoluble in water.

IRRITATING TO SKIN, EYES AND RESPIRATORY SYSTEM

Avoid breathing vapour. Avoid contact with skin and eyes. TLV 200 ppm (1050 mg m^{-3}).

Toxic effects The vapour irritates the respiratory system and the eyes. The liquid irritates the eyes severely. The liquid irritates the skin. Assumed to be irritant and narcotic if taken by mouth.

First aid *Vapour inhaled:* standard treatment (p 109).
Affected eyes: standard treatment (p 109).
Skin contact: standard treatment (p 108).
If swallowed: standard treatment (p 109).

Spillage disposal Instruct others to keep at a safe distance. Wear breathing apparatus and gloves. Apply dispersing agent if available and work to an emulsion with brush and water — run to waste diluting greatly with running water. If dispersant not available, absorb on sand, shovel into bucket and transport to safe open area for atmospheric evaporation. Site of spillage should be thoroughly ventilated.

4-Bromocyclopentene

Hazardous preparation In preparation from 3,5-dibromocyclopentene and lithium tetrahydro-aluminate (B442).

p-Bromo-*N*,*N*-dimethylaniline

Hazardous preparation Exploded during vacuum distillation (B550).

Bromoethane (ethyl bromide)

Colourless, volatile liquid with ethereal odour; bp 38 °C; sparingly soluble in water.

HARMFUL BY INHALATION
IRRITATING TO EYES

Avoid breathing vapour. Avoid contact with skin and eyes. TLV 200 ppm (890 mg m^{-3}).

Toxic effects   The vapour irritates the respiratory system; it has anaesthetic and narcotic effects. The liquid irritates the eyes. The liquid is poisonous if taken by mouth, causing damage to the kidneys. **Chronic effects** Can produce damage to the nervous system.

Hazardous preparation   In preparation from ethanol and bromide (B312).

First aid   *Vapour inhaled:* standard treatment (p 109).
Affected eyes: standard treatment (p 109).
Skin contact: standard treatment (p 108).
If swallowed: standard treatment (p 109).

Spillage disposal   Instruct others to keep at a safe distance. Wear breathing apparatus and gloves. Apply dispersing agent if available and work to an emulsion with brush and water — run this to waste diluting greatly with running water. If dispersant not available, absorb on sand, shovel into bucket(s) and transport to safe open area for atmospheric evaporation. Site of spillage should be thoroughly ventilated.

Bromoform (tribromomethane)

Heavy, colourless liquid with smell like chloroform; bp 150 °C; insoluble in water.

GIVES OFF VERY POISONOUS VAPOUR
CAUSES IRRITATION OF SKIN AND EYES

Avoid breathing vapour. Avoid contact with skin and eyes. TLV (skin) 0.5 ppm (5 mg m^{-3}).

Toxic effects   The vapour is lachrymatory and irritates the respiratory system. The liquid irritates the skin. Ingestion can cause respiratory difficulties, tremors and loss of consciousness.

Hazardous reactions   Reacts violently with acetone if catalysed by powdered KOH or other bases, even in presence of diluting solvents (B225).

First aid   *Vapour inhaled:* standard treatment (p 109).
Affected eyes: standard treatment (p 109).
Skin contact: standard treatment (p 108).
If swallowed: standard treatment (p 109).

Spillage disposal   Instruct others to keep at a safe distance. Wear breathing apparatus and gloves. Apply dispersing agent and work to an emulsion with brush and water — run this to waste diluting greatly with running water. If dispersant not available absorb on sand, shovel into bucket(s) and transport to safe open space for burial. Site of spillage should be washed thoroughly with water and soap or detergent.

Bromomethane (methyl bromide)

Colourless, volatile liquid (bp 4 °C) or gas with faint chloroform-like odour; liquid forms a crystalline hydrate with cold water and penetrates rubber.

TOXIC BY INHALATION
DANGER OF CUMULATIVE EFFECTS
CAUSES BURNS

Prevent inhalation of vapour. Prevent contact with skin and eyes. TLV (skin) 15 ppm (60 mg m^{-3}).

Toxic effects	Short exposures to high concentrations of vapour cause headache, dizziness, nausea, vomiting and weakness; this may be followed by mental excitement, convulsions and even acute mania. The longer inhalation of lower concentrations may lead to bronchitis and pneumonia. Both the vapour and liquid cause severe damage to the eyes. The liquid burns the skin, blisters appearing several hours after contact; itching and reddening of the skin may precede this. Assumed to be very poisonous if taken by the mouth.
Hazardous reactions	Forms pyrophoric Grignard-type compounds with zinc, aluminium and magnesium (B239).
First aid	*Vapour inhaled:* standard treatment (p 109). *Affected eyes:* standard treatment (p 109). *Skin contact:* standard treatment (p 108). *If swallowed:* standard treatment (p 109).
Spillage disposal	If ampoule is broken, instruct others to keep at a safe distance. Wear breathing apparatus and gloves. Organise ventilation of area to dispel vapour completely. Leaking cylinder should be placed in a well ventilated fume cupboard and vented slowly until discharged.

2-Bromomethylfuran

Hazardous reaction	Very unstable – will explode violently (B436).

3-Bromopropene – *see* Allyl bromide

3-Bromopropyne (1-bromoprop-2-yne; propargyl bromide)

Hazardous reactions	Classed as extremely shock-sensitive. Also danger of explosion in contact with copper, high-copper alloys, mercury and silver (B347).

N-Bromosuccinimide

White to pale buff crystalline solid smelling faintly of bromine; mp 177–181 °C with decomposition.

CAUSES BURNS

Avoid breathing dust. Prevent contact with skin or eyes.

Toxic effects	Irritates or burns the skin, eyes or respiratory system. Strongly irritant if taken by mouth.
Hazardous reactions	Reacts violently with aniline, diallyl sulphide, hydrazine hydrate (B384).
First aid	*Dust inhaled:* standard treatment (p 109). *Affected eyes:* standard treatment (p 109). *Skin contact:* standard treatment (p 109).
Spillage disposal	Wear face-shield or goggles, and gloves. Mop up with plenty of water and run to waste, diluting greatly with running water.

N-Bromotetramethylguanidine

Hazardous reaction	Unstable and explodes if heated above 50 °C (B456).

α-Bromotoluene – *see* Benzyl bromide

Buta-1,3-diene (vinylethylene)

Colourless gas; bp −4.7 °C; insoluble in water.

EXTREMELY FLAMMABLE

Avoid breathing gas. TLV 1000 ppm (2200 mg m^{-3}).

Toxic effects	The gas is of low toxicity but has narcotic effects in high concentration and can irritate the skin.
Hazardous reactions	May explode when heated under pressure. Peroxides formed on long contact with air are explosive but may also initiate polymerisation. Violent or explosive reactions with crotonaldehyde, nitrogen oxide, sodium nitrite (B389–390).
First aid	*Gas inhaled:* standard treatment (p 109).
Fire hazard	Flash point below −7 °C; explosive limits 2–11.5 %; ignition temp. 429 °C. Since the gas is supplied in a cylinder, turning off the valve will reduce any fire involving it; if possible cylinders should be removed quickly from an area in which a fire has developed.
Disposal	Surplus gas or leaking cylinder can be vented slowly to air in a safe open area or gas burnt off in a suitable burner.

Buta-1,3-diyne

Hazardous reactions	Polymerises rapidly above 0 °C, a gas above 10 °C. Potentially very explosive (B382).

Butane

Colourless gas; bp −0.5 °C; sparingly soluble in water.

EXTREMELY FLAMMABLE

Avoid breathing gas.

Toxic effects	The gas has an anaesthetic effect but is not toxic.
First aid	*Gas inhaled in quantity:* standard treatment (p 109).
Fire hazard	Flash point −60 °C; explosive limits 1.9–8.5%; ignition temp. 405 °C. Since the gas is supplied in a cylinder, turning off the valve will reduce any fire involving it; if possible, cylinders should be removed quickly from an area in which a fire has developed.
Disposal	Surplus gas or leaking cylinder can be vented slowly to air in a safe open area or gas burnt off in a suitable burner.

iso-Butane – *see* Isobutane

Butane-2,3-dione monoxime

Hazardous reaction	Has exploded during vacuum distillation (B404).

Butan-1-ol (n-butyl alcohol; 1-butanol)

Colourless liquid; bp 118 °C; 9 cm^3 dissolves in about 100 cm^3 water at 25 °C.

and

Butan-2-ol (s-butyl alcohol)

Colourless liquid; bp 99.5 °C; one part dissolves in 12 parts of water at about 25 °C.

FLAMMABLE
HARMFUL VAPOUR

Avoid breathing vapour. Avoid contact with skin and eyes. TLV 50 ppm (150 mg m^{-3}) and 150 ppm (450 mg m^{-3}) respectively.

Toxic effects   Vapour may irritate the respiratory system and the eyes. The liquid irritates the eyes and may irritate the skin causing dermatitis. If taken by mouth may cause headache, dizziness, drowsiness and narcosis.

First aid   *Vapour inhaled:* standard treatment (p 109).
Skin contact: standard treatment (p 109).
If swallowed: standard treatment (p 108).
Affected eyes: standard treatment (p 109).

Fire hazard   Flash points 24–29 °C; explosive limits 1.4–11%; ignition temps. 365–406 °C. *Extinguish fire with* water spray, dry powder, carbon dioxide or vaporising liquids.

Spillage disposal   Shut off all possible sources of ignition. Wear face-shield or goggles, and gloves. Mop up with plenty of water and run to waste, diluting greatly with running water. Ventilate area well to evaporate remaining liquid and dispel vapour.

Butanone (methyl ethyl ketone)

Colourless liquid with smell like acetone; bp 80 °C; one part dissolves in about 4 parts of water at 25 °C.

EXTREMELY FLAMMABLE

Avoid breathing vapour. Avoid contact with eyes. TLV 200 ppm (590 mg m^{-3}).

Toxic effects   Inhalation of vapour may cause dizziness, headache, nausea. The liquid irritates the eyes and may cause severe damage. If swallowed may cause gastric irritation and narcosis.

Hazardous reactions   Vigorous reaction with chloroform in presence of bases. Explosive peroxides formed by action of H_2O_2/HNO_3 (B407).

First aid   *Vapour inhaled:* standard treatment (p 109).
Affected eyes: standard treatment (p 109).
Skin contact: standard treatment (p 108).
If swallowed: standard treatment (p 109).

Fire hazard   Flash point −7 °C; explosive limits 2–10%; ignition temp. 515 °C. *Extinguish fire with* water spray, dry powder, carbon dioxide or vaporising liquids.

Spillage disposal   Shut off all possible sources of ignition. Wear face-shield and gloves. Mop up with plenty of water and run to waste diluting greatly with running water. Ventilate area well to evaporate remaining liquid and dispel vapour.

Butenes (butylenes)

Colourless gases; boil between −6 °C and 4 °C; all are insoluble in water.

EXTREMELY FLAMMABLE

Avoid breathing gases.

Toxic effects   The butenes are generally regarded as simple asphyxiants with some anaesthetic properties

First aid   *Gas inhaled in quantity:* standard treatment (p 109).

Fire hazards	Flash points are below −7 °C; explosive limits 1.6–9.7%; ignition temps. between 230 °C and 390 °C. Since the gases are supplied in cylinders, turning off the valve will reduce any fire involving them; if possible cylinders should be removed quickly from an area in which a fire has developed.
Disposal	Surplus gas or leaking cylinder can be vented slowly to air in a safe, open area or gas burnt off through a suitable burner.

But-1-en-3-yne

Hazardous reaction	Forms explosive compounds on contact with air (B384).

Butoxyacetylene

Hazardous reaction	Explodes on heating in sealed tubes (B497).

2-Butoxyethanol (ethylene glycol monobutyl ether)

Colourless liquid; bp 171 °C; one part dissolves in about 20 parts water at 25 °C

HARMFUL VAPOUR

Avoid contact with skin and eyes. TLV (skin) 50 ppm (240 mg m^{-3}).

Toxic effects	The liquid irritates the eyes and may irritate the skin. May have irritant and narcotic action if taken by mouth.
Hazardous reaction	Liable to form explosive peroxides on exposure to air and light which should be decomposed before the ether is distilled to small volume.
First aid	*Affected eyes:* standard treatment (p 109). *Skin contact:* standard treatment (p 108). *If swallowed:* standard treatment (p 109).
Fire hazard	Flash point 61 °C; explosive limits 1.1–12.7%. *Extinguish fire with* water spray, dry powder, carbon dioxide or vaporising liquids.
Spillage disposal	Wear face-shield or goggles, and gloves. Mop up with plenty of water and run to waste diluting greatly with running water. Ventilate area well to evaporate remaining liquid and dispel vapour.

Butyl acetate

Colourless liquid; bp 125 °C; slightly soluble in water.

FLAMMABLE
HARMFUL VAPOUR
IRRITATING TO SKIN, EYES AND RESPIRATORY SYSTEM

Avoid breathing vapour. Avoid contact with skin and eyes. TLV 150 ppm (710 mg m^{-3}).

Toxic effects   The vapour may irritate the respiratory system and cause headache and nausea. The liquid will irritate the eyes and may cause conjunctivitis. The liquid may irritate the skin and cause dermatitis. If taken by mouth, the liquid will cause irritation and act as a depressant of the central nervous system.

First aid   *Vapour inhaled:* standard treatment (p 109).
Affected eyes: standard treatment (p 109).
Skin contact: standard treatment (p 108).
If swallowed: standard treatment (p 109).

Fire hazard   Flash point 27 °C; explosive limits 1.4–7.6%; ignition temp. 399 °C. *Extinguish fire with* foam, dry powder, carbon dioxide or vaporising liquid.

Spillage disposal   Shut off all possible sources of ignition. Instruct others to keep at a safe distance. Wear goggles or face-shield, and gloves. Apply non-flammable dispersing agent if available and work to an emulsion with brush and water — run this to waste diluting greatly with running water. If dispersant not available, absorb on sand, shovel into bucket and transport to safe open area for atmospheric evaporation or burial. Ventilate area well to evaporate remaining liquid and dispel vapour.

Butyl acrylate

A colourless liquid; bp 145 °C; immiscible with water.

FLAMMABLE
HARMFUL BY SKIN ABSORPTION
IRRITATING TO SKIN, EYES AND RESPIRATORY SYSTEM

Avoid breathing vapour. Avoid contact with skin and eyes.

Toxic effects   The liquid irritates the skin and eyes. Assumed to be poisonous if taken by mouth.

First aid   *Affected eyes:* standard treatment (p 109).
Skin contact: standard treatment (p 108).
If swallowed: standard treatment (p 109).

Fire hazard   Flash point 49 °C. *Extinguish fire with* foam, dry powder, carbon dioxide or vaporising liquids.

Spillage disposal   Shut off all possible sources of ignition. Instruct others to keep at a safe distance. Wear breathing apparatus and gloves. Apply non-flammable dispersing agent and work to an emulsion with brush and water — run this to waste diluting greatly with running water. If dispersant not available, absorb on sand, shovel into bucket and transport to safe open area for burial. Site of spillage should be washed thoroughly with water and soap or detergent.

n-Butyl-alcohol – *see* Butan-1-ol

iso-Butyl alcohol – *see* Isobutyl alcohol

s-Butyl alcohol – *see* Butan-2-ol

t-Butyl alcohol – *see* 2-Methylpropan-2-ol

Butylamines

The butylamines (n-butylamine, isobutylamine, s-butylamine, butylamine, di-n-butylamine, di-isobutylamine, di-s-butylamine and tri-n-butylamine) are colourless liquids with an ammoniacal odour; miscible with water.

> EXTREMELY FLAMMABLE
> HARMFUL VAPOUR
> IRRITATING TO SKIN, EYES AND RESPIRATORY SYSTEM

Avoid breathing vapour. Avoid contact with skin and eyes.
TLV 5 ppm (15 mg m⁻³).

TLV 5 ppm (15 mg m^{-3}).

Toxic effects The vapours irritate the respiratory system. The vapours irritate and the liquids burn the eyes. The liquids may cause skin and eye burns. Assumed to be very irritant and poisonous if taken by mouth.

First aid *Vapour inhaled:* standard treatment (p 109).
Affected eyes: standard treatment (p 109).
Skin contact: standard treatment (p 108).
If swallowed: standard treatment (p 109).

Fire hazard Flash points −12 °C (n-), −9 °C (iso-); explosive limits 1.7–9.8% (n-) 1.7–9.8% (t); ignition temp. 312 °C (n-), 378 °C (iso-). *Extinguish fire with water spray, dry powder, carbon dioxide or vaporising liquid.*

Spillage disposal Shut off all possible sources of ignition. Instruct others to keep at a safe distance. Wear breathing apparatus and gloves. Mop up with plenty of water and run to waste diluting greatly with running water. Ventilate area well to evaporate remaining liquid and dispel vapour.

t-Butyl azidoformate

Hazardous reaction Has exploded during distillation at 74 °C/92mbar (92×10^2 Pa) (B447).

t-Butyl diazoacetate

Hazardous reaction Distillation under vacuum potentially hazardous (B496).

Butyldichloroborane

Hazardous reactions Ignites on prolonged exposure to air; hydrolysis may be explosive (B413).

iso-Butylene – *see* 2-Methylpropene

t-Butyl hydroperoxide

Colourless liquid; stable below 75 °C; slightly soluble in water.

FLAMMABLE
IRRITATING TO SKIN AND EYES

Avoid contact with skin, eyes and clothing. Avoid contamination with other materials. Store in a cool place.

Toxic effects	The liquid irritates the eyes and skin. Assumed to be toxic if taken by mouth.
Hazardous reaction	Liable to explode when distilled (B422).
First aid	*Affected eyes:* standard treatment (p 109). *Skin contact:* standard treatment (p 108). *If swallowed:* standard treatment (p 109).
Fire hazard	Flash point 27 °C. *Extinguish fire with* water spray, dry powder, carbon dioxide or vaporising liquids.
Spillage disposal	Wear face-shield or goggles, and gloves. Mix with sand and transport to safe open area for burial.

t-Butyl hypochlorite

Hazardous storage	Ampoules liable to burst unless stored cool and in dark (B415).

Butyl methacrylate

Colourless, mobile liquid, normally supplied containing a small amount of stabilising agent (*eg* 0.1% quinol); bp 163 °C; slightly soluble in water.

FLAMMABLE
IRRITATING TO SKIN, EYES AND RESPIRATORY SYSTEM

Avoid breathing vapour. Avoid contact with eyes and skin.

Toxic effects	The vapour irritates the eyes and respiratory system. The liquid irritates the eyes and may irritate the skin. Considered moderately toxic if taken by mouth.
First aid	*Vapour inhaled:* standard treatment (p 109). *Affected eyes:* standard treatment (p 109). *Skin contact:* standard treatment (p 108). *If swallowed:* standard treatment (p 109).
Fire hazard	Flash point 52 °C. *Extinguish fire with* foam, dry powder, carbon dioxide or vaporising liquids.
Spillage disposal	Shut off all possible sources of ignition. Wear breathing apparatus and gloves. Apply non-flammable dispersing agent and work to an emulsion with brush and water — run this to waste diluting greatly with running water. If dispersant not available, absorb on sand, shovel in to bucket(s) and transport to safe open area for burial. Site of spillage should be washed thoroughly with water and soap or detergent.

iso-Butyl methyl ketone
– see **4-methylpentan-2-one**

t-Butyl peracetate

Hazardous reaction Explodes violently when rapidly heated (B506).

t-Butyl perbenzoate

Hazardous reaction Exploded during interrupted vacuum distillation (B585).

But-1-yne (ethylacetylene)

Colourless liquid and gas; bp 8.1 °C; insoluble in water.

EXTREMELY FLAMMABLE

Avoid breathing gas.

Toxic effects The toxicity has not been fully investigated. It probably has some anaesthetic activity and can act as a simple asphyxiant.

First aid *Gas inhaled in quantity :* standard treatment (p 109).

Fire hazard No figures on flash point *etc* available. Since the gas is supplied in a cylinder, turning off the valve will reduce any fire involving it. If possible cylinders should be removed quickly from an area in which a fire has developed.

Disposal Surplus gas or leaking cylinder can be vented slowly to air in a safe open area or gas burnt off in a suitable burner.

But-2-ynedinitrile (dicyanoacetylene)

Hazardous reaction Potentially explosive in pure state or in concentrated solutions (B432).

But-2-yne-1,4-diol

Hazardous reactions Explodes on distillation in presence of traces of alkali or alkaline earth hydroxides or halides (B397).

But-2-yne-1-thiol

Hazardous reaction   Exposure to air results in polymer which may explode on heating (B401).

Butyraldehyde

Colourless liquid; bp 76 °C; 7 parts dissolve in about 100 parts water at 25 °C.

EXTREMELY FLAMMABLE
IRRITATING TO SKIN, AND EYES

Avoid breathing vapour. Avoid contact with skin and eyes.

Toxic effects   The vapour may irritate the eyes and respiratory system. The liquid will irritate the eyes and may irritate the skin. Assumed to be irritant and possibly narcotic if swallowed.

First aid   *Vapour inhaled:* standard treatment (p 109).
Affected eyes: standard treatment (p 109).
Skin contact: standard treatment (p 108).
If swallowed: standard treatment (p 109).

Fire hazard   Flash point −6.7 °C; ignition temp. 230 °C. *Extinguish fire with* water spray, foam, dry powder, carbon dioxide or vaporising liquids.

Spillage disposal   Shut off all possible sources of ignition. Wear face-shield or goggles, and gloves. Mop up with plenty of water and run to waste, diluting greatly with running water. Ventilate area well to evaporate remaining liquid and dispel vapour.

Butyraldehyde oxime

Hazardous reaction   Large batch exploded violently during vacuum distillation (B417).

Butyric acid

Colourless, oily liquid with very pungent smell; bp 163.5 °C; miscible with water.

CAUSES BURNS

Prevent contact with skin, eyes and clothing.

Toxic effects   Irritates or burns skin and eyes.

First aid   *Affected eyes:* standard treatment (p 109).
Skin contact: standard treatment (p 108).
If swallowed: standard treatment (p 109).

Spillage disposal   Wear face-shield or goggles and gloves. Mop up with plenty of water and run to waste diluting greatly with water. Ventilate area well to evaporate remaining liquid and dispel vapour.

iso-Butyric acid – *see* Isobutyric acid

Butyric *and* isobutyric anhydrides

Colourless liquids with pungent odour (bp 200 °C and 182 °C respectively) reacting with water to form the corresponding acids.

CAUSE BURNS
IRRITATING TO SKIN, EYES AND RESPIRATORY SYSTEM

Avoid contact with skin and eyes.

Toxic effects	The liquids burn the eyes and may burn the skin. They are irritant and corrosive if taken by mouth.
First aid	*Affected eyes:* standard treatment (p 109). *Skin contact:* standard treatment (p 108). *If swallowed:* standard treatment (p 109).
Spillage disposal	Wear face-shield or goggles, and gloves. Spread soda ash liberally over the spillage and mop up with plenty of water—run to waste, diluting greatly with running water.

Butyronitrile (propyl cyanide)

Colourless liquid; bp 117 °C; immiscible with water but tending to break down to cyanide by hydrolysis.

FLAMMABLE
SERIOUS RISK OF POISONING BY INHALATION OR SWALLOWING

Do not breathe vapour. Prevent contact with skin and eyes.

Toxic effects	Although there is no documented evidence of its toxicity to humans, it has been stated that rats exposed to its vapour rapidly develop weakness, laboured breathing and convulsions which usually result in death. Cases of poisoning seem to warrant the same urgent attention as those caused by hydrogen cyanide.
First aid	Special treatment for hydrogen cyanide (p 104).
Fire hazard	Flash point 26 °C. *Extinguish fire with* dry powder, carbon dioxide or vaporising liquids.
Spillage disposal	Shut off all possible sources of ignition. Instruct others to keep at a safe distance. Wear breathing apparatus and gloves. Apply non-flammable dispersing agent if available and work to emulsion with brush and water—run this waste diluting greatly with running water. If dispersant not available, absorb on sand, shovel into bucket(s) and transport to safe open area for burial. Site of spillage should be washed thoroughly with water and soap or detergent.

Butyryl nitrate

Hazardous reaction	Detonates on heating (B404).

Cadmium (metal)

Hazardous reactions   The powdered metal reacts violently or explosively with fused ammonium nitrate; it reacts vigorously with Se or Te on warming (B627).

Cadmium compounds

Chloride, nitrate and sulphate soluble in water; oxide and carbonate insoluble.

HARMFUL BY INHALATION
HARMFUL IF TAKEN INTERNALLY

Avoid inhaling dust. TLV (cadmium oxide fume) (0.05 mg m^{-3}).

Toxic effects   The inhalation of dust (usually the metal or oxide) irritates the lungs. The compounds cause increased salivation, choking, vomiting, stomach pains and diarrhoea if taken by mouth. **Chronic effects** Prolonged exposure to dust may cause damage to the lungs and kidneys and discoloration of teeth. (See references to toxicity of cadmium compounds in Ch 6.)

First aid   *Dust inhaled:* standard treatment (p 109).
If swallowed: standard treatment (p 109).

Spillage disposal   Cadmium compounds are not so toxic as to present serious disposal problems. The insoluble compounds can be mixed with wet sand, swept up and treated as normal waste. The soluble salts can be mopped up with water and run to waste, diluting greatly with water.

Cadmium diamide

Hazardous reactions   May explode when heated rapidly. Reacts violently with water (B627).

Cadmium diazide

Hazardous reactions   Dry solid explodes on heating or light friction; preparative solution exploded after standing for some hours (B628).

Cadmium propionate

Hazardous reaction   The salt exploded during drying in oven (B496).

Cadmium selenide

Hazardous preparation   Mixtures of powdered metal and selenium may explode (B628).

Caesium (metal)

Hazardous reactions	Ignites immediately in air and oxygen and on contact with water (B717).

Calcium (metal)

Grey metal with silver white surface when cut; mp 850 °C.

CONTACT WITH WATER LIBERATES HIGHLY FLAMMABLE GASES

Avoid contact with skin, eyes and clothing.

Toxic effects	Reaction with moisture on skin and eyes may cause irritation.
Hazardous reactions	Pyrophoric when finely divided; reacts explosively with $PbCl_2$, P_2O_5, S; reaction with water or dilute acids may be violent; ignites in fluorine (B618–619).
First aid	Standard treatments for skin and eye contacts (p 109).
Disposal	Allow to react in large excess of cold water and discharge to waste.

Calcium arsenate
– see **Arsenic compounds**

Calcium bis-*p*-iodylbenzoate

Hazardous reaction	Overdried formulated granules exploded (B599).

Calcium carbide (calcium acetylide)

Hazardous reactions	Incandesces with PbF_2 at room temperature, with HCl on warming, with Mg when heated in air, with Cl_2, Br_2 and I_2 at temperatures over 245 °C; very vigorous reaction with boiling methanol; forms highly sensitive explosive with silver nitrate solution; a mixture with sodium peroxide is explosive (B269–271).

Calcium cyanamide

Grey-black granules or powder; decomposed by water.

HARMFUL BY INHALATION
IRRITATING TO SKIN AND EYES

Avoid inhaling dust. Prevent contact with eyes and skin. TLV 0.5 mg m^{-3}.

Toxic effects	Inhalation of dust can cause severe irritation of the mucous membranes. The material irritates the eyes and can cause conjunctivitis. The material will burn the skin. If taken by mouth, there is severe internal irritation and damage which may result in death.
First aid	*Dust inhaled:* standard treatment (p 109). *Affected eyes:* standard treatment (p 109). *Skin contact:* standard treatment (p 108). *If swallowed:* standard treatment (p 109).
Spillage disposal	Wear face-shield or goggles, and gloves. Mop up with plenty of water and run to waste diluting greatly with water.

Calcium diazide

Hazardous reaction	Explodes on heating at about 150 °C (B623).

Calcium dihydride

Hazardous reactions	Mixtures with various bromates, chlorates, perchlorates explode on grinding; mixture with AgF becomes incandescent on grinding (B622).

Calcium disilicide

Hazardous reactions	Exploded when milled in CCl_4; ignites in close contact with alkali metal fluorides; mixture with potassium or sodium nitrate ignites readily (B625).

Calcium hydrogen di- and tri-fluorides
– *see* **Fluorides**

Calcium hypochlorite (bleaching powder)

White powder, smelling of chlorine; absorbs water and is decomposed by it.

CAUSES IRRITATION OF SKIN AND EYES
HARMFUL IF TAKEN INTERNALLY

Avoid breathing dust. Avoid contact with skin, eyes and clothing.

Toxic effects	Dust irritates the respiratory system; irritates the skin, eyes and alimentary system.
First aid	*Affected eyes:* standard treatment (p 109). *Skin contact:* standard treatment (p 108). *If swallowed:* standard treatment (p 109).
Spillage disposal	Wear face-shield or goggles, and gloves. Mop up with plenty of water and run to waste diluting greatly with running water.

Calcium oxalate – *see* Oxalates

Calcium oxide (lime)

White, amorphous lumps and powder; reacts vigorously with water forming calcium hydroxide.

HARMFUL BY INHALATION
IRRITATING TO SKIN, EYES AND RESPIRATORY SYSTEM

Avoid breathing dust. Avoid contact with skin and eyes.

Toxic effects   The dust irritates the skin, eyes and respiratory system.

Hazardous reactions   Incandesces in contact with liquid HF; mixture with P_2O_5 reacts violently if warmed or moistened; some mixtures with water develop enough heat to ignite combustible materials; glass bottles of the oxide may burst due to hydration expansion when the hydroxide is formed (B623–624).

First aid   *Affected eyes:* standard treatment (p 109).
Skin contact: standard treatment (p 108).
If swallowed: standard treatment (p 109).

Spillage disposal   Wear face-shield or goggles, and gloves. Shovel into dry bucket, transport to safe area and add, a little at a time, to a large quantity of water; after reaction complete, run suspension to waste, diluting greatly with running water.

Calcium peroxide

Hazardous reaction   Grinding with oxidisable materials may cause fire (B624).

Calcium peroxodisulphate

Hazardous reactions   Shock-sensitive; explodes violently (B624).

Calcium silicide

Hazardous reaction   Reacts vigorously with acid; the silanes evolved ignite (B625).

Calcium sulphate

Hazardous reactions   Reduced violently or explosively by Al powder. Contact with diazomethane vapour may result in detonation (B624).

Calcium sulphide

Hazardous reactions Reacts vigorously with chromyl chloride, lead dioxide; explodes with potassium chlorate and potassium nitrate (B625).

Caproic acid – *see* Hexanoic acid

ε-Caprolactam (2-oxohexamethylenimine, hexanolactam)

Hygroscopic leaflets; mp 70 °C; freely soluble in water.

CAUSES IRRITATION OF SKIN AND EYES

Avoid contact with skin, eyes and clothing.

Toxic effects Can cause local irritation.

First aid *Affected eyes:* standard treatment (p 109).
Skin contact: standard treatment (p 108).
If swallowed: standard treatment (p 109).

Spillage disposal Wear face-shield or goggles, and gloves. Mop up with plenty of water and run to waste diluting greatly with running water.

Carbon

Hazardous reactions Activated carbon is a potential fire hazard; contamination with drying oils or oxidising agents may ignite it spontaneously; numerous oxidants (O_2, oxides, peroxides, oxosalts, halogens, interhalogens, *etc*) in intimate contact with carbon, may cause ignition or explosion (B214–215).

Carbon dioxide (solid)

Whereas carbon dioxide presents negligible hazards in the laboratory as a gas, the solidified gas presents the risk of unpleasant skin burns resulting from the handling of the material without using adequately thick gloves. Burns received in this way are akin to frostbite and require medical attention.

Carbon disulphide (carbon bisulphide)

Colourless to yellow liquid, with unpleasant odour; bp 46 °C; immiscible with water.

EXTREMELY FLAMMABLE
GIVES OFF VERY POISONOUS VAPOUR

Avoid breathing vapour. TLV (skin) 20 ppm (60 mg m^{-3}).

Toxic effects High concentrations when inhaled produce narcotic effects and may result in unconsciousness. The liquid and vapour irritate the eyes. The liquid is poisonous if taken by mouth. **Chronic effects** Repeated inhalation of the vapour over a period may cause severe damage to the nervous system, including failure of vision, mental disturbance and paralysis. (*See* references to toxicity of carbon disulphide in Ch 6.)

Hazardous reactions Many fires and explosions have been caused by the ignition of the vapour from liquid poured down laboratory sinks; explosion may result from mixing with liquid chlorine in presence of iron; ignites on contact with fluorine; reacts with azide solutions to form explosive azidodithioformates; reacts with zinc dust with incandescence (B265).

First aid *Vapour inhaled :* standard treatment (p 109).
Affected eyes : standard treatment (p 109).
If swallowed : standard treatment (p 109).

Fire hazard Flash point $-30\,°C$; explosive limits 1–44%; ignition temp. 100 °C. *Extinguish fire with* foam, dry powder, carbon dioxide or vaporising liquids.

Spillage disposal Shut off all possible sources of ignition. Instruct others to keep at a safe distance. Wear breathing apparatus and gloves. Apply non-flammable dispersing agent if available and work to an emulsion with water and brush—run this to waste diluting greatly with running water. If dispersant not available, absorb on dry sand and transport to safe open area for atmospheric evaporation. Ventilate area of spillage thoroughly to dispel vapour.

Carbon monoxide (coal gas constituent)

Colourless, odourless gas, only slightly soluble in water.

EXTREMELY FLAMMABLE
VERY TOXIC BY INHALATION

Avoid breathing gas. TLV 50 ppm (55 mg m^{-3}).

Toxic effects Causes unconsciousness due to anoxia resulting from the combination of carbon monoxide with haemoglobin. Gas in lower concentrations causes headache, throbbing of temples, nausea, followed possibly by collapse. **Chronic effects** Headache, nausea and weakness. (*See* references to toxicity of carbon monoxide in Ch 6.)

Hazardous reactions May react explosively with F_2 and O_2 in the preparation of $C_2F_2O_4$ — also with BrF_3 and other interhalogens; reacts readily with K to form explosive carbonylpotassium; explosion occurred during reduction of Fe_2O_3 (B263).

First aid *Vapour inhaled :* standard treatment (p 109).

Fire hazard Explosive limits 12.5–74%; ignition temp. 609 °C. Since the gas is supplied in a cylinder, turning off the valve will reduce any fire involving it; if possible, cylinders should be removed quickly from an area in which a fire has developed.

Disposal Surplus gas or a leaking cylinder can be vented slowly to air in a safe open area or burnt off in a suitable gas burner.

Carbon tetrabromide

Colourless crystals; insoluble in water.

HARMFUL BY INHALATION

Avoid breathing vapour. Avoid contact with skin and eyes. TLV 0.1 ppm (1.1 mg m^{-3}).

Toxic effects	The vapour is narcotic in high concentrations. Assumed that the solid is poisonous by skin absorption and if taken by mouth.
First aid	*Vapour inhaled:* standard treatment (p 109). *Affected eyes:* standard treatment (p 109). *Skin contact:* standard treatment (p 108). *If swallowed:* standard treatment (p 109).
Spillage disposal	Wear face-shield or goggles, and gloves. Mix with sand and transport to safe open area for burial.

Carbon tetrachloride (tetrachloromethane)

Heavy, colourless liquid with a characteristic odour; bp 77 °C; immiscible with water. When used for extinguishing fires, phosgene *(qv)*, which is very poisonous, is liable to be formed.

GIVES OFF POISONOUS VAPOUR

Avoid breathing vapour. Avoid contact with skin and eyes. TLV 10 ppm (65 mg m^{-3}).

Toxic effects	Inhalation of high concentrations of vapour can cause headache, mental confusion, depression, fatigue, loss of appetite, nausea, vomiting and coma, these symptoms sometimes taking many hours to appear. The vapour and liquid irritate the eyes. It causes internal irritation, nausea and vomiting if taken by mouth; there is damage to the liver, kidneys, heart and nervous system and small doses have caused death. **Chronic effects** Prolonged inhalation of low concentrations may cause headache, nausea, stupor, vomiting, bronchitis and jaundice. Dermatitis may follow repeated contact with the liquid. (*See* references in Ch 6.)
Hazardous reactions	Exploded when milled with calcium disilicide; CCl_4 solutions of ClF_3 may detonate; initiated by dibenzoyl peroxide, mixtures with ethylene may explode; may react violently with dimethylformamide in presence of Fe; may react violently or explosively with F_2, Al, Ba, Be, K, Na, Zn. (B219–220).
First aid	*Vapour inhaled:* standard treatment (p 109). *Affected eyes:* standard treatment (p 109). *Skin contact:* standard treatment (p 108). *If swallowed:* standard treatment (p 109).
Spillage disposal	Instruct others to keep at a safe distance. Wear breathing apparatus and gloves. Apply dispersing agent if available and work to an emulsion with brush and water — run this to waste diluting greatly with running water. If dispersant not available, absorb on sand, shovel into bucket(s) and transport to safe open area for atmospheric evaporation. Ventilate area of spillage thoroughly to dispel vapour.

Carbonyl diazide

Hazardous reaction Violently explosive solid, usable only in solution (B262).

Carbonyllithium

Hazardous reaction Explodes on contact with water (B259).

Carbonylmetals Review of group (B27).

Carbonylpotassium

Hazardous reactions Reacts violently with oxygen; explodes on heating in air or on contact with water (B258).

Carbonylsodium

Hazardous reactions Explodes on heating in air and on contact with water (B262).

Catechol (pyrocatechol; 1,2-dihydroxybenzene)

Colourless crystalline powder; soluble in water.

HARMFUL BY SKIN ABSORPTION
CAUSES BURNS

Avoid contact with skin and eyes.

Toxic effects Irritates the eyes severely, causing burns. Irritates the skin and causes poisoning by absorption. Assumed to be irritant and poisonous if taken by mouth.

Hazardous reaction Explodes on contact with concentrated nitric acid (B488).

First aid *Affected eyes:* standard treatment (p 109).
Skin contact: standard treatment (p 108).
If swallowed: standard treatment (p 109).

Spillage disposal Wear goggles and gloves. Mop up with plenty of water and run to waste diluting greatly with running water.

Caustic potash – *see* Potassium hydroxide

Caustic soda – *see* Sodium hydroxide

Cellulose

Hazardous reactions
Reactions with calcium oxide and oxidants such as bleaching powder, perchlorates, perchloric acid, sodium chlorate, fluorine, nitric acid, $NaNO_2$, $NaNO_3$, Na_2O_2 are reviewed (B27).

Cellulose nitrate (nitrocellulose)

White or yellowish-white amorphous powder or matted filaments; ignites at 160–170 °C; usually supplied moistened with alcohol.

EXPLOSIVE WHEN DRY

Hazardous reactions
The combustion and explosion of cellulose nitrate are reviewed (B28). The hazards of cellulose nitrate (presented also as pyroxylin) centre on its flammability and explosive potential; the latter is classified as 'moderate' by one authority.

Spillage disposal
Shut off all sources of ignition. After damping with water transfer the nitrocellulose to an iron, steel or tinned container and add to it an equal volume of 10% sodium hydroxide solution. Allow to stand for an hour and then pour to waste, diluting greatly with running water.

Cerium (metal)

Hazardous reactions
Ignites and burns brightly at 160 °C; reaction with Zn is explosively violent, and with Sb or Bi very exothermic; Ce filings ignite in Cl_2 or Br_2 at about 215 °C; reacts violently with P above 400 °C (B629).

Cerium nitride

Hazardous reaction
Contact with limited amount of water or dilute acid causes rapid incandescence with ignition (B630).

Cerium trihydride

Hazardous reaction
May ignite in moist air (B630).

Chloral (trichloroacetaldehyde)

Colourless, oily liquid with pungent, irritating odour; bp 98 °C; soluble in water forming chloral hydrate.

IRRITATING TO SKIN, EYES AND RESPIRATORY SYSTEM

Avoid contact with skin, eyes and clothing.

Toxic effects	The vapour irritates the respiratory system and eyes. The liquid irritates the eyes severely. The liquid irritates the skin. If taken by mouth it will show the effects of chloral hydrate, namely nausea, vomiting, coldness of extremities and unconsciousness.
First aid	*Vapour inhaled:* standard treatment (p 109). *Affected eyes:* standard treatment (p 109). *Skin contact:* standard treatment (p 108). *If swallowed:* standard treatment (p 109).
Spillage disposal	Wear face-shield or goggles, and gloves. Mop up with plenty of water, and run to waste diluting greatly with water.

Chloral hydrate

Colourless crystals with acrid odour and bitter taste; soluble in water.

IRRITATING TO SKIN AND EYES

Avoid contact with skin and eyes.

Toxic effects	Irritates the skin and eyes. If taken by mouth it may cause nausea, vomiting, coldness of extremities and unconsciousness.
First aid	*Affected eyes:* standard treatment (p 109). *Skin contact:* standard treatment (p 108). *If swallowed:* standard treatment (p 109).
Spillage disposal	Wear face-shield or goggles, and gloves. Mop up with plenty of water and run to waste diluting greatly with running water.

Chloric acid

Hazardous reactions	Aqueous solution explodes if evaporated too far; it ignites filter paper and explodes with copper sulphide if concentrated; reactions with other oxidisable substances similar to those of chlorates (B640).

Chlorine

Greenish-yellow gas with irritating odour; soluble in water.

TOXIC BY INHALATION
IRRITATING TO SKIN, EYES AND RESPIRATORY SYSTEM

Avoid breathing gas. TLV 1 ppm (3 mg m^{-3}).

Toxic effects The gas causes severe lung irritation and damage. The gas irritates the eyes and can cause conjunctivitis. In high concentrations the gas irritates the skin.

Hazardous reactions Numerous reports of violent or explosive reactions with alcohols, BrF_3, CS_2, dibutyl phthalate, Cs_2O, diethyl ether, F_2O_2, F_2, glycerol, hexachlorodisilane, hydrocarbons, H_2; metal acetylides, carbides, hydrides and phosphides; Al, Bi, Ca, Cu, Fe, Ge, K, Mg, Mn, Na, Ni, Sb, Sn, Th, U, V, Zn; nitrogen compounds; AsH_3, PH_3, SiH_4, B_2H_6, SbH_3; P, B, C, As, Te; F_2O, P_2O_3, silicones, steel, metal sulphides, synthetic rubber, trialkylboranes (B665–672).

First aid *Gas inhaled:* standard treatment (p 109).
Affected eyes: standard treatment (p 109).
Skin contact: standard treatment (p 109).

Disposal Surplus gas or leaking cylinder can be vented slowly into water-fed scrubbing tower or column, or into a fume cupboard served by such a tower.

Chlorine azide

Hazardous reactions Extremely unstable, usually exploding violently without cause (B 659).

Chlorine dioxide

Hazardous reactions Explodes violently under slightest provocation (reference to guide on use provided); explodes on mixing with CO, on shaking with Hg, in contact with solid KOH or concentrated solution; phosphorus, sulphur, sugar or combustible materials ignite on contact and may cause explosion (B664).

Chlorine fluoride

Hazardous reactions Powerful oxidant reacting violently with a wide range of materials (reference given to information sheet) (B631).

Chlorine nitrate

Hazardous reactions Reacts explosively with alcohols, ethers and most organic materials (B657).

Chlorine perchlorate

Hazardous reaction Shock-sensitive and liable to explode (B683).

Chlorine trifluoride

Colourless gas or yellow-green liquid with somewhat sweet but highly irritant smell; bp 11.75 °C.

CONTACT WITH COMBUSTIBLE MATERIAL MAY CAUSE FIRE
REACTS VIOLENTLY WITH WATER
VERY TOXIC BY INHALATION
CAUSES SEVERE BURNS

Prevent inhalation of vapour. Prevent contact with skin, eyes and clothing. TLV 0.1 ppm (0.4 mg m^{-3}).

Toxic effects	The vapour severely irritates the eyes, skin and respiratory system, and may cause burns. The liquid severely burns all human tissue.
Hazardous reactions	Reacts violently or explosively with strong nitric and sulphuric acids, ammonium fluoride, carbon tetrachloride, ammonia, coal-gas, hydrogen, hydrogen sulphide, iodine; numerous metals and non-metals and their oxides; nitro compounds and organic materials generally; the reaction with water is violent and may be explosive even with ice. References to literature on the handling, *etc* are given (B634).
First aid	*Vapour inhaled:* standard treatment (p 109). *Affected eyes:* standard treatment (p 109). *Skin contact:* standard treatment (p 108). *Mouth contact:* standard treatment (p 109).
Spillage disposal	Instruct others to keep at a safe distance. Wear breathing apparatus and gloves. Spread soda ash liberally over the spillage and mop up cautiously with water — run to waste, diluting greatly with running water.

Chlorine pentafluoride

Hazardous reactions	Very vigorous reaction with water, anhydrous nitric acid (B638).

Chlorites Review of these unstable salts (B29).

Chloroacetamide

Colourless solid; 1 part is soluble in about 10 parts of cold water.

HARMFUL BY INHALATION
AND IF IN CONTACT WITH SKIN

Avoid inhaling dust. Avoid contact with skin and eyes.

Toxic effects	The toxicity of chloroacetamide is not well documented but, in view of the presence of an active chlorine atom, it is assumed that it may have harmful effects.

First aid	*Affected eyes:* standard treatment (p 109).
	Skin contact: standard treatment (p 108).
	If swallowed: standard treatment (p 109).
Spillage disposal	Wear face-shield or goggles, and gloves. Mop up with plenty of water and gloves. Mop up with plenty of water and run to waste diluting greatly with running water.

N-Chloroacetamide

Hazardous reactions	Exploded during desiccation of solid and during concentration of chloroform solution (B303).

Chloroacetic acid

Colourless to pale brown crystals; soluble in water.

CAUSES SEVERE BURNS

Prevent contact with skin and eyes.

Toxic effects	The solid or its solutions severely irritate or burn the eyes. The solid or its solutions produce severe skin burns which may only be apparent several hours after contact. Assumed to cause severe internal irritation and damage if taken by mouth.
First aid	*Affected eyes:* standard treatment (p 109).
	Skin contact: standard treatment (p 108).
	If swallowed: standard treatment (p 109).
Spillage disposal	Wear face-shield or goggles, and gloves. Spread soda ash liberally over the spillage and mop up cautiously with plenty of water — run to waste diluting readily with running water.

Chloroacetone

Hazardous reaction	May polymerise explosively on storage (B356).

ω-Chloroacetophenone
– *see* Phenacyl chloride

Chloroacetyl chloride

Colourless to pale yellow liquid; bp 106 °C; reacts with water.

CAUSES SEVERE BURNS
IRRITATING TO SKIN, EYES AND RESPIRATORY SYSTEM

Prevent inhalation of vapour. Prevent contact with skin and eyes.

Toxic effects   The vapour severely irritates all parts of the respiratory system. The vapour irritates and the liquid burns the eyes. The vapour irritates the skin and the liquid may produce blisters several hours after contact. Assumed to cause severe internal irritation and damage if taken by mouth.

First aid   *Vapour inhaled:* standard treatment (p 109).
Affected eyes: standard treatment (p 109).
Skin contact: standard treatment (p 108).
If swallowed: standard treatment (p 109).

Spillage disposal   Instruct others to keep at a safe distance. Wear breathing apparatus and gloves. Spread soda ash liberally over the spillage and mop up cautiously with water — run this to waste diluting greatly with running water.

Chloroacetylene

Hazardous reaction   May burn or explode in contact with air (B280).

Chloroamine

Hazardous reaction   Stable in ethereal solution, but solvent-free material decomposes violently or explosively (B648).

Chloroanilines

The *o*- and *m*-chloroanilines are yellow to brown liquids (bp 209 °C and 229 °C respectively); *p*-chloroaniline is an almost colourless crystalline solid or powder. All are insoluble in water.

TOXIC IN CONTACT WITH SKIN
GIVES OFF POISONOUS VAPOUR

Avoid breathing vapour. Avoid contact with skin and eyes.

Toxic effects   The inhalation of vapour and absorption through the skin may cause cyanosis, and damage to the liver and kidneys. Assumed that similar poisoning will result from ingestion.

First aid   *Vapour inhaled:* standard treatment (p 109).
Affected eyes: standard treatment (p 109).
Skin contact: standard treatment (p 108).
If swallowed: standard treatment (p 109).

Spillage disposal   Wear breathing apparatus (or face-shield if amount is small) and gloves. Mix with sand and shovel into suitable vessel (glass or enamel) for dispersion in an excess of dilute hydrochloric acid (1 volume concentrated hydrochloric acid in 2 volumes of water). Allow to stand, with occasional stirring, for 24 hours and then run acid extract to waste, diluting greatly with running water and washing the sand. The sand can be treated as normal refuse.

1-Chloroaziridine

Hazardous property   Liable to explode on long storage (B303).

2-Chlorobenzaldehyde

Colourless liquid or crystals; mp 11 °C; very slightly soluble in water.

CAUSES BURNS

Avoid contact with skin, eyes and clothing.

Toxic effects   Contact with skin or eyes may cause irritation or burns. There may be severe irritation and damage if the substance is swallowed.

First aid   *Affected eyes:* standard treatment (p 109).
Skin contact: standard treatment (p 108).
If swallowed: standard treatment (p 109).

Spillage disposal   Wear face-shield or goggles, and gloves. Apply dispersing agent if available and work to an emulsion with brush and water — run this to waste, diluting greatly with running water. If dispersant not available, absorb on sand, shovel into bucket(s) and transport to safe, open area for burial. Site of spillage should be washed thoroughly with water and soap or detergent.

Chlorobenzene (monochlorobenzene)

Clear colourless liquid with a faint, not unpleasant almond-like odour; bp 132 °C; immiscible with water.

FLAMMABLE
TOXIC BY INHALATION
IRRITATING TO SKIN, EYES AND RESPIRATORY SYSTEM

Avoid breathing vapour. Avoid contact with skin, eyes and clothing. TLV 75 ppm (350 mg m^{-3}).

Toxic effects   The vapour may cause drowsiness and unconsciousness. The liquid irritates the skin. The liquid may cause stupor and unconsciousness after a few hours if taken by mouth.

First aid   *Vapour inhaled:* standard treatment (p 109).
Affected eyes: standard treatment (p 109).
Skin contact: standard treatment (p 108).
If swallowed: standard treatment (p 109).

Fire hazard   Flash point 29 °C; explosive mixture 1.3–7.1%; ignition temp. 630 °C. *Extinguish fire with* foam, dry powder carbon dioxide or vaporising liquids.

Spillage disposal   Shut off all possible sources of ignition. Instruct others to keep at a safe distance. Wear breathing apparatus and gloves. Apply non-flammable dispersing agent if available and work to an emulsion with brush and water — run this to waste diluting greatly with running water. If dispersant not available, absorb on sand, shovel into bucket(s) and transport to safe, open area for atmospheric evaporation or burial. Site of spillage should be washed thoroughly with water and soap or detergent.

1-Chlorobenzotriazole

Hazardous reaction   May ignite spontaneously (B470).

1-Chlorobutan-2-one

Hazardous reaction   Bottle of stabilised material exploded spontaneously (B402).

1-Chlorobut-1-en-3-one

Hazardous reaction   Liable to explode soon after preparation (B387).

Chlorocyanoacetylene
– *see* **3-Chloropropiolonitrile**

2-Chloro-1-cyanoethanol
– *see* **3-Chloro-2-hydroxypropiononitrile**

4-Chloro-2,6-dinitroaniline

Hazardous reactions   Dangerous decomposition set in during large-scale manufacture from 4-chloro-2-nitroaniline. Explosion occurred during large-scale diazotisation of the amine (B470).

1-Chloro-2,4-dinitrobenzene (2,4-dinitrochlorobenzene)

Pale yellow crystals; insoluble in water.

SERIOUS RISK OF POISONING
BY INHALATION, SWALLOWING OR SKIN CONTACT

Avoid breathing dust and vapour. Prevent contact with skin and eyes.

Toxic effects   The dust, or vapour from the molten compound, irritates the respiratory system. Irritates the skin and may cause dermatitis. Cyanosis and liver injury may follow inhalation or skin absorption. Assumed to be poisonous if taken by mouth.

First aid   *Dust or vapour inhaled:* standard treatment (p 109).
Affected eyes: standard treatment (p 109).
Skin contact: standard treatment (p 108).
If swallowed: standard treatment (p 109).

Spillage disposal   Wear face-shield or goggles, and gloves. Mix with sand and transport to a safe open area for burial.

1-Chloro-2,3-epoxypropane (epichlorhydrin)

Colourless liquid with irritating chloroform-like odour; bp 118 °C; immiscible with water.

FLAMMABLE
TOXIC IN CONTACT WITH SKIN
GIVES OFF POISONOUS VAPOUR
IRRITATING TO SKIN, EYES AND RESPIRATORY SYSTEM

Prevent inhalation of vapour. Prevent contact with skin and eyes. TLV (skin) 5 ppm (19 mg m^{-3}).

Toxic effects	The vapour irritates the respiratory system and in severe cases can cause respiratory paralysis. The vapour and liquid irritate the eyes and may cause conjunctivitis. Poisonous by skin absorption and ingestion. Skin blistering and severe pain may develop after latent period. **Chronic effects** Prolonged exposure to low concentrations of vapour may cause conjunctivitis, chronic weariness and stomach upset.
Hazardous reactions	Reacts violently with isopropylamine; also with trichloroethylene (B356).
First aid	*Vapour inhaled:* standard treatment (p 109). *Affected eyes:* standard treatment (p 109). *Skin contact:* standard treatment (p 108). *If swallowed:* standard treatment (p 109).
Fire hazard	Flash point 41 °C; *Extinguish fire with* water spray, dry powder, carbon dioxide or vaporising liquids.
Spillage disposal	Shut off all possible sources of ignition. Instruct others to keep at a safe distance. Wear breathing apparatus and gloves. Apply non-flammable dispersing agent if available and work to an emulsion with brush and water — run this to waste diluting greatly with running water. If dispersant not available, absorb on sand, shovel into bucket(s) and transport to safe open area for burial. Site of spillage should be washed thoroughly with water and soap or detergent.

Chloroethane (ethyl chloride)

Colourless gas and liquid with pungent, ethereal odour, which is used as a local anaesthetic and refrigerant; bp 12.4 °C; sparingly soluble in water.

EXTREMELY FLAMMABLE

Avoid breathing vapour. TLV 1000 ppm (2600 mg m^{-3}).

Toxic effects	The vapour is mildly irritating to the mucous membranes; at high concentrations it is narcotic.
First aid	*Vapour inhaled:* standard treatment (p 109). **Fire hazard** Flash point −50 °C; explosive limits 3.6–15.4%; ignition temp. 519 °C. Since the gas is supplied in a cylinder, turning off the valve will reduce any fire involving it; if possible, cylinders should be removed quickly from an area in which a fire has developed.
Disposal	Surplus gas or leaking cylinder can be vented slowly to air in a safe open area or gas burnt off in a suitable burner.

2-Chloroethanol (ethylene chlorohydrin: chloroethyl alcohol)

Colourless liquid with a faint ethereal odour; bp 129 °C; miscible with water.

GIVES OFF VERY POISONOUS VAPOUR

Prevent inhalation of vapour. Prevent contact with skin and eyes.
TLV (skin) 1 ppm (3 mg m^{-3}).

Toxic effects	The vapour causes nausea, headaches, vomiting, stupefaction and unconsciousness. It irritates the mucous membranes. The liquid is rapidly absorbed by the skin, producing similar effects to inhalation. Assumed to be extremely poisonous if taken by mouth.
First aid	*Vapour inhaled:* standard treatment (p 109). *Affected eyes:* standard treatment (p 109). *Skin contact:* standard treatment (p 108). *If swallowed:* standard treatment (p 109).
Spillage disposal	Instruct others to keep at a safe distance. Wear breathing apparatus and gloves. Mop up with plenty of water and run to waste, diluting greatly with running water. Ventilate area well to evaporate remaining liquid and dispel vapour.

2-Chloroethylamine

Hazardous reaction	May polymerise explosively (B319).

Chlorofluoroalkanes

The commonly available compounds in this series, all of which are colourless gases or low-boiling liquids supplied in cylinders, are: chlorodifluoroethane*; chlorodifluoromethane; chloropentafluoroethane; chlorotrifluoromethane; dichlorodifluoromethane; dichlorofluoromethane; dichlorotetrafluoroethane; dichlorofluoroethane; trichlorofluoromethane. They are all relatively innocuous gases, their toxicity being of the same order as nitrogen or carbon dioxide. Only one, marked*, is flammable, the remainder are non-flammable.

Fire hazard	(chlorodifluoroethane) Explosive limits 9–14.8%; ignition temp. 632 °C. Since the gas is supplied in a cylinder, turning off the valve will reduce any fire involving it; cylinders should be removed quickly from an area in which a fire has developed.
Disposal	Surplus gas or a leaking cylinder can be vented slowly in air in a safe open area or gas burnt off in a suitable burner.

Chloroform (trichloromethane)

Colourless volatile liquid with a characteristic odour; bp 61 °C; immiscible with water.

HARMFUL VAPOUR

Avoid breathing vapour. Avoid contact with eyes. TLV 25 ppm (120 mg m^{-3}).

Toxic effects   The vapour has anaesthetic properties, causing drowsiness, giddiness, headache, nausea, vomiting and unconsciousness. The vapour and liquid irritate the eyes causing conjunctivitis. The liquid is poisonous if taken by mouth.

Hazardous reactions   Vigorous reaction with acetone in the presence of KOH or $Ca(OH)_2$; may react explosively with fluorine, N_2O_4, Al, Li, Na, Na/methanol, NaOH/ methanol, sodium methoxide (B226–227).

First aid   *Vapour inhaled:* standard treatment (p 109).
Affected eyes: standard treatment (p 109).
If swallowed: standard treatment (p 109).

Spillage disposal   Instruct others to keep at a safe distance. Wear breathing apparatus and gloves. Apply dispersing agent if available and work to an emulsion with brush and water — run this to waste diluting greatly with running water. If dispersant not available, absorb on sand, shovel into bucket(s) and transport to safe, open area for atmospheric evaporation. Site of spillage should be washed thoroughly with water and soap or detergent.

Chlorogermane

Hazardous reaction   Reacts with ammonia to form explosive product (B638).

3-Chloro-2-hydroxypropiononitrile
(2-chloro-1-cyanoethanol)

Hazardous reaction   May explode during vacuum distillation (B351).

3-Chloro-2-hydroxypropyl perchlorate

Hazardous reaction   Explodes violently on heating (B360).

Chloromethane (methyl chloride)

Colourless gas; bp −24 °C; 2.2 volumes dissolve in 1 volume of water at 20 °C.

EXTREMELY FLAMMABLE
TOXIC BY INHALATION

Avoid breathing vapour. TLV 100 ppm (210 mg m^{-3}).

Toxic effects   Inhalation of vapour may cause dizziness, drowsiness, nausea, stomach pains, visual disturbances, mental confusion and unconsciousness; heavy exposure can be fatal and some symptoms may be delayed. (*See* note in Ch 6, p 84.)

Hazardous reactions	Ignites or explodes on contact with BrF_3 or BrF_5. May react explosively with Mg, K, Na and, probably, Zn (B239).
First aid	*Vapour inhaled:* standard treatment (p 109).
Fire hazard	Flash point below 0 °C; explosive limits 10.7–17.4%; ignition temp. 632 °C. Since the gas is supplied in a cylinder, turning off the valve will reduce any fire involving it; if possible cylinders should be removed quickly from an area in which a fire has developed.
Disposal	Surplus gas or a leaking cylinder can be vented slowly to air in a safe open area or gas burnt off in a suitable burner.

2-Chloromethylfuran

Hazardous reaction	Liable to explode violently owing to polymerisation or decomposition (B436).

4-Chloro-2-methylphenol

Hazardous reaction	Vigorous reaction followed by explosion when large quantity was left in contact with concentrated NaOH solution (B528).

2-Chloromethylthiophen

Hazardous reactions	Unstable and gradually decomposes: closed containers may explode (B437).

N-Chloronitroamines   Review of group (B30).

Chloronitroanilines

2-Chloro-4-nitro- and 4-chloro-2-nitro-anilines are yellow to brownish-yellow powders or crystals; insoluble in water.

> SERIOUS RISK OF POISONING
> BY INHALATION, SWALLOWING
> OR SKIN CONTACT

Prevent contact with skin and eyes.

Toxic effects	Absorption through the skin may cause dermatitis, cyanosis and damage to the liver and kidneys. Assumed that similar poisoning will result from ingestion.
First aid	*Affected eyes:* standard treatment (p 109). *Skin contact:* standard treatment (p 108). *If swallowed:* standard treatment (p 109).
Spillage disposal	Wear face-shield or goggles, and gloves. Mix with sand and transport to a safe open area for burial. Site of spillage should be washed thoroughly with water and soap or detergent.

Chloronitrobenzenes

o-, *m-* and *p-*Chloronitrobenzenes are yellow solids of low melting point; insoluble in water.

> SERIOUS RISK OF POISONING
> BY INHALATION, SWALLOWING OR SKIN CONTACT
> DANGER OF CUMULATIVE EFFECTS
> CAUSES IRRITATION OF SKIN AND EYES

Prevent contact with skin and eyes. Avoid breathing dust and vapour. TLV (*p-*, skin) (1 mg m^{-3}).

Toxic effects The dust or vapour from the molten compounds irritates the respiratory system. Contact with the skin may cause dermatitis. Cyanosis and liver injury may follow inhalation or skin absorption. Assumed to be poisonous if taken by mouth.

Hazardous reaction *p-*Chloronitrobenzene reacted violently and finally explosively when added to a solution of sodium methoxide in methanol (B469).

First aid *Dust or vapour inhaled:* standard treatment (p 109).
Affected eyes: standard treatment (p 109).
Skin contact: standard treatment (p 108).
If swallowed: standard treatment (p 109).

Spillage disposal Wear face-shield or goggles, and gloves. Mix with sand and transport to a safe, open area for burial. Site of spillage should be washed thoroughly with water and soap or detergent.

Chloronitromethane

Hazardous preparation Product of chlorination of nitromethane decomposed explosively during vacuum distillation (B231).

2-Chloro-4-nitrotoluene

Hazardous reaction Residue from vacuum distillation of crude material exploded (B526).

Chlorophenols

*o-*Chlorophenol is a pale brown to brown liquid; *p-*chlorophenol is a colourless to pale brown crystalline solid with a phenolic (carbolic) odour. Both compounds are sparingly soluble in water.

> HARMFUL SUBSTANCE IF TAKEN INTERNALLY
> OR IF IN CONTACT WITH SKIN
> CAUSE BURNS
> IRRITATING TO SKIN, EYES AND RESPIRATORY SYSTEM

Avoid breathing vapour. Prevent contact with skin and eyes.

Toxic effects   The vapour when inhaled irritates the respiratory system. In contact with the eyes they cause irritation or burning. They irritate and may burn the skin and must be assumed to be very poisonous and irritant if taken by mouth.

First aid   *Vapour inhaled:* standard treatment (p 109).
Affected eyes: standard treatment (p 109).
Skin contact: standard treatment (p 108). But see also note on phenol p 107.
If swallowed: standard treatment (p 109).

Spillage disposal   Wear face-shield or goggles, and gloves. Mix with sand and transport to a safe open area for burial. Site of spillage should be washed thoroughly with water and soap or detergent.

Chlorophenyldiazirine
– see Phenylchlorodiazirine

p-Chlorophenyllithium

Hazardous preparation   Traces of O_2 in reaction atmosphere may cause explosion (B469).

Chloropicrin *– see* Trichloronitromethane

Chloropropanes

Colourless, volatile liquids; bps:1-isomer, 47 °C; 2-isomer, 35 °C; immiscible with water.

EXTREMELY FLAMMABLE
HARMFUL BY INHALATION
IRRITATING TO EYES

Avoid breathing vapour. Avoid contact with skin and eyes.

Toxic effects   Inhalation of 1-chloropropane irritates the respiratory system: high concentrations of both isomers cause narcosis. The liquids irritate the eyes. Assumed to be poisonous if taken by mouth.

First aid   *Vapour inhaled:* standard treatment (p 109).
Affected eyes: standard treatment (p 109).
If swallowed: standard treatment (p 109).

Fire hazard   Flash point below −18 °C for 1- isomer, −32 °C for 2-; explosive limits 2.6–11.1% for 1- isomer, 2.8–10.7% for 2-; ignition temp. 520 °C for 1-isomer, 592 °C for 2-. *Extinguish fire with* water spray, foam, dry powder, carbon dioxide or vaporising liquids.

Spillage disposal   Shut off all possible sources of ignition. Instruct others to keep at a safe distance. Wear breathing apparatus and gloves. Apply non-flammable dispersing agent if available and work to an emulsion with brush and water — run this to waste, diluting greatly with running water. If dispersant not available, absorb on sand, shovel into bucket(s) and transport to safe open area for atmospheric evaporation. Site of spillage should be washed thoroughly with water and soap or detergent.

3-Chloropropene – *see* Allyl chloride

3-Chloropropiolonitrile (chlorocyanoacetylene)

Hazardous reaction   Explosion hazard when heated in nearly closed vessel (B342).

3-Chloropropionyl chloride

Colourless liquid with acrid odour; reacts with water, forming chloropropionic acid and hydrogen chloride.

TOXIC BY INHALATION
CAUSES BURNS
IRRITATING TO SKIN AND EYES

Prevent inhalation of vapour. Prevent contact with skin and eyes.

Toxic effects   The vapour irritates the eyes and respiratory system severely. The liquid burns the eyes and skin. Causes severe internal irritation and damage if taken by mouth.

First aid   *Vapour inhaled:* standard treatment (p 109).
Affected eyes: standard treatment (p 109).
Skin contact: standard treatment (p 108).
If swallowed: standard treatment (p 109).

Spillage disposal   Shut off all possible sources of ignition. Instruct others to keep at a safe distance. Wear breathing apparatus and gloves. Absorb on sand, shovel into bucket(s), transport to safe, open area and tip into large volume of water; leave to decompose before decanting the water to waste, diluting greatly with running water. Site of spillage should be ventilated after washing thoroughly with water and soap or detergent.

3-Chloropropyne (1-chloro-2-propyne)

Hazardous reaction   Violent reaction with ammonia under pressure followed by explosion (B348).

N-Chlorosuccinimide

Hazardous reactions   Violent or explosive reaction with aliphatic alcohols, benzylamine, hydrazine hydrate (B384).

Chlorosulphuric acid (chlorosulphonic acid)

Colourless to brown fuming liquid; bp 151 °C; decomposing with explosive violence when mixed with water.

CAUSES SEVERE BURNS
IRRITATING TO SKIN, EYES AND RESPIRATORY SYSTEM

Prevent inhalation of vapour. Prevent contact with eyes and skin.

Toxic effects	The fumes are very irritant to the lungs and mucous membranes. The fumes irritate the eyes severely. The liquid burns the skin and eyes. If taken by mouth there would be severe local and internal corrosive effects.
Hazardous reactions	Powerful oxidising agent; explosive reaction with P; violent reaction with water (B641).
First aid	*Vapour inhaled:* standard treatment (p 109). *Affected eyes:* standard treatment (p 109). *Skin contact:* standard treatment (p 108). *If swallowed:* standard treatment (p 109).
Spillage disposal	Instruct others to keep at a safe distance. Wear breathing apparatus and gloves (and rubber boots or overshoes if spillage is large). Spread soda ash liberally over the spillage and mop up cautiously with plenty of water — run to waste diluting greatly with running water.

N-Chlorotetramethylguanidine

Hazardous reaction	Unstable; explodes if heated above 50 °C (B456).

Chlorotoluenes

All three isomers are liquids at normal temperatures (*o*- bp 159 °C; *m*- bp 162 °C; *p*- bp 162 °C; mp 7 °C); they are insoluble in water.

FLAMMABLE
HARMFUL BY INHALATION

Avoid breathing vapour.

Toxic effects	The vapour of the chlorotoluenes is considered potentially toxic in moderate concentrations and is known to be narcotic at high concentrations in the case of the 3-isomer.
First aid	*Vapour inhaled:* standard treatment (p 109). *Skin contact:* standard treatment (p 108). *If swallowed:* standard treatment (p 109).
Fire hazard	Flash point 47–50 °C. *Extinguish fire with* foam, dry powder, carbon dioxide or vaporising liquids.
Spillage disposal	Shut off all possible sources of ignition. Instruct others to keep at a safe distance. Wear breathing apparatus and gloves. Apply non-flammable dispersing agent if available and work to an emulsion with brush and water — run this to waste diluting greatly with running water. If dispersant not available, absorb on sand, shovel into bucket(s) and transport to safe, open area for burial. Site of spillage should be washed thoroughly with water and soap or detergent.

α-Chlorotoluene – *see* Benzyl chloride

Chloryl hypofluorite

Hazardous reaction   Explosive (B632).

Chloryl perchlorate

Hazardous reactions   Very powerful oxidant; reacts violently or explosively with ethanol, stopcock grease, wood and organic matter generally. Liable to explode on contact with water, thionyl chloride (B684).

Chromates and dichromates

Generally yellow or orange-red crystals or powder; usually soluble in water.

**CONTACT WITH COMBUSTIBLE MATERIAL MAY CAUSE FIRE
HARMFUL IF TAKEN INTERNALLY**

Avoid inhaling dust. Avoid contact with eyes and skin. TLV (as CrO_3) (0.1 mg m^{-3}).

Toxic effects   The dust irritates the respiratory tract. The dust irritates the eyes severely. If taken by mouth there is irritation and internal damage. **Chronic effects** Frequent exposure of skin to dust can cause ulceration. Long-continued absorption can cause liver and kidney disease and even cancer.

Hazardous reactions   Ammonium dichromate decomposes thermally at 190 °C, the flame spreading rapidly with emission of green Cr_2O_3; if confined it will explode; hydroxylamine reacts explosively with both potassium and sodium dichromates; the dihydrated sodium salt reacts violently and finally explosively with acetic anhydride (B715–716).

First aid   *Dust inhaled:* standard treatment (p 109).
Affected eyes: standard treatment (p 109).
Skin contact: standard treatment (p 108).
If swallowed: standard treatment (p 109).

Spillage disposal   Shovel into bucket of water and run solution or suspension to waste diluting greatly with running water. Site of spillage should be washed thoroughly to remove all oxidant, which is liable to render any organic matter (particularly wood, paper and textiles) with which it comes into contact, dangerously combustible when dry. Clothing wetted with the solution should be washed thoroughly.

Chromic acid – *see* Chromium trioxide

Chromium diacetate

Hazardous reaction   Anhydrous salt is pyrophoric in air (B391).

Chromium trioxide (chromic acid)

Dark red crystalline masses or flakes; soluble in water.

**CONTACT WITH COMBUSTIBLE MATERIAL MAY CAUSE FIRE
CAUSES SEVERE BURNS**

Avoid inhaling dust. Prevent contact with eyes and skin. TLV 0.1 mg m^{-3}.

Toxic effects   The dust irritates all parts of the respiratory system. The solid and its solutions cause severe eye burns. The solid and its solutions burn the skin. If taken by mouth there would be severe internal irritation and damage. **Chronic effects** Frequent exposure of skin to the material may result in ulceration.

Hazardous reactions   Very powerful oxidant; violent or explosive reactions with acetic acid, acetic anhydride, P, Se; reacts with incandescence with K, Na, NH_3, As, butyric acid, H_2S; may ignite acetone, methanol, ethanol, propan-2-ol, butanol, cyclohexanol, *N,N*-dimethylformamide, glycerol, pyridine, sulphur. (B710–714).

First aid   *Dust inhaled:* standard treatment (p 109).
Affected eyes: standard treatment (p 109).
Skin contact: standard treatment (p 108).
If swallowed: standard treatment (p 109).

Spillage disposal   Wear face-shield or goggles, and gloves (and rubber boots or overshoes if spillage is large). Spread soda ash liberally over the spillage to neutralise and mop up cautiously with plenty of water—run to waste diluting greatly with running water.

Chromyl acetate

Hazardous preparation   Explosion occurred when being prepared from chromium trioxide and acetic anhydride (B392).

Chromyl azide chloride

Hazardous reaction   Explosive solid (B630).

Chromyl chloride

Red fuming liquid with pungent musty odour; bp 117 °C; decomposed vigorously by water; can ignite organic matter on contact.

CONTACT WITH COMBUSTIBLE MATERIAL MAY CAUSE FIRE
CAUSES SEVERE BURNS

Avoid breathing vapour. Prevent contact with eyes and skin.

Toxic effects   The vapour irritates all parts of the respiratory system. The vapour irritates the eyes severely. The liquid burns the skin and eyes. If taken by mouth there would be severe local and internal irritation and damage. **Chronic effects** Frequent exposure of skin to the material may result in ulceration.

Hazardous reactions   Explosions may occur with alkyl aromatics, liquid chlorine, PCl_3, sodium azide; ignites S_2Cl_2, acetone, ethanol, diethyl ether, turpentine, sulphur; incandescent reaction with ammonia (B673–674).

First aid   *Vapour inhaled:* standard treatment (p 109).
Affected eyes: standard treatment (p 109).
Skin contact: standard treatment (p 108).
If swallowed: standard treatment (p 109).

Spillage disposal   Instruct others to keep at a safe distance. Wear breathing apparatus and gloves. Spread soda ash liberally over the spillage and mop up cautiously with water — run this to waste diluting greatly with running water.

Chromyl nitrate

Hazardous reactions   Powerful oxidant and nitrating agent which ignites many hydrocarbons and organic solvents on contact (B710).

Chromyl perchlorate

Hazardous reactions   Explodes violently above 80 °C; ignites organic solvents (B674).

Cinnamaldehyde

Hazardous reaction   Rags soaked in NaOH solution and the aldehyde ignited in waste bin (B568).

Coal gas – *see* Carbon monoxide

Cobalt nitride

Hazardous reaction   A pyrophoric powder (B707).

Cobalt trifluoride

Hazardous reactions Fluorinating agent: reacts violently with hydrocarbons, water (B704).

Copper compounds

Blue or greenish-blue crystals or powder.

HARMFUL IF TAKEN INTERNALLY

Avoid inhaling dust. Avoid contact with eyes and skin.

Toxic effects The dust irritates the mucous membranes. The dust and solutions of salts irritate the eyes. Ingestion may cause violent vomiting and diarrhoea with intense abdominal pain and collapse.

First aid *Dust inhaled:* standard treatment (p 109).
Affected eyes: standard treatment (p 109).
If swallowed: standard treatment (p 109).

Spillage disposal Wear face-shield or goggles, and gloves. Mop up with plenty of water and run to waste, diluting greatly with running water.

Copper(I) hydride

Hazardous reaction Impure material may decompose explosively on heating (B719).

Copper(I) nitride

Hazardous reaction May explode on heating in air (B723).

Copper(II) phosphinate

Hazardous reaction Explodes at about 90 °C (B719).

Cresols (cresylic acid)

o-Cresol—colourless to pale brown crystals. *m*-Cresol—a colourless to yellow liquid. *p*-Cresol —colourless to pink crystals. Cresylic acid (mixed isomers)—colourless to brown liquid. All cresols have a phenolic (carbolic) odour and are sparingly soluble in water.

**SERIOUS RISK OF POISONING BY INHALATION, SWALLOWING OR SKIN CONTACT
CAUSES BURNS**

Avoid breathing vapour. Prevent contact with skin and eyes. TLV (skin) 5 ppm (22 mg m^{-3}).

Toxic effects The vapour from heated cresols is irritant to the respiratory system. Cresols burn the eyes severely and irritate or burn the skin. Considerable absorption through the skin may give effects similar to those caused by ingestion, namely headache, dizziness, nausea, vomiting, stomach pain, exhaustion and possibly coma. **Chronic effects** Repeated inhalation or absorption of small amounts may cause damage to liver or kidneys; repeated contact with skin may cause dermatitis.

First aid *Vapour inhaled:* standard treatment (p 109).
Affected eyes: standard treatment (p 109).
Skin contact: standard treatment (p 108). But see also note on phenol (p 107).
If swallowed: standard treatment (p 109).

Spillage disposal Wear face-shield or goggles, and gloves. Apply dispersing agent if available and work to an emulsion with brush and water — run this to waste, diluting greatly with running water. If dispersant not available, absorb on sand, shovel into bucket(s) and transport to safe, open area for burial. Site of spillage should be washed thoroughly with water and soap or detergent.

Crotonaldehyde

Colourless liquid with a pungent, suffocating odour; bp 104 °C; immiscible with water.

HIGHLY FLAMMABLE
GIVES OFF POISONOUS VAPOUR
CAUSES IRRITATION OF SKIN AND EYES

Prevent of inhalation of vapour. Prevent contact with skin and eyes. TLV 2 ppm (6 mg m^{-3}).

Toxic effects The vapour irritates the respiratory system. The vapour and liquid irritate the eyes severely, causing lachrymation and burns. The liquid irritates the skin. Assumed to be irritant and poisonous if taken by mouth.

Hazardous reactions Exploded when heated with butadiene in autoclave at 180 °C; explodes on contact with concentrated nitric acid (B394).

First aid *Vapour inhaled:* standard treatment (p 109).
Affected eyes: standard treatment (p 109).
Skin contact: standard treatment (p 108).
If swallowed: standard treatment (p 109).

Fire hazard Flash point 13 °C; explosive limits 2.1–15.5%; ignition temp. 207 °C. *Extinguish fire with* water spray, foam, dry powder, carbon dioxide or vaporising liquids.

Spillage disposal Shut off all possible sources of ignition. Instruct others to keep at a safe distance. Wear breathing apparatus and gloves. Apply non-flammable dispersing agent if available and work to an emulsion with brush and water — run this to waste diluting greatly with running water. If dispersant not available, absorb on sand, shovel into bucket(s) and transport to safe open area for atmospheric evaporation or burial. Site of spillage should be washed thoroughly with water and soap or detergent.

Cryogenic liquids Review of two manuals on safe handling (B31).

Cumene (isopropylbenzene)

Colourless liquid; bp 152 °C; insoluble in water.

> FLAMMABLE
> HARMFUL VAPOUR
> IRRITATING TO SKIN, EYES AND RESPIRATORY SYSTEM

Avoid breathing vapour. Avoid contact with skin, eyes and clothing. TLV (skin) 50 ppm (245 mg m⁻³).

Toxic effects	The vapour irritates the respiratory system. The vapour irritates the eyes and may cause conjunctivitis. The liquid irritates the skin and, by absorption and slow elimination, may cause serious poisoning. Assumed to be harmful if taken by mouth.
First aid	*Vapour inhaled:* standard treatment (p 109).
	Affected eyes: standard treatment (p 109).
	Skin contact: standard treatment (p 108).
	If swallowed: standard treatment (p 109).
Fire hazard	Flash point 44 °C; explosive limits 0.9–6.5 %; ignition temp. 424 °C. *Extinguish fire with* foam, dry powder, carbon dioxide or vaporising liquid.
Spillage disposal	Shut off all possible sources of ignition. Wear face-shield or goggles, and gloves. Apply non-flammable dispersing agent if available and work to an emulsion with brush and water — run this to waste diluting greatly with running water. If dispersant not available, absorb on sand, shovel into bucket(s) and transport to safe, open area for burial. Ventilate site of spillage well to evaporate remaining liquid and dispel vapour.

Cyanamide

Hazardous reactions	Acids or alkalis and moisture speed decomposition which may become violent above 49 °C; explosive polymerisation may occur on evaporating aqueous solutions to dryness (B233).

isoCyanatomethane – *see* Isocyanatomethane

Cyanides (water soluble)

Colourless crystals or powders which are soluble in water and react with acids to generate hydrogen cyanide.

> POISONOUS
> SERIOUS RISK OF POISONING BY
> INHALATION, SWALLOWING OR SKIN CONTACT
> CONTACT WITH ACIDS LIBERATES A TOXIC GAS

TLV (skin) (as CN) 5 mg m⁻³.

Toxic effects   Cyanides and their solutions, and hydrogen cyanide liberated from these by the action of acids, are extremely poisonous. Both the cyanide solutions and the gas can be absorbed through the skin. Whatever the route of absorption, severe poisoning may result. The early warning symptoms of poisoning are general weakness and heaviness of the arms and legs, increased difficulty in breathing, headache, dizziness, nausea, vomiting, and these may be rapidly followed by pallor, unconsciousness, cessation of breathing and death.

First aid   Special treatment for soluble cyanides (p 105).
Spillage   Wear breathing apparatus and gloves. Instruct others to keep at a safe
disposal   distance. When cyanide solutions have been spilt, bleaching powder should be scattered liberally over the spillage, or an excess of sodium hypochlorite solution added. The treated spillage should then be mopped up into a bucket and allowed to stand for 24 hours before running to waste, diluting greatly with running water. Solid cyanides should be swept up and placed in a large volume of water in which they can be rendered innocuous by adding an excess of sodium hypochlorite solution and allowing to stand for 24 hours before running to waste, diluting greatly with running water.

Cyano compounds Review of class (B32).

Cyanogen

Colourless gas with almond-like odour; bp −21 °C; 4 volumes dissolve in 1 volume of water at 15 °C and 760 mmHg.

    EXTREMELY FLAMMABLE
    TOXIC BY INHALATION
    IRRITATES THE EYE AND RESPIRATORY SYSTEM

Prevent inhalation of gas. TLV 10 ppm (20 mg m^{-3}).

Toxic effects   The gas irritates the respiratory system, leading to headache, dizziness, rapid pulse, nausea, vomiting, unconsciousness, convulsions and death, depending upon exposure. The gas irritates the eyes causing lachrymation.

First aid   Special treatment for hydrogen cyanide (p 104).
Fire hazard   Explosive limits 6–32%. Since the gas is supplied in a cylinder, turning off the valve will reduce any fire involving it; if possible, cylinders should be removed quickly from an area in which a fire has developed.
Disposal   Surplus gas or leaking cylinder can be vented slowly into water-fed scrubbing tower or column, or into fume cupboard served by such a tower.

2-Cyanopropan-2-ol
−*see* 2-Hydroxy-2-methylpropiononitrile

1-Cyano-2-propen-1-ol
−*see* 2-Hydroxybut-3-enonitrile

3-Cyanotriazenes Review of group (B34).

Cyanuric chloride
– *see* 2,4,6-Trichloro-s-triazine

Cyclic peroxides Review of class (B34).

Cyclohexa-1,3-diene

Hazardous reaction	Slowly forms explosive peroxide on contact with air (B491).

Cyclohexane

Colourless, mobile liquid with pungent odour when impure; bp 81 °C; insoluble in water.

EXTREMELY FLAMMABLE
IRRITATING TO SKIN, EYES AND RESPIRATORY SYSTEM

Avoid breathing vapour. Avoid contact with skin and eyes. TLV 300 ppm (1050 mg m^{-3}).

Toxic effects   Irritates the eyes, skin and respiratory system; the inhalation of high concentrations may cause narcosis. Assumed to be irritant and narcotic if taken by mouth.

Hazardous reaction   Addition of liquid dinitrogen tetraoxide to hot cyclohexane caused explosion (B500).

First aid   *Vapour inhaled:* standard treatment (p 109).
Affected eyes: standard treatment (p 109).
Skin contact: standard treatment (p 108).
If swallowed: standard treatment (p 109).

Fire hazard   Flash point −20 °C; explosive limits 1.3–8.4%; ignition temp. 260 °C. *Extinguish fire with* foam, dry powder, carbon dioxide or vaporising liquid.

Spillage disposal   Shut off all possible sources of ignition. Instruct others to keep at a safe distance. Wear breathing apparatus and gloves. Apply non-flammable dispersing agent if available and work to an emulsion with brush and water – run this to waste, diluting greatly with running water. If dispersant not available, absorb on sand, shovel into bucket(s) and transport to safe open area for atmospheric evaporation. Site of spillage should be washed thoroughly with water and soap or detergent.

Cyclohexane-1,2-dione

Hazardous preparation   Explosion when being prepared by HNO_3 oxidation of cyclohexanol (B493).

Cyclohexanol

Colourless hygroscopic crystals or viscous liquid with camphorlike odour; bp 161 °C; mp 23–25 °C; sparingly soluble in water.

> HARMFUL VAPOUR
> IRRITATING TO SKIN, EYES AND RESPIRATORY SYSTEM

Avoid breathing vapour. Avoid contact with skin and eyes. TLV 50 ppm (200 mg m^{-3}).

Toxic effects	The vapour may irritate the eyes, skin and respiratory system. The liquid irritates the eyes and may cause conjunctivitis and more serious damage. The liquid irritates the skin; major absorption may lead to tremors and kidney or liver damage. Assumed to be irritant and damaging to the alimentary system if taken by mouth.
Hazardous reactions	Ignited by CrO_3; explosion with HNO_3 (B502).
First aid	*Vapour inhaled:* standard treatment (p 109). *Affected eyes:* standard treatment (p 109). *Skin contact:* standard treatment (p 108). *If swallowed:* standard treatment (p 109).
Spillage disposal	Shut off all possible sources of ignition. Wear face-shield or goggles, and gloves. Apply non-flammable dispersing agent if available and work to an emulsion with brush and water — run this to waste, diluting greatly with running water. If dispersant not available, absorb on sand, shovel into bucket(s) and transport to safe open area for burial. Ventilate site of spillage well to evaporate remaining liquid and dispel vapour.

Cyclohexanone

Colourless, oily liquid with odour somewhat similar to that of acetone; bp 156 °C; sparingly soluble in water.

> FLAMMABLE
> HARMFUL VAPOUR
> IRRITATING TO SKIN, EYES AND RESPIRATORY SYSTEM

Avoid breathing vapour. Avoid contact with skin and eyes. TLV 50 ppm (200 mg m^{-3}).

Toxic effects	The vapour may irritate the eyes, skin and respiratory system. The liquid irritates the eyes and may cause conjunctivitis. The liquid may irritate the skin. Assumed to be irritant and damaging to the alimentary system if taken by mouth.
Hazardous reactions	Forms explosive peroxide with H_2O_2; may explode when added to HNO_3 at about 75 °C (B497).
First aid	*Vapour inhaled:* standard treatment (p 109). *Affected eyes:* standard treatment (p 109). *Skin contact:* standard treatment (p 108). *If swallowed:* standard treatment (p 109).
Fire hazard	Flash point 44 °C; explosive limits 1.1–8.1%; ignition temp. 420 °C. *Extinguish fire with* water spray, foam, dry powder, carbon dioxide or vaporising liquid.

Spillage disposal	Shut off all possible sources of ignition. Wear face-shield or goggles, and gloves. Apply non-flammable dispersing agent if available and work to an emulsion with brush and water — run this to waste, diluting greatly with running water. If dispersant not available, absorb on sand, shovel into bucket(s) and transport to safe open area for burial. Ventilate site of spillage well to evaporate remaining liquid and dispel vapour.

Cyclohexene

Colourless liquid; bp 83 °C; insoluble in water.

EXTREMELY FLAMMABLE
IRRITATING TO RESPIRATORY SYSTEM

Avoid breathing vapour. TLV 300 ppm (1015 mg m^{-3}).

Toxic effects	The vapour irritates the respiratory system and may irritate the eyes and skin. Assumed to be irritant if taken by mouth.
First aid	*Vapour inhaled:* standard treatment (p 109). *Affected eyes:* standard treatment (p 109). *Skin contact:* standard treatment (p 108). *If swallowed:* standard treatment p 109).
Fire hazard	Flash point −60 °C. *Extinguish fire with* foam, dry powder, carbon dioxide or vaporising liquid.
Spillage disposal	Shut off all possible sources of ignition. Instruct others to keep at a safe distance. Wear breathing apparatus and gloves. Apply non-flammable dispersing agent if available and work to an emulsion with brush and water — run this to waste, diluting greatly with running water. If dispersant not available absorb on sand, shovel into bucket(s) and transport to safe open area for atmospheric evaporation. Site of spillage should be washed thoroughly with water and soap or detergent.

Cyclohexylamine (hexahydroaniline)

Colourless liquid with an ammoniacal, fishy odour; bp 134 °C; miscible with water.

FLAMMABLE
HARMFUL VAPOUR
IRRITATING TO SKIN, EYE AND RESPIRATORY SYSTEM

Avoid breathing vapour. Prevent contact with skin and eyes. TLV (skin) 10 ppm (40 mg m^{-3}).

Toxic effects	The vapour may irritate the eyes and respiratory system, causing difficulty in breathing. The liquid can burn the eyes and skin; skin absorption may cause nausea and vomiting. Assumed to be poisonous if taken by mouth.
First aid	*Vapour inhaled:* standard treatment (p 109). *Affected eyes:* standard treatment (p 109). *Skin contact:* standard treatment (p 108). *If swallowed:* standard treatment (p 109).
Fire hazard	Flash point 32 °C; ignition temp. 293 °C. *Extinguish fire with* water spray, foam, dry powder, carbon dioxide or vaporising liquids.
Spillage disposal	Shut off all possible sources of ignition. Instruct others to keep at a safe distance. Wear breathing apparatus and gloves. Mop up with plenty of water and run to waste diluting greatly with running water. Ventilate area well to evaporate remaining liquid and dispel vapour.

Cyclopenta-1,3-diene

Colourless liquid; bp 42 °C; insoluble in water.

HIGHLY FLAMMABLE

Avoid breathing vapour. Avoid contact with skin and eyes. TLV 75 ppm (200 mg m^{-3}).

Toxic effects	The main toxic effects arise from inhalation which leads to depression of central nervous system, liver damage and a benzene-like effect upon the blood. **Chronic effects** include headache, abdominal pains, jaundice and anaemia. An acute effect may be narcosis.
Hazardous reactions	Dimerisation is exothermic and may cause rupture of closed, uncooled containers; heat sensitive explosive peroxides formed on exposure to O_2; explosive reaction with N_2O_4 (B438).
First aid	*Vapour inhaled:* standard treatment (p 109). *Affected eyes:* standard treatment (p 109). *Skin contact:* standard treatment (p 108). *If swallowed:* standard treatment (p 109).
Fire hazard	No data about flash point, *etc* was traced. *Extinguish fire with* foam, dry powder, carbon dioxide or vaporising liquid.
Spillage disposal	Shut off all possible sources of ignition. Instruct others to keep at a safe distance. Wear breathing apparatus and gloves. Apply non-flammable dispersing agent if available and work to an emulsion with brush and water — run this to waste, diluting greatly with running water. If dispersant not available, absorb on sand, shovel into bucket(s) and transport to safe open area for atmospheric evaporation. Site of spillage should be washed thoroughly with water and soap or detergent.

Cyclopentadienylsodium

Hazardous reaction	Pyrophoric in air (B438).

Cyclopentane

Colourless, mobile liquid; bp 49.3 °C; insoluble in water.

EXTREMELY FLAMMABLE

Avoid breathing vapour.

Toxic effects	May act as mild narcotic in high concentrations.
First aid	*Vapour inhaled:* standard treatment (p 109).
Fire hazard	Flash point below −6.7 °C *Extinguish fire with* foam, dry powder, carbon dioxide or vaporising liquid.
Spillage disposal	Shut off all possible sources of ignition. Instruct others to keep at a safe distance. Wear breathing apparatus and gloves. Apply non-flammable dispersing agent if available and work to an emulsion with brush and water — run this to waste, diluting greatly with running water. If dispersant not available, absorb on sand, shovel into bucket(s) and transport to safe open area for atmospheric evaporation. Site of spillage should be washed thoroughly with water and soap or detergent.

Cyclopentanone

Hazardous reaction — Mixtures with nitric acid and hydrogen peroxide react vigorously and may become explosive (B444).

Cyclopropane

Colourless gas with smell like that of petroleum spirit; bp −33 °C; 1 volume of gas dissolves in 2.7 volumes of water at 15 °C and 760 mmHg.

EXTREMELY FLAMMABLE

Avoid breathing gas.

Toxic effects — The gas is an aneasthetic and is employed for this purpose.

First aid — *Gas inhaled:* remove from exposure, rest and keep warm.

Fire hazard — Explosive limits 2.4–10.4%; ignition temp. 498 °C. Since the gas is supplied in a cylinder, turning off the valve will reduce any fire involving it; if possible, cylinders should be removed quickly from an area in which a fire has developed.

Disposal — Surplus gas or a leaking cylinder can be vented slowly to air in a safe open area or burnt off in a suitable gas burner.

Decaborane

Hazardous reactions — Forms impact-sensitive mixtures with ethers and halocarbons; ignites in oxygen at 100 °C (B192).

Decanedioyl dichloride
– *see* Sebacoyl dichloride

Devarda's alloy Safe usage discussed (B34).

Diacetone alcohol
– *see* 4-Hydroxy-4-methylpentan-2-one

Diacetyl peroxide

Hazardous reaction — When dry is a shock-sensitive explosive; use in ethereal solution (B400).

Diacyl peroxides Review of group (B35).

Dialkyl hyponitrites Review of group (B36).

Dialkylmagnesiums Review of group (B36).

Dialkyl peroxides Review of group (B36).

Dialkylzincs Review of group (B37).

Diallyl ether

Hazardous reaction   Peroxidises readily in air and sunlight to explosive peroxide (B498).

Diallyl phosphite

Hazardous reaction   Liable to explode during distillation (B500).

Diallyl phthalate

Colourless, oily liquid; bp 157 °C: insoluble in water.

HARMFUL IN CONTACT WITH SKIN

Avoid contact with skin, eyes and clothing.

Toxic effects   The liquid irritates the eyes and skin and can cause internal disorders by continued skin absorption. Assumed to be harmful if taken by mouth.

First aid   *Affected eyes:* standard treatment (p 109).
Skin contact: standard treatment (p 108).
If swallowed: standard treatment (p 109).

Spillage disposal   Wear face-shield or goggles, and gloves. Apply dispersing agent if available and work to an emulsion with brush and water—run this to waste, diluting greatly with running water. If dispersant not available, absorb on sand, shovel into bucket(s) and transport to safe open area for burial. Ventilate site of spillage well to evaporate remaining liquid and dispel vapour.

Diallyl sulphate

Hazardous reaction Exploded during distillation (B499).

4,4'-Diaminodiphenylmethane
– see Bis(4-aminophenyl)methane

1,2-Diaminoethane (ethylenediamine)

Clear, colourless liquid with ammoniacal odour, bp 117 °C; miscible with water.

> FLAMMABLE
> HARMFUL VAPOUR
> IRRITATING TO SKIN, EYES AND RESPIRATORY SYSTEM

Avoid breathing vapour. Avoid contact with skin and eyes. TLV 10 ppm (25 mg m⁻³).

Toxic effects The vapour irritates the respiratory system. The liquid and vapour cause irritation of skin and eyes. If swallowed, may cause digestive disturbance and possibly damage to the kidneys. **Chronic effects** Repeated inhalation of vapour or skin contact may cause sensitisation of skin or respiratory system.

Hazardous reactions May ignite on contact with cellulose nitrate; dangerous reactions with nitromethane and diisopropyl peroxydicarbonate (B330).

First aid *Vapour inhaled:* standard treatment (p 109).
Affected eyes: standard treatment (p 109).
Skin contact: standard treatment (p 108).
If swallowed: standard treatment (p 109).

Fire hazard Flash point 43 °C. *Extinguish fire with* dry powder, carbon dioxide or vaporising liquid.

Spillage disposal Shut off all possible sources of ignition. Instruct others to keep at a safe distance. Wear breathing apparatus and gloves. Mop up with plenty of water and run to waste, diluting greatly with running water. Ventilate area well to evaporate remaining liquid and dispel vapour.

1,6-Diaminohexane (hexane-1,6-diamine)

Colourless leaflets; mp 42 °C; freely soluble in water.

> IRRITATING TO SKIN, EYES AND RESPIRATORY SYSTEM

Toxic effects Indicated in above warning.

First aid *Affected eyes:* standard treatment (p 109).
Skin contact: standard treatment (p 108).
If swallowed: standard treatment (p 109).

Spillage disposal Wearing gloves, brush up and dissolve in bucket of water. Run solution to waste diluting with running water. Mop up residual amine with water.

Dianilinium dichromate

Hazardous reaction Unstable on storage (B594).

o-Dianisidine and hydrochloride – *see* 3,3′-Dimethoxybenzidine and hydrochloride

Diazidodimethylsilane

Hazardous reaction Old sample exploded (B322).

Diazidoethanes

Hazardous reaction Explosives (B306).

Diazidomalononitrile

Hazardous reaction Shock-sensitive explosive but may explode without warning (B380).

1,3-Diazidopropene

Hazardous reaction Exploded while being weighed (B353).

Diazirines Review of group (B37).

Diazoacetonitrile

Hazardous reaction Liable to explode through friction (B284).

Diazo compounds Review of group (B37).

2-Diazocyclohexanone

Hazardous   May explode on heating (B492).
reaction

Diazocyclopentadiene

Hazardous   Exploded violently during vacuum distillation (B435).
reaction

Diazoindene

Hazardous   Distillation residue may explode (B566).
preparation

Diazomalononitrile (dicyanodiazomethane)

Hazardous   May explode at 75 °C (B379).
reaction

Diazomethane

A yellow gas generally employed for organic synthesis in chloroformic or ethereal solution and prepared as part of the process.

EXTREME RISK OF EXPLOSION BY SHOCK, FRICTION, FIRE OR OTHER SOURCES OF IGNITION
TOXIC BY INHALATION

Prevent inhalation of vapour and contact with skin and eyes. TLV 0.2 ppm (0.4 mg m^{-3}).

Toxic effects   Irritates the respiratory system, eyes and skin. Inhalation may result in chest discomfort, headache, weakness and, in severe cases, collapse.

Hazardous   Gaseous diazomethane may explode on ground glass surfaces and when
reactions   heated to about 100 °C; concentrated solutions may also explode especially if impurities are present; explosions occur on contact with alkali metals and the exothermic reaction with calcium sulphate is also dangerous (B234).

First aid   *Vapour inhaled :* standard treatment (p 109).
                       Affected eyes : standard treatment (p 109).

Diazonium salts Review of class (B38).

2-Diazo-2-phenylacetophenone
(1,1-benzoylphenyldiazomethane)

Hazardous reaction   May explode if heated above 40 °C (B600).

Dibenzoyl peroxide (benzoyl peroxide)

White granular crystals, normally supplied moistened with about 30 % water; liable to explode when heated above melting point (mp 103–105 °C) or when subjected to friction or shock when dry. Insoluble in water.

> EXTREME RISK OF EXPLOSION BY SHOCK, FRICTION, FIRE OR
> OTHER SOURCE OF IGNITION
> IRRITATING TO SKIN AND EYES

Avoid contact with skin and eyes. TLV 5 mg m^{-3}.

Toxic effects   Contact with skin or eyes causes irritation.

Hazardous reactions   Dry material burns and is sensitive to heat (explodes above mp), shock, friction or contact with combustible materials; has exploded on heating with ethylene and CCl_4 under pressure; explosions have resulted from contact with dimethylaniline or dimethyl sulphide; ignition occurs with methyl methacrylate (B601–602).

First aid   *Affected eyes:* standard treatment (p 109).
Skin contact: standard treatment (p 108).

Spillage disposal   Moisten well with water and mix with plenty of sand. Disposal in any quantity depends upon local conditions and regulations. Deep burial in an isolated area or consignment to deep sea water in a heavy container can be considered if quantities are large.

Dibenzylamine

Colourless, oily liquid with ammoniacal odour; bp 300 °C with decomposition; immiscible with water.

> CAUSES BURNS

Prevent contact with skin and eyes.

Toxic effects   The liquid irritates and burns the skin and eyes. Corrosive and highly irritant if taken by mouth.

First aid   *Affected eyes:* standard treatment (p 109).
Skin contact: standard treatment (p 108).
If swallowed: standard treatment (p 109).

Spillage disposal   Wear face-shield or goggles, and gloves. Apply dispersing agent if available and work to an emulsion with brush and water—run this to waste, diluting greatly with running water. If dispersant not available, absorb on sand, shovel into bucket(s) and transport to safe open area for burial. Site of spillage should be washed thoroughly with water and soap or detergent.

Dibenzyl ether

Hazardous reaction   Exploded with aluminium dichloride hydride diethyl etherate (B604).

Dibenzyl phosphite

Hazardous reaction   Decomposes at 160 °C (B604).

Diborane

Hazardous reactions   Usually ignites in air and delayed ignition may be followed by violent explosions; reacts explosively with chlorine and forms explosive compound with dimethylsulphoxide; reacts violently with halocarbon liquids (B188).

Diboron tetrachloride

Hazardous reaction   Exposure to air may cause explosion (B187).

Diboron tetrafluoride

Hazardous reaction   Extremely explosive in presence of oxygen (B188).

Dibromoacetylene

Hazardous reactions   Ignites in air and explodes on heating (B269).

2,6-Dibromobenzoquinone 4-chloroimine

Hazardous reaction   Liable to decompose explosively on heating (B462).

1,2-Dibromoethane (ethylene dibromide)

Colourless liquid with sweetish chloroform-like odour, bp 131 °C; immiscible with water.

HARMFUL VAPOUR
TOXIC IN CONTACT WITH SKIN

Avoid breathing vapour. Avoid contact with skin and eyes. TLV (skin) 20 ppm (145 mg m^{-3}).

Toxic effects   The vapour irritates the respiratory system and may have narcotic action. The vapour and liquid irritate the eyes. The liquid irritates the skin and may cause dermatitis. Poisonous by skin absorption and ingestion, effects being nausea, vomiting, pain and jaundice resulting from liver and kidney damage.

First aid   *Vapour inhaled:* standard treatment (p 109).
Affected eyes: standard treatment (p 109).
Skin contact: standard treatment (p 108).
If swallowed: standard treatment (p 109).

Spillage disposal   Instruct others to keep at a safe distance. Wear breathing apparatus and gloves. Apply dispersing agent if available and work to an emulsion with brush and water—run this to waste, diluting greatly with running water. If dispersant not available, absorb on sand, shovel into bucket(s) and transport to safe open area for burial. Site of spillage should be washed thoroughly with water and soap or detergent.

Dibromomethane

Hazardous reaction   Forms shock-sensitive explosive with potassium (B231).

Dibutylamines – *see* Butylamines

Di-t-butyl chromate

Hazardous preparation   Addition of t-butanol to CrO_3 resulted in explosion (B561).

Dibutyl ether (n-butyl ether)

Colourless liquid; bp 142 °C; insoluble in water. May form explosive peroxides on exposure to light and air which should be decomposed before distillation to small volume.

FLAMMABLE

Avoid breathing vapour. Avoid contact with skin and eyes.

Toxic effects   The vapour is somewhat irritating to the respiratory system. The liquid irritates the eyes, and is considered to present some hazard by skin absorption.

First aid   *Vapour inhaled:* standard treatment (p 109).
Affected eyes: standard treatment (p 109).
Skin contact: standard treatment (p 108).
If swallowed: standard treatment (p 109).

Fire hazard   Flash point 25 °C; explosive limits 1.5–7.6%; ignition temp. 194 °C. *Extinguish fire with* dry powder, carbon dioxide, or vaporising liquid.

Spillage disposal   Shut off all possible sources of ignition. Instruct others to keep at a safe distance. Wear breathing apparatus and gloves. Apply non-flammable dispersing agent if available and work to an emulsion with brush and water—run this to waste diluting greatly with running water. If dispersant not available, absorb on sand, shovel into bucket(s) and transport to safe open area for burial. Ventilate site of spillage well to evaporate remaining liquid and dispel vapour.

2,6-Di-t-butyl-4-nitrophenol

Hazardous reaction   Exploded on heating to 100 °C (B604).

Di-t-butyl peroxide

Hazardous preparation   Addition of t-butanol to 50% H_2O_2/78% H_2SO_4 mixtures may result in explosions (B562).

Dibutyl phthalate

Hazardous reaction   Mixture with liquid chlorine in S.S. bomb reacted explosively at 118 °C (B607).

Dichlorine oxide

Hazardous reactions   The liquid explodes on pouring, the gas on heating or sparking; explodes with charcoal, dicyanogen, NO, K, NH_3, As, Sb, S, PH_3, H_2S, paper, cork, rubber, turpentine and many other oxidisable materials (B680).

Dichlorine trioxide

Hazardous reaction   Vapour explodes well below 0 °C (B683).

Dichloroacetic acid

Colourless liquid with pungent odour; bp 194 °C; miscible with water.

CAUSES SEVERE BURNS

Avoid breathing vapour. Prevent contact with skin and eyes.

Toxic effects   The vapour irritates the eyes and respiratory system. The liquid burns the skin and eyes. Assumed to cause severe irritation and damage if taken by mouth.

First aid	*Vapour inhaled:* standard treatment (p 109). *Affected eyes:* standard treatment (p 109). *Skin contact:* standard treatment (p 108). *If swallowed:* standard treatment (p 109).
Spillage disposal	Wear goggles and gloves (and rubber boots or overshoes if spillage is large). Spread soda ash liberally over the spillage and mop up cautiously with plenty of water—run this to waste, diluting greatly with running water.

Dichloroacetyl chloride

Colourless, fuming liquid with acrid penetrating odour; bp 107 °C; reacts with water forming dichloroacetic and hydrochloric acids.

CAUSES SEVERE BURNS

Prevent inhalation of vapour. Prevent contact with skin and eyes.

Toxic effects	The vapour irritates the eyes and respiratory system. The liquid burns the skin and eyes. Assumed to cause severe irritation and damage if taken by mouth.
First aid	*Vapour inhaled:* standard treatment (p 109). *Affected eyes:* standard treatment (p 109). *Skin contact:* standard treatment (p 108). *If swallowed:* standard treatment (p 109).
Spillage disposal	Wear goggles or face-shield, and gloves (and rubber boots or overshoes if spillage is large). Spread soda ash liberally over the spillage to neutralise and mop up cautiously with plenty of water—run to waste diluting greatly with running water.

Dichloroacetylene

Hazardous reaction	Heat-sensitive explosive gas which ignites in contact with air (B273).

N,N-Dichloroaniline

Hazardous reaction	Explosive (B477).

Dichlorobenzenes (*o-* and *p-*)

*o-*Isomer is colourless liquid with a pleasant aromatic smell; bp 185 °C; *p-*isomer consists of colourless, volatile crystals with characteristic disinfectant smell, mp 53 °C; both are insoluble in water.

HARMFUL VAPOUR
IRRITATING TO SKIN, EYES AND RESPIRATORY SYSTEM

Avoid breathing vapour. Prevent contact with skin and eyes. TLVs *o-* 50 ppm (300 mg m^{-3}), *p-* 75 ppm (450 mg m^{-3}).

Toxic effects   Inhalation of vapours may cause drowsiness and irritation of nose; both isomers irritate the eyes; the *o*-isomer is more irritating to the skin and may cause dermatitis; long exposure to either isomer may result in liver damage.

First aid   *Vapour inhaled:* standard treatment (p 109).
Affected eyes: standard treatment (p 109).
Skin contact: standard treatment (p 108).
If swallowed: standard treatment (p 109).

Spillage disposal   The solid *p*-isomer may be mixed with dry sand, swept up and placed in waste bin. The liquid *o*-isomer: wear breathing apparatus and gloves. Apply dispersing agent if available, and work to an emulsion with brush and water—run this to waste diluting greatly with running water. If dispersant not available, absorb on sand, shovel into bucket(s) and transport to approved open area for burial. Site of spillage should be washed thoroughly with water and soap or detergent.

1,4-Dichlorobut-2-yne

Hazardous preparation   A preferred method using dichloromethane as diluent is indicated (B385).

1,2-Dichloroethane (ethylene dichloride)

Colourless liquid with a chloroform-like odour; bp 84 °C; immiscible with water.

HIGHLY FLAMMABLE
HARMFUL VAPOUR
IRRITATING TO SKIN, EYES AND RESPIRATORY SYSTEM

Avoid breathing vapour. Avoid contact with skin and eyes. TLV 50 ppm (200 mg m^{-3}).

Toxic effects   In high concentrations, the vapour irritates the eyes and respiratory system; it may also cause drowsiness, headache, vomiting and mental confusion. The liquid may cause serious damage to the eyes. Poisonous if taken by mouth. The liquid irritates the skin. **Chronic effects** Continued exposure to low concentrations may result in dizziness, nausea and abdominal pain, and there may be damage to the eyes and liver. Dermatitis may follow repeated contact with the skin.

Hazardous reactions   Mixtures with N_2O_4 or potassium are explosive when subjected to shock; reaction with aluminium powder may be violent or explosive (B304).

First aid   *Vapour inhaled:* standard treatment (p 109).
Affected eyes: standard treatment (p 109).
Skin contact: standard treatment (p 108).
If swallowed: standard treatment (p 109).

Fire hazard   Flash point 13 °C; explosive limits 6.2–15.9%; ignition temp. 413 °C. *Extinguish fire with* water spray, foam, dry powder, carbon dioxide or vaporising liquid.

Spillage disposal
Shut off all possible sources of ignition. Instruct others to keep at a safe distance. Wear breathing apparatus and gloves. Apply non-flammable dispersing agent if available and work to an emulsion with brush and water —run this to waste, diluting greatly with running water. If dispersant not available, absorb on sand, shovel into bucket(s) and transport to safe open area for atmospheric evaporation. Site of spillage should be washed thoroughly with water and soap or detergent.

1,1-Dichloroethylene (vinylidene chloride)

Hazardous reactions
Rapidly absorbs oxygen forming a violently explosive peroxide; reaction products with ozone are particularly dangerous; reaction under pressure with chlorotrifluoroethylene may develop into explosive polymerisation (B289).

1,2-Dichloroethylene (acetylene dichloride)

Colourless liquid with a slight chloroform-like odour; bp 60 °C (cis-isomer); immiscible with water.

> HIGHLY FLAMMABLE
> MAY FORM EXPLOSIVE PEROXIDES
> HARMFUL VAPOUR

Avoid breathing vapour. Avoid contact with skin and eyes. TLV 200 ppm (790 mg m^{-3}).

Toxic effects
The vapour irritates the eyes and mucous membranes; in high concentrations it may cause drowsiness and unconsciousness. Poisonous if taken by mouth. **Chronic effects** Continued exposure to low concentrations of vapour may cause drowsiness and digestive disturbance.

Hazardous reactions
Contact with solid caustic alkalies or their concentrated solution will form chloroacetylene which ignites in air; forms explosive mixtures with N_2O_4 (289).

First aid
Vapour inhaled: standard treatment (p 109).
Affected eyes: standard treatment (p 109).
Skin contact: standard treatment (p 108).
If swallowed: standard treatment (p 109).

Fire hazard
Flash point 2–4 °C; explosive limits 9.7–12.8%. *Extinguish fire with* water spray, foam, dry powder, carbon dioxide or vaporising liquid.

Spillage disposal
Shut off all possible sources of ignition. Instruct others to keep at a safe distance. Wear breathing apparatus and gloves. Apply non-flammable dispersing agent if available and work to an emulsion with brush and water—run this to waste, diluting greatly with running water. If dispersant not available, absorb on sand, shovel into bucket(s) and transport to safe open area for atmospheric evaporation. Site of spillage should be washed thoroughly with water and soap or detergent.

Di(2-chloroethyl) ether
- *see* Bis (2-chloroethyl) ether

1,6-Dichloro-2,4-hexadiyne

Hazardous reaction   Extremely shock-sensitive explosive (B471).

1,3-Dichlorohydrin
- see **1,3-Dichloropropan-2-ol**

Dichloromethane (methylene chloride)

Colourless volatile liquid with chloroform-like odour; bp 40 °C; immiscible with water.

> HARMFUL VAPOUR
> IRRITATING TO EYES

Avoid breathing vapour. Avoid contact with skin and eyes. TLV 100 ppm (360 mg m^{-3}).

Toxic effects   The vapour irritates the eyes and respiratory system and may cause head-ache and nausea; high concentrations may result in cyanosis and un-consciousness. The liquid irritates the eyes. Assumed to be poisonous if taken by mouth. *CYANOSIS - BLUISH DISCOLORATION dueTO DEFICIENT OXYGENATION OF THE BLOOD*

Hazardous reactions   Solution of dinitrogen pentaoxide in dichloromethane liable to explode; mixtures with Li, Na, N_2O_4, HNO_3 are liable to explode (B231–232).

First aid   *Vapour inhaled:* standard treatment (p 109).
Affected eyes: standard treatment (p 109).
If swallowed: standard treatment (p 109).

Spillage disposal   Wear face-shield or goggles, and gloves. Apply dispersing agent if available and work to an emulsion with brush and water—run this to waste, diluting greatly with running water. If dispersant not available, absorb on sand, shovel into bucket(s) and transport to safe open area for atmospheric evaporation. Site of spillage should be washed thoroughly with water and soap or detergent.

N,N-Dichloromethylamine

Hazardous reactions   Exploded on warming with water or on distillation over calcium hypo-chlorite; exploded on contact with solid sodium sulphide (B241).

1,4-Dichloro-5-nitrobenzene (2,5-dichloronitrobenzene)

Colourless crystals; mp 33 °C; insoluble in water.

> HARMFUL VAPOUR
> HARMFUL IN CONTACT WITH SKIN

Avoid breathing vapour. Avoid contact with skin, eyes and clothing.

Toxic effects   These are not well documented but it is assumed that, in common with other substituted benzene compounds of this type, it is irritant to the eyes, skin and respiratory system in vapour form. The liquid or solid would irritate the skin, eyes and respiratory system.

First aid   *Vapour inhaled:* standard treatment (p 109).
Affected eyes: standard treatment (p 109).
Skin contact: standard treatment (p 108).
If swallowed: standard treatment (p 109).

Spillage disposal   Wear face-shield or goggles, and gloves. Mix with sand, shovel into bucket(s) and transport to safe open area for burial. Site of spillage should be washed thoroughly with water and soap or detergent.

2,4-Dichlorophenol

Colourless crystals; almost insoluble in water.

> HARMFUL SUBSTANCE IF TAKEN INTERNALLY OR
> IF IN CONTACT WITH SKIN
> IRRITATING TO SKIN AND EYES
> CAUSES BURNS

Toxic effects   Causes severe irritation or burns in contact with the eyes and skin. Assumed to be very poisonous and irritant if taken by mouth.

First aid   *Affected eyes:* standard treatment (p 109).
Skin contact: drench with water and swab contaminated skin with glycerol for at least 10 minutes (use water if glycerol is not available); remove and wash contaminated clothing before re-use; if contamination has been other than slight, obtain medical attention.
If swallowed: standard treatment (p 109).

Spillage disposal   Wear face-shield or goggles, and gloves. Mix with sand and transport to safe, open area for burial. Site of spillage should be washed thoroughly with water and soap or detergent.

1,3-Dichloropropan-2-ol (1,3-dichlorohydrin)

Colourless liquid with ethereal odour; bp 174 °C; sparingly soluble in water.

> HARMFUL VAPOUR
> IRRITATING TO SKIN AND EYES

Avoid breathing vapour. Avoid contact with skin, eyes and clothing.

Toxic effects   The vapour irritates the eyes and respiratory system. Inhalation may cause headache, vertigo, nausea, vomiting and pulmonary oedema. The liquid irritates the skin and eyes. Nausea, vomiting, coma and liver damage may result if it is taken by mouth.

First aid   *Vapour inhaled:* standard treatment (p 109).
Affected eyes: standard treatment (p 109).
Skin contact: standard treatment (p 108).
If swallowed: standard treatment (p 109).

Spillage disposal Wear face-shield or goggles, and gloves. Apply dispersing agent if available and work to an emulsion with brush and water—run this to waste, diluting greatly with running water. If dispersant not available absorb on sand, shovel into bucket(s) and transport to safe, open area for burial. Site of spillage should be washed thoroughly with water and soap or detergent.

Dicrotonoyl peroxide

Hazardous reaction Very shock-sensitive explosive (B551).

Dicyanoacetylene – see But-2-ynedinitrile

1,4-Dicyano-2-butene – see Hex-3-enedinitrile

Dicyanodiazomethane – see Diazomalononitrile

Dicyanogen

Hazardous reactions Mixtures with oxidants may explode (B335).

Dicyanogen *N,N'*-dioxide

Hazardous reaction Decomposes at −45 °C under vacuum before exploding (B335).

Dicyanomethane – see Malononitrile

Dicyclohexylcarbonyl peroxide

Hazardous reaction In bulk, may explode without apparent reason (B605).

Dicyclopentadiene (3a,4,7,7a-tetrahydro-4,7-methanoindene)

The commercial product that is commonly used is a colourless liquid with a camphor-like odour; bp about 167 °C; insoluble in water. The pure compound is in the form of colourless crystals, mp 33 °C.

FLAMMABLE
IRRITATING TO SKIN AND EYES

Avoid breathing vapour. Avoid contact with skin, eyes and clothing. TLV 5 ppm

Toxic effects   These are not well documented, but suppliers advise against the inhalation of vapour and contact with the skin.

First aid   *Vapour inhaled :* standard treatment (p 109).
Affected eyes : standard treatment (p 109).
Skin contact : standard treatment (p 108).
If swallowed : standard treatment (p 109).

Fire hazard   Flash point 35 °C. *Extinguish fire with* foam, dry powder, carbon dioxide or vaporising liquid.

Spillage disposal   Shut off all possible sources of ignition. Wear face-shield or goggles, and gloves. Apply non-flammable dispersing agent if available and work to an emulsion with brush and water — run this to waste, diluting greatly with running water. If dispersant not available, absorb on sand, shovel into bucket(s) and transport to safe open area for burial. Ventilate site of spillage well to evaporate remaining liquid and dispel vapour.

Dienes Review of class (B44).

1,1-Diethoxyethane (acetal)

Colourless, volatile liquid with a pleasant odour; bp 102 °C; sparingly soluble in water; liable to form explosive peroxides on exposure to light and air, which requires that these be decomposed before the ether is distilled to small volume.

EXTREMELY FLAMMABLE
HARMFUL VAPOUR

Avoid breathing vapour.

Toxic effects   These are not well documented. High concentrations of vapour are liable to cause narcosis when inhaled.

Hazardous reaction   Peroxidised material exploded during distillation (B511).

First aid   *Vapour inhaled in quantity :* standard treatment (p 109).
Affected eyes : standard treatment (p 109).
Skin contact : standard treatment (p 108).
If swallowed : standard treatment (p 109).

Fire hazard   Flash point −20 °C; explosive limits 1.7–10.4%, ignition temp. 230 °C. *Extinguish fire with* water spray, foam, dry powder, carbon dioxide or vaporising liquid.

Spillage disposal   Shut off all possible sources of ignition. Wear face-shield or goggles, and gloves. Apply non-flammable dispersing agent if available and work to an emulsion with brush and water — run this to waste, diluting greatly with running water. If dispersant not available, absorb on sand, shovel into bucket(s) and transport to safe open area for atmospheric evaporation or burial. Ventilate site of spillage well to evaporate remaining liquid and dispel vapour.

Diethylamine

Colourless liquid with an ammoniacal odour; bp 56 °C; miscible with water.

EXTREMELY FLAMMABLE
IRRITATING TO SKIN, EYES AND RESPIRATORY SYSTEM

Avoid breathing vapour. Avoid contact with skin and eyes. TLV 25 ppm (75 mg m^{-3}).

Toxic effects The vapour irritates the eyes and respiratory system. The liquid irritates the skin and eyes. Assumed to be poisonous if taken by mouth.

First aid *Vapour inhaled:* standard treatment (p 109).
Affected eyes: standard treatment (p 109).
Skin contact: standard treatment (p 108).
If swallowed: standard treatment (p 109).

Fire hazard Flash point below −26 °C; explosive limits 1.8–10.1%; ignition temp. 312 °C. *Extinguish fire with* water spray, foam, dry powder, carbon dioxide or vaporising liquid.

Spillage disposal Shut off all possible sources of ignition. Instruct others to keep at a safe distance. Wear breathing apparatus and gloves. Mop up with plenty of water and run to waste, diluting greatly with running water. Ventilate area well to evaporate remaining liquid and dispel vapour.

2-Diethylaminoethanol (*N,N*-diethylethanolamine)

Colourless hygroscopic liquid; bp 163 °C; miscible with water.

IRRITATING TO SKIN, EYES AND RESPIRATORY SYSTEM

Avoid breathing vapour. Avoid contact with skin, eyes and clothing. TLV (skin) 10 ppm (50 mg m^{-3}).

Toxic effects The vapour irritates the eyes and respiratory system. The liquid injures the eyes and is absorbed by the skin which may be irritated. The liquid is moderately toxic if taken by mouth.

First aid *Vapour inhaled:* standard treatment (p 109).
Affected eyes: standard treatment (p 109).
Skin contact: standard treatment (p 108).
If swallowed: standard treatment (p 109).

Spillage disposal Shut off all possible sources of ignition. Wear face-shield or goggles, and gloves. Mop up with plenty of water and run to waste, diluting greatly with running water. Ventilate area well to evaporate remaining liquid and dispel vapour.

N,N-Diethylaniline

Colourless to brown liquid; bp 216 °C; sparingly soluble in water.

HARMFUL VAPOUR
HARMFUL IN CONTACT WITH SKIN

Avoid breathing vapour. Prevent contact with skin and eyes.

Toxic effects   Excessive breathing of the vapour or absorption of the liquid through the skin can cause headache, drowsiness, cyanosis, mental confusion and, in severe cases, convulsions. The liquid is dangerous to the eyes and the above effects can also be experienced if it is swallowed. **Chronic effects** Continued exposure to the vapour or slight skin exposure to the liquid over a period may affect the nervous system and the blood, causing fatigue, loss of appetite, headache and dizziness.

First aid   *Vapour inhaled:* standard treatment (p 109).
Affected eyes: standard treatment (p 109).
Skin contact: standard treatment (p 108).
If swallowed: standard treatment (p 109).

Spillage disposal   Wear face-shield or goggles, and gloves. Mix with sand and shovel mixture into glass, enamel or polythene vessel for dispersion in an excess of dilute hydrochloric acid (1 volume of concentrated acid diluted with 2 volumes of water). Allow to stand, with occasional stirring, for 24 hours and then run acid extract to waste, diluting greatly with running water and washing the sand. The sand can be treated as normal refuse.

Diethylarsine

Hazardous reaction   Inflames in air (B425).

Diethyl azodiformate

Hazardous reaction   Shock-sensitive explosive (B497).

Diethylberyllium

Hazardous reactions   Ignites in air; reacts explosively with water (B419).

Diethylcadmium

Hazardous reaction   Liable to explode under differing circumstances (B419).

Diethyl carbonate

Colourless liquid with pleasant, ethereal smell; bp 126 °C; practically insoluble in water.

FLAMMABLE
IRRITATING TO EYES AND RESPIRATORY SYSTEM

Avoid breathing vapour. Avoid contact with eyes.

Toxic effects   The vapour irritates the eyes and respiratory system. The liquid irritates the eyes and is assumed to be irritant and harmful if taken by mouth.

First aid	*Vapour inhaled:* standard treatment (p 109). *Affected eyes:* standard treatment (p 109). *Skin contact:* standard treatment (p 108). *If swallowed:* standard treatment (p 109).
Fire hazard	Flash point 25 °C. *Extinguish fire with* water spray, foam, dry powder, carbon dioxide or vaporising liquid.
Spillage disposal	Shut off all possible sources of ignition. Wear face-shield or goggles, and gloves. Apply non-flammable dispersing agent if available and work to an emulsion with brush and water—run this to waste, diluting greatly with running water. If dispersant not available, absorb on sand, shovel into bucket(s) and transport to safe open area for atmospheric evaporation or burial. Ventilate site of spillage well to evaporate remaining liquid and dispel vapour.

Diethylene dioxide – *see* Dioxan

Diethylene oximide – *see* Morpholine

Diethylenetriamine

Yellow, viscous liquid with ammoniacal smell; bp 208 °C; miscible with water.

CAUSES BURNS
IRRITATING TO SKIN, EYES AND RESPIRATORY SYSTEM

Prevent contact with skin and eyes. Avoid breathing vapour. TLV 10 ppm (42 mg m^{-3}).

Toxic effects	As indicated in warning phrases.
First aid	*Vapour inhaled:* standard treatment (p 109). *Affected eyes:* standard treatment (p 109). *Skin contact:* standard treatment (p 108). *If swallowed:* standard treatment (p 109).
Spillage disposal	Instruct others to keep at a safe distance. Wear breathing apparatus and gloves. Mop up with plenty of water and run to waste, diluting greatly with running water.

N,*N*-Diethylethanolamine – *see* 2-Diethylaminoethanol

Diethyl ether (ether; ethyl ether; sulphuric ether)

Colourless, highly volatile liquid with characteristic odour, bp 34 °C; immiscible with water; liable to form explosive peroxides on exposure to air and light, which should be decomposed before the ether is distilled to small volume.

EXTREMELY FLAMMABLE
MAY FORM EXPLOSIVE PEROXIDES
HARMFUL VAPOUR

Avoid breathing vapour. TLV 400 ppm (1200 mg m^{-3}).

Toxic effects Inhalation of vapour may cause drowsiness, dizziness, mental confusion, faintness and, in high concentrations, unconsciousness. Ingestion may also produce these effects. **Chronic effects** Continued inhalation of low concentrations may cause loss of appetite, dizziness, fatigue and nausea. Repeated inhalation or swallowing may lead to 'ether habit', with symptoms resembling chronic alcoholism.

Hazardous reactions Peroxide formation and subsequent explosion extensively reviewed; powerful oxidants also produce explosive reactions readily; reacts vigorously with sulphuryl chloride (B421–422).

First aid *Vapour inhaled:* standard treatment (p 109).
Affected eyes: standard treatment (p 109).
If swallowed: standard treatment (p 109).

Fire hazard Flash point −45 °C; explosive limits 1.85–48%; ignition temp. 180 °C. *Extinguish fire with* dry powder, carbon dioxide or vaporising liquid.

Spillage disposal Shut off all possible sources of ignition. Instruct others to keep at a safe distance. Wear breathing apparatus and gloves. Apply non-flammable dispersing agent if available and work to an emulsion with brush and water—run this to waste, diluting greatly with running water. If dispersant not available, organise effective ventilation of area until the liquid and vapour have been dispersed.

Diethyl ketone – *see* Pentan-3-one

Diethylmagnesium

Hazardous reactions Water ignites solid or ethereal solution (B420).

Diethylmethylphosphine

Hazardous reaction May ignite on long exposure to air (B459).

Diethyl peroxide

Hazardous reaction Explosive (B423).

N,N-Diethyl-*p*-phenylenediamine

(4-amino-*N,N*-diethylaniline) **and salts**

The free base is a reddish-brown liquid which darkens on exposure to light and air; bp 261 °C; insoluble in water. The hydrochloride and sulphate are buff or grey crystalline powders which darken on exposure to light and air; they are soluble in water.

HARMFUL IN CONTACT WITH SKIN

Avoid contact with skin and eyes.

Toxic effects	These are not recorded but may be expected to be similar to those resulting from contact with phenylenediamine, namely eye and skin irritation; dermatitis and more serious eye injury may also result from major contact.
First aid	*Affected eyes:* standard treatment (p 109). *Skin contact:* standard treatment (p 108). *If swallowed:* standard treatment (p 109).
Spillage disposal	Wear face-shield or goggles, and gloves. Mix with sand and shovel mixture into a glass, polythene or enamel vessel for dispersion in an excess of dilute hydrochloric acid (1 volume of concentrated acid diluted with 2 volumes of water). Allow to stand, with occasional stirring, for 24 hours and then run acid extract to waste, diluting greatly with running water and washing the sand. The sand can be treated as normal refuse.

Diethylphosphine

Hazardous reaction	Readily ignites in air (B426).

Diethyl sulphate (ethyl sulphate)

Colourless liquid with faint ethereal odour; bp 209 °C with decomposition; insoluble in water.

HARMFUL VAPOUR
HARMFUL IN CONTACT WITH SKIN

Avoid breathing vapour. Avoid contact with skin and eyes.

Toxic effects	Its effect on humans has not been recorded, but authorities consider that animal experiments justify its classification as a dangerous chemical. It is assumed to be poisonous or irritant by inhalation, eye and skin contact and ingestion.
Hazardous reactions	Violent reactions with 2,7-dinitro-9-phenylphenanthridine/water and potassium *t*-butoxide (B424).
First aid	*Vapour inhaled:* standard treatment (p 109). *Affected eyes:* standard treatment (p 109). *Skin contact:* standard treatment (p 108). *If swallowed:* standard treatment (p 109).
Spillage disposal	Instruct others to keep at a safe distance. Wear breathing apparatus and gloves. Apply dispersing agent if available and work to an emulsion with brush and water—run this to waste diluting greatly with running water. If dispersant not available, absorb on sand, shovel into bucket(s) and transport to safe open area for burial. Site of spillage should be washed thoroughly with water and soap or detergent.

Diethylzinc

Hazardous reactions	Pyrophoric in air; reacts violently with water (B425).

Difluoroamine

**Hazardous
reaction**   A dangerous explosive (B735).

Difluoroamino compounds Review of group (B45).

Difluorodiazene

**Hazardous
reaction**   Reacts explosively with hydrogen above 90 °C. (B735).

1,1-Difluoroethylene

Colourless gas with faint ethereal odour; bp −83 °C; slightly soluble in water.

FLAMMABLE

Toxic effects   The gas, according to current evidence, has no substantial toxicity but shows the asphyxiant properties of non-toxic gases such as nitrogen.

Fire hazard   Explosive limits 5.5–21.3%. Since the gas is supplied in a cylinder, turning off the valve will reduce any fire involving it; if possible, cylinders should be removed quickly from an area in which a fire has developed.

Disposal   Surplus gas or leaking cylinder can be vented slowly to air in an open area.

Di-2-furoyl peroxide

**Hazardous
reaction**   Explodes violently on heating and friction (B576).

Dihexanoyl peroxide

**Hazardous
reaction**   Explodes at 85 °C (B595).

1,2-Dihydroxybenzene – *see* Catechol

Di-iodoacetylene

**Hazardous
reaction**   Explodes on friction, impact, and on heating to 84 °C (B333).

Di-iodoamine

Hazardous reaction   Explosive (B756).

Di-isobutylene – *see* 2,4,4-Trimethylpentene

Di-isobutyl ketone

Colourless liquid; bp 169 °C; sparingly soluble in water.

**HARMFUL VAPOUR
IRRITATING TO SKIN AND EYES**

Avoid breathing vapour. Avoid contact with eyes. TLV 25 ppm (250 mg m^{-3}).

Toxic effects   The vapour may irritate the respiratory system and is narcotic in high concentrations. The liquid irritates the eyes and may irritate the skin; it is assumed to be harmful if taken by mouth.

First aid   *Vapour inhaled in high concentrations:* standard treatment (p 109).
Affected eyes: standard treatment (p 109).
Skin contact: standard treatment (p 108).
If swallowed: standard treatment (p 109).

Spillage disposal   Shut off all possible sources of ignition. Wear face-shield or goggles, and gloves. Apply non-flammable dispersing agent if available and work to an emulsion with brush and water — run this to waste, diluting greatly with running water. If dispersant not available, absorb on sand, shovel into bucket(s) and transport to safe open area for burial. Ventilate site of spillage well to evaporate remaining liquid and dispel vapour.

Di-isobutyryl peroxide

Hazardous reaction   Explodes on standing at room temperature; solution in ether exploded during evaporation (B556, B562).

Di(4-isocyanatophenyl)methane
– *see* Bis (4-isocyanatophenyl)methane

2,4-Di-isocyanatotoluene (toluene 2,4-di-isocyanate)

Pale yellow liquid with sharp pungent smell; bp 251 °C; reacts with water with evolution of carbon dioxide; commonly contains about 20% of the 2,6-isomer.

**HARMFUL VAPOUR
IRRITATING TO SKIN, EYES AND RESPIRATORY SYSTEM**

Avoid breathing vapour. Avoid contact with skin, eyes and clothing. TLV 0.02 ppm (0.14 mg m^{-3}).

Toxic effects   The vapour irritates the respiratory system and may cause bronchial asthma. The vapour and liquid are very irritating to the eyes. The liquid irritates the skin and may cause severe dermatitis. Assumed to be highly irritant and poisonous if taken by mouth.

Hazardous reactions   May polymerise vigorously on contact with bases and acyl chlorides (B566).

First aid
Vapour inhaled: standard treatment (p 109).
Affected eyes: standard treatment (p 109).
Skin contact: standard treatment (p 108).
If swallowed: standard treatment (p 109).

Spillage disposal   Instruct others to keep at a safe distance. Wear breathing apparatus and gloves. Absorb on sand, shovel into bucket(s), transport to safe open area and tip into large volume of water; leave to decompose before decanting the water to waste, diluting greatly with running water. Site of spillage should be ventilated after washing thoroughly with water and soap or detergent.

Di-isopropylamine

Colourless, strongly alkaline liquid; bp 84 °C; miscible with water.

EXTREMELY FLAMMABLE
HARMFUL VAPOUR
IRRITATING TO SKIN, EYES AND RESPIRATORY SYSTEM

Avoid breathing vapour. Avoid contact with skin, eyes and clothing. TLV 5 ppm (20 mg m^{-3}) (skin).

Toxic effects   The vapour irritates eyes and respiratory system. The liquid irritates the skin and eyes and may cause burns to the eyes. Assumed to be irritant and poisonous if taken internally.

First aid
Vapour inhaled: standard treatment (p 109).
Affected eyes: standard treatment (p 109).
Skin contact: standard treatment (p 108).
If swallowed: standard treatment (p 109).

Fire hazard   Flash point −1 °C. *Extinguish fire with* dry powder, carbon dioxide or vaporising liquid.

Spillage disposal   Shut off all possible sources of ignition. Instruct others to keep at a safe distance. Wear breathing apparatus and gloves. Mop up with plenty of water and run to waste diluting greatly with running water. Ventilate area well to evaporate remaining liquid and dispel vapour.

Di-isopropyl ether (isopropyl ether)

Colourless liquid with ethereal odour; bp 69 °C; sparingly soluble in water. The unstabilised ether readily forms explosive peroxides on exposure to light and air; these must be destroyed before the ether is distilled.

EXTREMELY FLAMMABLE
MAY FORM EXPLOSIVE PEROXIDES
HARMFUL VAPOUR
IRRITATING TO EYES AND RESPIRATORY SYSTEM

Avoid breathing vapour. Avoid contact with skin and eyes. TLV 500 ppm (2100 mg m^{-3}).

Toxic effects   The vapour irritates the respiratory system and eyes and inhalation may lead to headache, dizziness, nausea, vomiting and narcosis. The liquid irritates the eyes causing conjunctivitis; it will defat the skin and may lead to dermatitis. If taken internally, it gives effects similar to those indicated for inhalation of the vapour.

Hazardous reaction   Formation of peroxides responsible for numerous explosions; methods of inhibiting this are reviewed (B510).

First aid   *Vapour inhaled:* standard treatment (p 109).
Affected eyes: standard treatment (p 109).
Skin contact: standard treatment (p 108).
If swallowed: standard treatment (p 109).

Fire hazard   Flash point −28 °C; explosive limits 1.4–21%; ignition temp. 443 °C. *Extinguish fire with* dry powder, carbon dioxide or vaporising liquid.

Spillage disposal   Shut off all possible sources of ignition. Instruct others to keep at a safe distance. Wear breathing apparatus and gloves. Apply non-flammable dispersing agent if available and work to an emulsion with brush and water—run this to waste, diluting greatly with running water. If dispersant not available, absorb on sand, shovel into bucket(s) and transport to safe open area for atmospheric evaporation. Site of spillage should be washed thoroughly with water and soap or detergent.

Di-isopropyl peroxydicarbonate

Hazardous reaction   Undergoes self-accelerating decomposition when warmed above its melting point (10 °C) which may become dangerously violent (B556).

Diketen (4-methyleneoxetan-2-one, diketene)

Colourless liquid with pungent odour; bp 127 °C; decomposed by water; reacts violently with acids and alkalies.

FLAMMABLE
CAUSES BURNS
IRRITATING TO SKIN, EYES AND RESPIRATORY SYSTEM

Prevent inhalation of vapour. Prevent contact with skin, eyes and clothing.

Toxic effects   The vapour irritates the respiratory system and the eyes severely, causing lachrymation. The liquid irritates the skin and may cause burns; it is assumed to cause severe damage if taken internally.

Hazardous reactions   Violent polymerisation is catalysed by acids or bases (B386).

First aid   *Vapour inhaled:* standard treatment (p 109).
Affected eyes: standard treatment (p 109).
Skin contact: standard treatment (p 108).
If swallowed: standard treatment (p 109).

Fire hazard   Flash point 46 °C (open cup). *Extinguish fire with* water spray, dry powder, carbon dioxide or vaporising liquids.

Spillage disposal
Shut off all possible sources of ignition. Instruct others to keep at a safe distance. Wear breathing apparatus and gloves. Apply non-flammable dispersing agent if available and work to an emulsion with brush and water—run this to waste, diluting greatly with running water. If dispersant not available, absorb on sand, shovel into bucket(s) and transport to safe open area for atmospheric evaporation or burial. Site of spillage should be washed thoroughly with water and soap or detergent.

Dilithium acetylide

Hazardous reactions
Burns brilliantly in chlorine or fluorine; burns vigorously in P, S and Se vapours (B334).

Dimercury dicyanide oxide

Hazardous reaction
Heat- and impact-sensitive explosive (B333).

3,3′-Dimethoxybenzidine (o-dianisidine) and its dihydrochloride

o-Dianisidine is a colourless to grey-mauve powder; insoluble in water. Its dihydrochloride is a colourless to grey powder, sparingly soluble in water. The use of o-dianisidine and its salts is controlled in the United Kingdom by the Carcinogenic Substances Regulations 1967.

SERIOUS RISK OF POISONING BY INHALATION, SKIN CONTACT OR SWALLOWING
DANGER OF VERY SERIOUS IRREVERSIBLE EFFECTS
CAUSES IRRITATION OF SKIN AND EYES

Prevent inhalation of dust. Prevent contact with skin and eyes.

Toxic effects
The dust irritates the nose severely, causing sneezing. Solutions of the dihydrochloride irritate the eyes. The effects of ingestion are not recorded. **Chronic effects** There is evidence that o-dianisidine, through continued absorption, can cause cancer of the bladder.

First aid
Dust inhaled: standard treatment (p 109).
Affected eyes: standard treatment (p 109).
Skin contact: standard treatment (p 108).
If swallowed: standard treatment (p 109).

Spillage disposal
Wear breathing apparatus and gloves. Mix spillage with moist sand and shovel mixture into a glass, enamel or polythene vessel for dispersion in an excess of dilute hydrochloric acid (1 volume of concentrated acid diluted with 2 volumes of water). Allow to stand, with occasional stirring, for 24 hours and then run extract to waste, diluting greatly with running water and washing the sand. The residual sand can be treated as normal refuse. The site of the spillage should be washed with water and soap or detergent.

1,2-Dimethoxyethane (ethylene glycol dimethyl ether)

Colourless liquid with sharp, ethereal odour; bp 85 °C; miscible with water; liable to form explosive peroxides on exposure to air and light which should be decomposed before the ether is distilled to small volume.

HIGHLY FLAMMABLE
HARMFUL VAPOUR
MAY FORM EXPLOSIVE PEROXIDES

Avoid breathing vapour. Avoid contact with skin and eyes.

Toxic effects   These are not well documented apart from animal experiments which indicate that the vapour is irritant to the respiratory system and that skin and eyes are liable to be affected by contact with the liquid.

First aid   *Vapour inhaled:* standard treatment (p 109).
Affected eyes: standard treatment (p 109).
Skin contact: standard treatment (p 108).
If swallowed: standard treatment (p 109).

Fire hazard   Flash point 4.5 °C; *Extinguish fire with* water spray, foam, dry powder, carbon dioxide or vaporising liquids.

Spillage disposal   Shut off all possible sources of ignition. Instruct others to keep at a safe distance. Wear breathing apparatus and gloves. Mop up with plenty of water and run to waste diluting greatly with running water. Ventilate area well to evaporate remaining liquid and dispel vapour.

Dimethoxymethane (methylal)

Colourless, volatile liquid; bp 42 °C; 1 part dissolves in 3 parts water at about 25 °C; may form explosive peroxides on exposure to air and light which should be decomposed before the ether is distilled to small volume.

EXTREMELY FLAMMABLE
MAY FORM EXPLOSIVE PEROXIDES

Avoid breathing vapour. Avoid contact with eyes. TLV 1000 ppm (3100 mg m^{-3}).

Toxic effects   Considered to be of low toxicity though high concentrations may cause narcosis. It has produced injury to lungs, liver, kidneys and heart in experiments on animals.

First aid   *Vapour inhaled:* standard treatment (p 109).
Affected eyes: standard treatment (p 109).
Skin contact: standard treatment (p 108).
If swallowed: standard treatment (p 109).

Fire hazard   Flash point −18 °C (open cup); ignition temp. 237 °C. *Extinguish fire with* dry powder, carbon dioxide or vaporising liquid.

Spillage disposal   Shut off all possible sources of ignition. Instruct others to keep at a safe distance. Wear breathing apparatus and gloves. Mop up with plenty of water and run to waste, diluting greatly with running water. Ventilate area well to evaporate remaining liquid and dispel vapour.

2,2-Dimethoxypropane

Hazardous reactions	Violent explosions when dehydration of hydrated manganese and nickel perchlorates was attempted using the ether (B457).

Dimethylamine and solutions

Colourless gas at ordinary temperatures (bp 7 °C); readily soluble in water. Commonly available in aqueous and ethanolic solution.

> EXTREMELY FLAMMABLE
> IRRITATING TO SKIN, EYES AND RESPIRATORY SYSTEM

Avoid breathing vapour. Avoid contact with skin and eyes. TLV 10 ppm (18 mg m⁻³).

Toxic effects	The vapour irritates the mucous membranes and respiratory system; in high concentrations it may affect the nervous system. The vapour and solutions irritate the eyes. The solutions may irritate the skin. Assumed to be poisonous if taken by mouth.
Hazardous reactions	The gas incandesces on contact with fluorine; causes maleic anhydride to decompose exothermically above 150 °C (B328).
First aid	*Vapour inhaled:* standard treatment (p 109). *Affected eyes:* standard treatment (p 109). *Skin contact:* standard treatment (p 108). *If swallowed:* standard treatment (p 109).
Fire hazard (Gas)	Flash point −50 °C; explosive limits 2.8–14.4%; ignition temp. 402 °C. *Extinguish fire with* water spray, foam, dry powder, carbon dioxide or vaporising liquid. Latter applies also to solutions in water or alcohol.
Spillage disposal	(liquid and solutions) Shut off all possible sources of ignition. Instruct others to keep at a safe distance. Wear breathing apparatus and gloves. Mop up with plenty of water and run to waste, diluting greatly with running water. Ventilate area well to evaporate remaining liquid and dispel vapour.

2-Dimethylaminoethanol (*N,N*-dimethylethanolamine)

Colourless liquid; bp 135 °C; miscible with water.

> FLAMMABLE
> IRRITATING TO EYES

Avoid contact with skin, eyes and clothing.

Toxic effects	Little is recorded about these, but it is suggested that splashing of the eyes could cause serious injury.
First aid	*Affected eyes:* standard treatment (p 109). *Skin contact:* standard treatment (p 108). *If swallowed:* standard treatment (p 109).
Fire hazard	Flash point 41 °C (open cup). *Extinguish fire with* water spray, foam, dry powder, carbon dioxide or vaporising liquid.

Spillage disposal	Shut off all possible sources of ignition. Wear face-shield or goggles, and gloves. Mop up with plenty of water and run to waste diluting greatly with running water. Ventilate area well to evaporate remaining liquid and dispel vapour.

N,N-Dimethylaniline

Colourless to brown liquid; bp 193 °C; sparingly soluble in water.

> HARMFUL VAPOUR
> HARMFUL IN CONTACT WITH SKIN

Avoid breathing vapour. Prevent contact with skin and eyes. TLV (skin) 5 ppm (25 mg m⁻³).

Toxic effects	Excessive breathing of the vapour or absorption of the liquid through the skin can cause headache, drowsiness, cyanosis, mental confusion and, in severe cases, convulsions. The liquid is dangerous to the eyes and the above effects can also be experienced if it is swallowed. **Chronic effects** Continued exposure to the vapour or slight skin exposure to the liquid over a period may affect the nervous system and the blood causing fatigue, loss of appetite, headache and dizziness.
Hazardous reaction	Contact with a drop causes dibenzoyl peroxide to explode (B552).
First aid	*Vapour inhaled:* standard treatment (p 109). *Affected eyes:* standard treatment (p 109). *Skin contact:* standard treatment (p 108). *If swallowed:* standard treatment (p 109).
Spillage disposal	Wear face-shield or goggles, and gloves. Mix with sand and shovel mixture into glass, enamel or polythene vessel for dispersion into an excess of dilute hydrochloric acid (1 volume of concentrated acid diluted with 2 volumes of water). Allow to stand, with occasional stirring, for 24 hours and then run acid extract to waste, diluting greatly with running water and washing the sand. The sand can be treated as normal refuse.

Dimethylantimony chloride

Hazardous reaction	Ignites in air at 40 °C (B320).

Dimethylarsine

Hazardous reaction	Inflames in air (B328).

2,2´-Dimethylbenzidine – *see* *O*-Tolidine

3,5-Dimethylbenzoic acid

Hazardous preparation   Explosion when mesitylene being oxidised with nitric acid in autoclave (B569).

Dimethylberyllium

Hazardous reactions   Ignites in moist air or in CO_2; reacts explosively with water (B318).

2,3-Dimethylbuta-1,3-diene

Hazardous reaction   Polymeric peroxide autoxidation residue exploded violently on ignition (B495).

Dimethylcadmium

Hazardous reaction   Peroxide formed on exposure to air is explosive (B319).

Dimethyl carbonate (methyl carbonate)

Colourless liquid with pleasant odour; bp 90 °C; insoluble in water.

> HIGHLY FLAMMABLE
> HARMFUL VAPOUR
> IRRITATES EYES AND RESPIRATORY SYSTEM

Avoid breathing vapour. Avoid contact with skin, eyes and clothing.

Toxic effects   The vapour irritates the eyes and respiratory system. The liquid irritates the eyes and is assumed to be poisonous if taken internally.

First aid   *Vapour inhaled:* standard treatment (p 109).
Affected eyes: standard treatment (p 109).
Skin contact: standard treatment (p 108).
If swallowed: standard treatment (p 109).

Fire hazard   Flash point 19 °C (open cup). *Extinguish fire with* water spray, foam, dry powder, carbon dioxide or vaporising liquid.

Spillage disposal   Shut off all possible sources of ignition. Instruct others to keep at a safe distance. Wear breathing apparatus and gloves. Apply non-flammable dispersing agent if available and work to an emulsion with brush and water — run this to waste, diluting greatly with running water. If dispersant not available, absorb on sand, shovel into bucket(s) and transport to safe open area for atmospheric evaporation. Site of spillage should be washed thoroughly with water and soap or detergent.

Dimethyldichlorosilane

Colourless, fuming liquid which reacts violently with water; bp 70 °C.

> **EXTREMELY FLAMMABLE**
> **CAUSES BURNS**
> **IRRITATING TO SKIN, EYES AND RESPIRATORY SYSTEM**

Avoid breathing vapour. Avoid contact with skin, eyes and clothing.

Toxic effects The vapour irritates the eyes and respiratory system. The liquid burns the skin and eyes. The liquid will burn the mouth and alimentary system if taken by mouth.

First aid *Vapour inhaled:* standard treatment (p 109).
Affected eyes: standard treatment (p 109).
Skin contact: standard treatment (p 108).
If swallowed: standard treatment (p 109).

Fire hazard Flash point −9 °C; explosive limits 3.4–9.5%. *Extinguish fire with* dry powder, carbon dioxide, dry sand or earth.

Spillage disposal Shut off all possible sources of ignition. Instruct others to keep at safe distance. Wear breathing apparatus and gloves. Absorb on sand, shovel into bucket(s), transport to safe open area and tip into large volume of water; leave to decompose before decanting the water to waste, diluting greatly with running water. Site of spillage should be ventilated after washing thoroughly with water and soap or detergent.

N,N-Dimethylethanolamine
-see 2-Dimethylaminoethanol

Dimethyl ether (methyl ether)

Colourless gas with slight ethereal odour; bp −25 °C; slightly soluble (7% by weight) in water.

> **EXTREMELY FLAMMABLE**
> **HARMFUL BY INHALATION**

Avoid breathing gas.

Toxic effects The gas is about one fourth as potent as diethyl ether as an anaesthetic but is not used for this purpose because of other toxic effects, notably rushing of blood through the head and sickness.

First aid *Vapour inhaled:* standard treatment (p 109).

Fire hazard Flash point −41 °C; explosive limits 3.4–18%; ignition temp. 350 °C. Since the gas is supplied in a cylinder, turning off the valve will reduce any fire involving it; if possible, cylinders should be removed quickly from an area in which a fire has developed.

Disposal Surplus gas or leaking cylinder can be vented slowly to air in a safe open area or gas burnt off in a suitable burner.

Dimethylformamide (formdimethylamide)

Colourless liquid with faint amine-like odour; bp 153 °C; miscible with water.

> HARMFUL VAPOUR
> IRRITATING TO SKIN, EYES AND RESPIRATORY SYSTEM

Avoid breathing vapour. Avoid contact with skin and eyes. TLV (skin) 10 ppm (30 mg m^{-3}).

Toxic effects The vapour from the hot liquid irritates the eyes and respiratory system. The liquid irritates the skin and eyes. Assumed to be poisonous if taken by mouth. **Chronic effects** Prolonged inhalation of vapour has resulted in liver damage in experimental animals.

Hazardous reactions Reacts vigorously or violently with a range of materials including Br$_2$, CCl$_4$, CrO$_3$, Na, magnesium nitrate (B369).

First aid *Vapour inhaled:* standard treatment (p 109).
Affected eyes: standard treatment (p 109).
Skin contact: standard treatment (p 108).
If swallowed: standard treatment (p 109).

Spillage disposal Shut off all possible sources of ignition. Wear face-shield or goggles, and gloves. Mop up with plenty of water and run to waste, diluting greatly with running water.

1,1-Dimethylhydrazine

Hazardous reactions Ignites violently with oxidants such as N$_2$O$_4$, H$_2$O$_2$, HNO$_3$ (B330).

Dimethylketen

Hazardous reaction Forms extremely explosive peroxide when exposed to air (B395).

Dimethylmagnesium

Hazardous reaction Water ignites the solid or its ethereal solution (B321).

Dimethylmercury

Hazardous reaction Reacts explosively with diboron tetrachloride (B321).

N,N-Dimethyl-*p*-nitrosoaniline (nitrosodimethylaniline)

Green powder, insoluble in water.

HARMFUL BY INHALATION
IRRITATING TO SKIN, EYES AND RESPIRATORY SYSTEM

Avoid breathing dust. Avoid contact with skin and eyes.

Toxic effects   The dust irritates the respiratory system. The dust irritates the eyes and skin and may cause dermatitis. Assumed to be irritant and poisonous if taken by mouth.

First aid   *Dust inhaled:* standard treatment (p 109).
Affected eyes: standard treatment (p 109).
Skin contact: standard treatment (p 108).
If swallowed: standard treatment (p 109).

Spillage disposal   Wear face-shield or goggles, and gloves. Mix with sand and transport to a safe, open area for burial. Site of spillage should be washed thoroughly with water and soap or detergent.

1,2-Dimethyl-1-nitrosohydrazine

Hazardous reaction   Deflagrates on heating (B329).

Dimethyl peroxide

Hazardous reaction   Heat- and shock-sensitive explosive as liquid or vapour (B325).

3,3-Dimethyl-1-phenyltriazene

Hazardous reaction   Exploded on attempted distillation at atmospheric pressure (B553).

Dimethylphosphine

Hazardous reaction   Readily ignites in air (B329).

2,2-Dimethylpropane (neopentane)

Colourless liquid or gas; bp 9.5 °C; insoluble in water.

EXTREMELY FLAMMABLE

Avoid breathing gas.

Toxic effects This is classed as a simple asphyxiant, anaesthetic gas of low toxicity which may show irritant and narcotic effects in high concentrations.

First aid *Vapour inhaled:* standard treatment (p 109).

Fire hazard Flash point below −7 °C; explosive limits 1.4–7.5%; ignition temp. 450 °C. Since the gas is supplied in a cylinder, turning off the valve will reduce any fire involving it; if possible, cylinders should be removed from an area in which a fire has developed.

Disposal Surplus gas or leaking cylinder can be vented slowly to air in a safe, open area or gas burnt off in a suitable gas burner.

Dimethyl selenate

Hazardous reaction Explodes at about 150 °C when distilled at atmospheric pressure (B327).

Dimethyl sulphate (methyl sulphate)

Colourless, odourless liquid; bp 189 °C with decomposition; somewhat soluble in water.

GIVES OFF POISONOUS VAPOUR
CAUSES BURNS
IRRITATING TO SKIN, EYES AND RESPIRATORY SYSTEM

Prevent inhalation of vapour. Prevent contact with skin and eyes. TLV (skin) 1 ppm (5 mg m^{-3}).

Toxic effects Vapour causes severe irritation of respiratory system, with possible severe lung injury after a latent period. Vapour and liquid irritate or burn the eyes severely after a latent period, resulting in temporary or permanent dimming of vision. The vapour or liquid may blister the skin and skin absorption may result in severe poisoning after a latent period. Extremely poisonous and irritant if taken by mouth.

Hazardous reaction Reacts violently with concentrated aqueous ammonia (B327).

First aid *Vapour inhaled:* standard treatment (p 109).
Affected eyes: standard treatment (p 109).
Skin contact: standard treatment (p 108).
If swallowed: standard treatment (p 109).

Spillage disposal Instruct others to keep at a safe distance. Wear breathing apparatus and gloves. Apply dispersing agent if available and work to an emulsion with brush and water—run this to waste, diluting greatly with running water. If dispersant not available, absorb on sand, shovel into bucket(s) and transport to safe open area for burial. Site of spillage should be washed thoroughly with water and soap or detergent.

Dimethyl sulphoxide

Colourless, hygroscopic liquid; bp 189 °C; miscible with water.

HARMFUL SUBSTANCE IF TAKEN INTERNALLY
IRRITATING TO EYES

Avoid contact with skin or eyes.

Toxic effects	May cause redness, itching and scaling of skin and damage to eyes. Absorbed readily by skin and volunteers have reported nausea, vomiting, cramps, chills and drowsiness from applications.
Hazardous reactions	Reacts violently or explosively with acetyl chloride, benzenesulphonyl chloride, cyanuric chloride, PCl_3, $POCl_3$, $SiCl_4$, SCl_2, S_2Cl_2, $SOCl_2$, N_2O_4, IF_5, magnesium perchlorate, $HClO_4$, HIO_4, sodium hydride, SO_3 or AgF_2 (B323–325).
First aid	*Affected eyes:* standard treatment (p 109). *Skin contact:* standard treatment (p 108). *If swallowed:* standard treatment (p 109).
Spillage disposal	Wear goggles and gloves. Mop up with plenty of water and run to waste diluting greatly with running water.

Dimethylzinc

Hazardous reactions	Ignites in air and explodes in oxygen (B328).

Di-1-naphthoyl peroxide

Hazardous reaction	Explodes on friction (B615).

2,4-Dinitroaniline

Yellow granules or powder; mp 188 °C; insoluble in water.

SERIOUS RISK OF POISONING BY INHALATION, SWALLOWING OR SKIN CONTACT

Prevent contact with skin and eyes.

Toxic effects	Records have not been traced, but it can be assumed that skin absorption is liable to cause dermatitis and cyanosis, and that the eyes would be damaged by contact. It must also be assumed to be poisonous if taken by mouth.
First aid	*Affected eyes:* standard treatment (p 109). *Skin contact:* standard treatment (p 108). *If swallowed:* standard treatment (p 109).
Spillage disposal	Wear face-shield or goggles, and gloves. Mix with sand and shovel mixture into glass, enamel or polythene vessel for dispersion in an excess of dilute hydrochloric acid (1 volume of concentrated acid diluted with 2 volumes of water). Allow to stand, with occasional stirring, for 24 hours and then run acid extract to waste, diluting greatly with running water and washing the sand. The sand can be disposed of as normal refuse.

1,3-Dinitrobenzene

Colourless to yellow crystals; mp 89–90 °C; insoluble in water.

TOXIC IF TAKEN INTERNALLY
DANGER OF CUMULATIVE EFFECTS

Avoid breathing vapour. Prevent contact with skin and eyes. TLV (skin) 0.15 ppm (1 mg m⁻³).

Toxic effects   The vapour may cause headache, vertigo and vomiting; in severe cases this may be followed by exhaustion, cyanosis, drowsiness, and unconsciousness. Contact will damage the eyes, and skin absorption may lead to the above symptoms. Assumed to be poisonous if taken by mouth.

First aid   *Vapour inhaled:* standard treatment (p 109).
Affected eyes: standard treatment (p 109).
Skin contact: standard treatment (p 108).
If swallowed: standard treatment (p 109).

Spillage disposal   Wear face-shield or goggles, and gloves. Mix with sand and transport to a safe, open area for burial. Site of spillage should be washed thoroughly with water and soap or detergent.

Dinitrobenzenes

Hazardous reaction   Mixtures with concentrated nitric acid possess high explosive properties (B474).

2,4-Dinitrobenzenesulphenyl chloride

Hazardous preparation   Must not be overheated as it may explode (B465).

Dinitrobutenes

Hazardous reaction   Liable to violent decomposition or explosion when heated (B393).

2,4-Dinitrochlorobenzene
– *see* 1-Chloro-2,4-dinitrobenzene

Dinitro-o-cresol

Yellow crystals or powder; mp 85 °C almost insoluble in water.

SERIOUS RISK OF POISONING
BY INHALATION, SWALLOWING OR CONTACT WITH SKIN
DANGER OF CUMULATIVE EFFECTS

Prevent inhalation of dust. Prevent contact with skin and eyes. TLV (skin) (0.2 mg m^{-3}.)

Toxic effects The inhalation of dust may cause profuse sweating, fever, shortness of breath and yellow coloration of skin of hands and feet. Similar symptoms may follow ingestion and absorption through the skin. Skin contact may cause dermatitis.

First aid *Dust inhaled:* standard treatment (p 109).
Skin contact: standard treatment (p 109).
If swallowed: standard treatment (p 108).
Affected eyes: standard treatment (p 109).

Spillage disposal Wear face-shield or goggles, and gloves. Mix with sand and transport to a safe, open area for burial. Site of spillage should be washed thoroughly with water and soap or detergent.

Dinitrogen oxide (nitrous oxide)

Hazardous reactions Amorphous boron ignites when heated in the gas; a mixture of phosphine with excess of the oxide can be exploded by sparking (B874).

Dinitrogen pentoxide

Hazardous reactions K and Na burn in gas, Hg and As are vigorously oxidised; explodes with naphthalene, other organic materials react vigorously; reacts explosively with sulphur dichloride and sulphuryl chloride (B879).

Dinitrogen tetraoxide (nitrogen dioxide; nitrous fumes)

Nitrogen dioxide is a red-brown gas which is a common by-product of the reaction of nitric acid with metals and organic materials; it is also available in cylinders as the pure liquefied gas (bp 20 °C).

TOXIC BY INHALATION
IRRITATING TO SKIN, EYES AND RESPIRATORY SYSTEM

Avoid breathing gas. TLV 5 ppm (9 mg m^{-3}).

Toxic effects Although nitrogen dioxide has some irritant effect upon the respiratory system, its danger lies in the delay before its full effects upon the lungs are shown by feelings of weakness and coldness, headache, nausea, dizziness, abdominal pain and cyanosis; in severe cases, convulsions and asphyxia may follow. (*See* note in Ch 6, p 83.)

Hazardous reactions	Reacts violently or explosively with wide range of materials including acetonitrile, alcohols, liquid ammonia, carbonylmetals, dimethyl sulphoxide, halocarbons, hydrazine and derivatives, hydrocarbons, nitrobenzene, organic compounds, Mg filings, potassium, reduced iron, pyrophoric manganese (B875–879).
First aid	It is advisable in all cases where appreciable inhalation of nitrogen dioxide is believed to have occurred to obtain medical attention immediately, even if the person exposed is not complaining of discomfort; removal from exposure, rest and warmth are essential until under professional care. In severe cases, administer oxygen through a face mask. If breathing has stopped, apply artificial respiration.
Disposal	Surplus gas or leaking cylinder can be vented slowly into a water-fed scrubbing tower or column in a fume cupboard, or into a fume cupboard served by such a tower.

Dinitrophenols

Yellow crystals or powder; sparingly soluble in water.

> HARMFUL SUBSTANCE IF TAKEN INTERNALLY
> OR IF IN CONTACT WITH SKIN
> DANGER OF CUMULATIVE EFFECTS

Prevent inhalation of dust. Prevent contact with skin and eyes.

Toxic effects	The inhalation of dust may cause profuse sweating, fever, shortness of breath and yellow coloration of skin of hands and feet. Similar symptoms may follow ingestion and absorption through the skin. Skin contact may cause dermatitis. (*See* note in Ch 6, p 84.)
First aid	*Dust inhaled :* standard treatment (p 109). *Affected eyes :* standard treatment (p 109). *Skin contact :* drench with water and swab contaminated skin with glycerol for at least 10 minutes (use water if glycerol is not available); remove and wash contaminated clothing before re-use; if contamination has been other than slight, obtain medical attention. *If swallowed :* standard treatment (p 109).
Spillage disposal	Wear face-shield or goggles, and gloves. Mix with sand and transport to a safe, open area for burial. Site of spillage should be washed thoroughly with water and soap or detergent.

2,4-Dinitrophenylhydrazine

Red crystalline powder; mp about 200 °C; slightly soluble in water which is normally added in storage to reduce explosion risk.

> RISK OF EXPLOSION BY SHOCK, FRICTION, HEAT
> OR OTHER SOURCES OF IGNITION
> HARMFUL SUBSTANCE BY INHALATION,
> SWALLOWING OR SKIN CONTACT

Avoid inhalation of dust. Avoid contact with skin and eyes.

Toxic effects   No effects have been recorded, nor testing carried out, but the relationship with phenylhydrazine and presence of two nitro groups indicate that the above warning is likely to be justified.

First aid   *Dust inhaled:* standard treatment (p 109).
Affected eyes: standard treatment (p 109).
Skin contact: standard treatment (p 108).
If swallowed: standard treatment (p 109).

Spillage disposal   Wear face-shield or goggles and gloves. Mix with sand and transport to a safe, open area for burial. Site of spillage should be washed thoroughly with water and detergent.

2,4-Dinitrotoluene

Yellow crystals; insoluble in water.

HARMFUL VAPOUR
HARMFUL BY SKIN ABSORPTION

Prevent contact with skin and eyes. TLV (skin) (1.5 mg m^{-3}.)

Toxic effects   The vapour and crystals irritate the eyes. Contact with the skin may cause dermatitis. Assumed to be poisonous if taken by mouth.

First aid   *Affected eyes:* standard treatment (p 109).
Skin contact: standard treatment (p 108).
If swallowed: standard treatment (p 109).

Spillage disposal   Wear face-shield or goggles, and gloves. Mix with sand and transport to a safe, open area for burial. Site of spillage should be washed thoroughly with water and soap or detergent.

Dioxan (diethylene dioxide; diethylene oxide; dioxane)

Colourless, almost odourless, liquid; bp 101 °C; miscible with water; liable to form explosive peroxides on exposure to light and air which should be decomposed before the ether is distilled to small volume.

HIGHLY FLAMMABLE
MAY FORM EXPLOSIVE PEROXIDES
HARMFUL VAPOUR

Avoid breathing vapour. TLV (skin) 100 ppm (360 mg m^{-3}).

Toxic effects   The vapour irritates nose and eyes and this may be followed by headache and drowsiness. High concentrations may also cause nausea and vomiting, while injury to the kidney and liver are possible. Shows some of the above effects when taken by mouth.

Hazardous reactions   Forms explosive peroxides on exposure to air; reacts almost explosively with Raney nickel above 210 °C; addition complex with sulphur trioxide decomposes violently on storage (B410).

First aid	*Vapour inhaled:* standard treatment (p 109). *Affected eyes:* standard treatment (p 109). *If swallowed:* standard treatment (p 109).
Fire hazard	Flash point 12 °C; explosive limits 2–22%; ignition temp. 180 °C. *Extinguish fire with* water spray, dry powder, carbon dioxide or vaporising liquid.
Spillage disposal	Shut off all possible sources of ignition. Instruct others to keep at a safe distance. Wear breathing apparatus and gloves. Mop up with plenty of water and run to waste, diluting greatly with running water. Ventilate area well to evaporate remaining liquid and dispel vapour.

Dioxygen difluoride

Hazardous reactions	Very powerful oxidant reacting with many materials at low temperatures (B738).

Dipentene (DL-limonene; *p*-mentha-1,8-diene)

Colourless liquid with pleasant, lemon-like odour; bp 178 °C; insoluble in water.

FLAMMABLE
IRRITATING TO SKIN AND EYES

Avoid contact with skin, eyes and clothing.

Toxic effects	It is considered to be a moderate skin irritant and sensitiser, but its effects are not well documented.
First aid	*Affected eyes:* standard treatment (p 109). *Skin contact:* standard treatment (p 108). *If swallowed:* standard treatment (p 109).
Fire hazard	Flash point 45 °C; ignition temp. 237 °C. *Extinguish fire with* foam, dry powder, carbon dioxide or vaporising liquid.
Spillage disposal	Shut off all possible sources of ignition. Wear face-shield or goggles, and gloves. Apply non-flammable dispersing agent if available and work to an emulsion with brush and water—run this to waste diluting greatly with running water. If dispersant not available, absorb on sand, shovel into bucket(s) and transport to safe open area for burial. Ventilate site of spillage well to evaporate remaining liquid and dispel vapour.

1,1-Diphenylethylene

Hazardous reaction	Forms explosive peroxide with oxygen (B602).

Diphenyldistibene

Hazardous reactions	Ignites in air and is oxidised explosively by nitric acid (B592).

Diphenylmagnesium

Hazardous reactions   Ignites in moist air; reacts violently with water (B591).

Diphenylmercury

Hazardous reactions   Reacts violently with sulphur trioxide or dichlorine oxide (B590).

Diphenyltin

Hazardous reaction   Ignites on contact with fuming HNO_3 (B592).

1,3-Diphenyltriazene

Hazardous reactions   Decomposes explosively at mp 98 °C; mixture with acetic anhydride exploded on warming (B592).

Diphosphane

Hazardous reaction   Ignites in air; when present 0.2% v/v causes other flammable gases to ignite in air (B815).

Dipotassium acetylide

Hazardous reaction   Contact with water may cause ignition and explosion of evolved acetylene (B334).

Dipropionyl peroxide

Hazardous reaction   Explodes on standing at room temperature (B498).

Dipyridinesodium

Hazardous reaction   This addition product of sodium and pyridine ignites in air (B579).

Dirubidium acetylide

Hazardous reactions   Ignited by concentrated hydrochloric acid; burns in halogens; may react violently or explosively with some metal oxides; ignites on warming in CO_2, NO or SO_2; ignites with As, burns in S or Se vapour; reacts vigorously with B or Si (B339).

Disilane

Hazardous reactions   Explodes on contact with Br_2, CCl_4 or SF_6; ignites spontaneously in air (B819).

Disilver acetylide

Hazardous reaction   Powerful detonator which will initiate explosive acetylene-containing gas mixtures (B267).

Disodium acetylide

Hazardous reactions   Burns in chlorine and probably in fluorine, and in contact with bromine and iodine on warming; violent reactions when ground with finely divided lead, aluminium, iron or mercury; incandesces in CO_2 or SO_2; ignites on warming in oxygen; burns vigorously in P vapour; rubbing in mortar with some salts (chlorides, iodides and sulphates—and probably nitrates) results in vigorous reactions or explosions (B338).

Disulphur dichloride

Yellow-brown, fuming liquid with a suffocating odour; bp 136 °C; decomposed by water with formation of hydrochloric acid, sulphur dioxide and sulphur.

CAUSES BURNS
IRRITATING TO SKIN, EYES AND RESPIRATORY SYSTEM

Avoid breathing vapour. Prevent contact with skin and eyes. TLV 1 ppm (6 mg m^{-3}).

Toxic effects   The vapour irritates the respiratory system. The vapour and liquid irritate the eyes and skin severely and the liquid may cause burns. Its ingestion would result in severe internal irritation and damage.

Hazardous reactions   Vigorous or violent reactions with Sb, Sb or As sulphides, chromyl chloride, dimethyl sulphoxide, P_2O_3, HgO, Na_2O_2, water and a number of organic compounds; a mixture with potassium is shock-sensitive and explodes on heating (B686).

First aid *Vapour inhaled:* standard treatment (p 109).
Affected eyes: standard treatment (p 109).
Skin contact: standard treatment (p 108).
If swallowed: standard treatment (p 109).

Spillage disposal Instruct others to keep at a safe distance. Wear breathing apparatus and gloves. Spread soda ash liberally over the spillage and mop up cautiously with water — run this to waste, diluting greatly with running water.

Disulphur dinitride

Hazardous reactions Explosion initiated by shock, friction, pressure and heat (B881).

Disulphur heptaoxide

Hazardous reaction Explodes in moist air (B926).

Disulphuryl diazide

Hazardous reaction Explodes below 80 °C; with dilute alkali at 0 °C an explosive deposit is formed (B885).

Disulphuryl dichloride

Hazardous reactions Reacts vigorously with red phosphorus; reaction with water can be violent (B684).

Disulphuryl difluoride

Hazardous reaction Reacts violently with ethanol (B739).

Dodecanoyl peroxide (lauroyl peroxide)

White powder; mp 55 °C; decomposes slowly above 60 °C, especially in sunlight; insoluble in water.

HIGHLY FLAMMABLE
IRRITATING TO SKIN, EYES AND RESPIRATORY SYSTEM

Avoid contact with skin, eyes and clothing.

Toxic effects It irritates and may burn the skin and eyes; assumed to be irritant and harmful if taken by mouth.

First aid *Affected eyes:* standard treatment (p 109).
Skin contact: standard treatment (p 108).
If swallowed: standard treatment (p 109).

Spillage disposal Wear face-shield or goggles, and gloves. Mix with sand and transport to a safe, open area for burial. Site of spillage should be washed thoroughly with water.

Epichlorohydrin
– see **1-Chloro-2,3-epoxypropane**

1,2-Epoxypropane *– see* **Propylene oxide**

Ethane

Colourless gas; bp −89 °C; insoluble in water.

EXTREMELY FLAMMABLE

Avoid breathing gas.

Toxic effects Considered to be a simple asphyxiant gas which can act as an anaesthetic at high concentrations.

First aid *Gas inhaled in quantity:* standard treatment (p 109).

Fire hazard Explosive limits 3–12.5%; ignition temp. 515 °C. Since the gas is supplied in a cylinder, turning off the valve will reduce any fire involving it; if possible, cylinders should be removed quickly from an area in which a fire has developed.

Disposal Surplus gas or leaking cylinder can be vented slowly to air in a safe, open area or gas burnt off through a suitable burner.

Ethane-1,2-diol (ethylene glycol)

Colourless syrupy liquid with sweetish taste; miscible with water; bp 198 °C. It is the main constituent of 'anti-freeze' mixtures.

HARMFUL IF TAKEN INTERNALLY

Toxic effects Death has followed the drinking of ethanediol as a substitute for spirits (100 cm³ may prove fatal). Smaller quantities may result in restlessness, unsteady gait, drowsiness and coma and injury to the kidneys.

Hazardous reactions Product of reaction with perchloric acid explodes on addition of water; explosion occurred when it was heated with P_2S_5 in hexane (B325).

First aid *If swallowed:* standard treatment (p 109).
Spillage disposal Mop up with plenty of water and run to waste, diluting greatly with running water.

Ethanethiol (ethyl mercaptan)

Colourless liquid with penetrating garlic-like odour; bp 35 °C; sparingly soluble in water (6.8 g in one litre at 20 °C).

HIGHLY FLAMMABLE
HARMFUL VAPOUR

Avoid breathing vapour. Avoid contact with skin and eyes. TLV 0.5 ppm (1 mg m⁻³).

Toxic effects The vapour irritates the respiratory system and may be narcotic in high concentrations. The liquid irritates the eyes and mucous membranes; it is assumed to be poisonous if taken by mouth.

First aid *Vapour inhaled:* standard treatment (p 109).
Affected eyes: standard treatment (p 109).
Skin contact: standard treatment (p 108).
If swallowed: standard treatment (p 109).

Fire hazard Flash point below 21°C; explosive limits 2.8–18%; ignition temp. 299 °C. *Extinguish fire with* water spray, foam, dry powder, carbon dioxide or vaporising liquid.

Spillage disposal Shut off all possible sources of ignition. Instruct others to keep at a safe distance. Wear breathing apparatus and gloves. Apply non-flammable dispersing agent if available and work to an emulsion with brush and water—run this to waste, diluting greatly with running water. If dispersant not available, absorb on sand, shovel into bucket(s) and transport to safe open area for atmospheric evaporation or burial. Site of spillage should be washed thoroughly with water and soap or detergent.

Ethanol (ethyl alcohol)

Colourless, mobile liquid with characteristic smell, bp 79 °C; miscible with water. Ethanol is supplied to laboratories in a variety of forms mainly because of Excise control. In the United Kingdom, 'duty paid' material comes as 90% (rectified spirit), 95% and 99/100% (absolute) grades; denatured ethanol is sold as industrial methylated spirit (supplied against Excise requisitions) and methylated spirit (mineralised). The materials used for denaturing ethanol add substantially to its toxicity and are not taken into account in this monograph which deals with pure (99/100%) material.

HIGHLY FLAMMABLE

Avoid breathing vapour in high concentrations. TLV 1000 ppm (1900 mg m⁻³).

Toxic effects   The intoxicating qualities of ethanol are so well appreciated that a stark summary of them is superfluous.

Hazardous reactions   Reacts with varying degrees of violence with a wide range of oxidants, including silver nitrate with which an explosion was reported (B322).

First aid   *Affected eyes:* standard treatment (p 109).
Fire hazard   Flash point 12 °C; explosive limits 3.3–19%; ignition temp. 423 °C. *Extinguish fire with* water spray, dry powder, carbon dioxide or vaporising liquid.
Spillage disposal   Shut off all possible sources of ignition. Wear face-shield or goggles, and gloves. Mop up with plenty of water and run to waste diluting greatly with running water. Ventilate area well to evaporate remaining liquid and dispel vapour.

Ethanolamine – *see* 2-Aminoethanol

Ethene – *see* Ethylene

Ether – *see* Diethyl ether

Ethoxyacetylene

Hazardous reaction   Explodes when heated at around 100 °C in sealed tubes (B395).

Ethoxyanilines (phenetidines)

o- and *p*-Phenetidines are red to brown liquids; bp 229 °C and 254 °C respectively; immiscible with water.

SERIOUS RISK OF POISONING
BY INHALATION, SWALLOWING
OR SKIN CONTACT

Avoid breathing vapour. Avoid contact with skin and eyes.

Toxic effects   These are not recorded, but are assumed to be similar to those of aniline poisoning, *ie* headache, drowsiness and cyanosis.

First aid   *Vapour inhaled:* standard treatment (p 109).
Affected eyes: standard treatment (p 109).
Skin contact: standard treatment (p 108).
If swallowed: standard treatment (p 109).
Spillage disposal   Wear face-shield or goggles, and gloves. Mix with sand and shovel mixture into glass, enamel or polythene vessel for dispersion in an excess of dilute hydrochloric acid (1 volume of concentrated acid diluted with 2 volumes of water). Allow to stand, with occasional stirring, for 24 hours and then run acid extract to waste, diluting greatly with running water and washing the sand. The sand can be disposed of as normal refuse.

2-Ethoxyethanol (ethylene glycol monoethyl ether; cellosolve)

Colourless liquid; bp 135 °C; miscible with water; liable to form explosive peroxides on exposure to light and air which should be decomposed before the ether is distilled to small volume.

> FLAMMABLE
> MAY FORM EXPLOSIVE PEROXIDES
> HARMFUL VAPOUR

Avoid breathing vapour. Avoid contact with skin, eyes and clothing. TLV (skin) 100 ppm (370 mg m⁻³).

Toxic effects	The vapour irritates the eyes and respiratory system when in high concentrations. The liquid irritates the eyes. The liquid is poisonous if taken by mouth and may cause blood and kidney damage.
First aid	*Vapour inhaled:* standard treatment (p 109). *Affected eyes:* standard treatment (p 109). *Skin contact:* standard treatment (p 108). *If swallowed:* standard treatment (p 109).
Fire hazard	Flash point 40 °C; explosive limits 1.7–15.6%; ignition temp. 238 °C. *Extinguish fire with* water spray, foam, dry powder, carbon dioxide or vaporising liquid.
Spillage disposal	Shut off all possible sources of ignition. Wear face-shield or goggles, and gloves. Mop up with plenty of water and run to waste, diluting greatly with running water. Ventilate area well to evaporate remaining liquid and dispel vapour.

2-Ethoxyethyl acetate

(ethylene glycol monoethyl ether monoacetate)

Colourless liquid with pleasant ester-like odour; bp 156 °C; sparingly soluble in water; liable to form explosive peroxides on exposure to air and light which should be decomposed before the ether ester is distilled to small volume.

> FLAMMABLE
> MAY FORM EXPLOSIVE PEROXIDES
> HARMFUL VAPOUR
> CAUSES IRRITATION OF SKIN AND EYES

Avoid breathing vapour. Avoid contact with skin, eyes and clothing. TLV 100 ppm (540 mg m⁻³).

Toxic effects	The vapour irritates the eyes and respiratory system in high concentrations. The liquid irritates the eyes. The liquid is poisonous if taken by mouth and may cause blood and kidney damage.
First aid	*Vapour inhaled:* standard treatment (p 109). *Affected eyes:* standard treatment (p 109). *Skin contact:* standard treatment (p 108). *If swallowed:* standard treatment (p 109).
Fire hazard	Flash point 51 °C; explosive limits 1.2–13%; ignition temp. 379 °C. *Extinguish fire with* foam, dry powder, carbon dioxide or vaporising liquid.

Spillage **disposal**	Shut off all possible sources of ignition. Wear face-shield or goggles, and gloves. Apply non-flammable dispersant and work to an emulsion with brush and water—run this to waste, diluting greatly with running water. If dispersant not available, absorb on sand, shovel into bucket(s) and transport to safe, open area for atmospheric evaporation or burial.

3-Ethoxypropyne (1-ethoxy-2-propyne)

Hazardous **reaction**	Exploded during distillation under vacuum (B444).

Ethyl acetate

Colourless, volatile liquid with fragrant odour; bp 77 °C; 1 part soluble in about 35 parts water at 25 °C.

> EXTREMELY FLAMMABLE
> IRRITATING TO EYES AND RESPIRATORY SYSTEM

Avoid breathing vapour. Avoid contact with eyes. TLV 400 ppm (1400 mg m^{-3}).

Toxic effects	The vapour may irritate the eyes and respiratory system. The liquid irritates the eyes and mucous surfaces. Prolonged inhalation may cause kidney and liver damage.
First aid	*Vapour inhaled:* standard treatment (p 109). *Affected eyes:* standard treatment (p 109). *Skin contact:* standard treatment (p 108). *If swallowed:* standard treatment (p 109).
Fire hazard	Flash point −4.4 °C; explosive limits 2.5–11.5%; ignition temp. 427 °C. *Extinguish fire with* water spray, foam, dry powder, carbon dioxide or vaporising liquid.
Spillage **disposal**	Shut off all possible sources of ignition. Wear face-shield or goggles, and gloves. Apply non-flammable dispersing agent if available and work to an emulsion with brush and water — run this to waste, diluting greatly with running water. If dispersant not available, absorb on sand, shovel into bucket(s) and transport to safe open area for atmospheric evaporation. Ventilate site of spillage well to evaporate remaining liquid and dispel vapour.

Ethylacetylene – *see* But-1-yne

Ethyl acrylate

Colourless liquid with acrid, penetrating odour; bp 100 °C; immiscible with water; normally supplied containing stabiliser.

> HIGHLY FLAMMABLE
> HARMFUL VAPOUR
> IRRITATING TO SKIN, EYES AND RESPIRATORY SYSTEM

Avoid breathing vapour. Avoid contact with skin and eyes. TLV (skin) 25 ppm (100 mg m^{-3}).

Toxic effects   The vapour irritates the eyes and respiratory system. High concentrations may result in lethargy and convulsions. The liquid irritates the skin and eyes. Animal experiments indicate high degree of toxicity if taken by mouth.

First aid   *Vapour inhaled:* standard treatment (p 109).
Affected eyes: standard treatment (p 109).
Skin contact: standard treatment (p 108).
If swallowed: standard treatment (p 109).

Fire hazard   Flash point 16 °C; ignition temp. 273 °C. *Extinguish fire with* water spray, foam, dry powder, carbon dioxide or vaporising liquid.

Spillage disposal   Shut off all possible sources of ignition. Instruct others to keep at a safe distance. Wear breathing apparatus and gloves. Apply non-flammable dispersing agent if available and work to an emulsion with brush and water — run this to waste diluting greatly with running water. If dispersant not available, absorb on sand, shovel into bucket(s) and transport to safe, open area for burial. Site of spillage should be washed thoroughly with water and soap or detergent.

Ethyl alcohol – *see* Ethanol

Ethylamine and solutions

Colourless volatile liquid with a fishy ammoniacal odour; bp 17 °C. The solutions in water and ethanol are commonly used.

EXTREMELY FLAMMABLE
IRRITATING TO SKIN, EYES AND RESPIRATORY SYSTEM

Avoid breathing vapour. Avoid contact with skin and eyes. TLV 10 ppm (18 mg m^{-3}).

Toxic effects   The vapour irritates the mucous membranes and respiratory system; in high concentrations it may affect the nervous system. The vapour and solutions irritate the eyes. The solutions may irritate the skin. Assumed to be irritant and poisonous if taken by mouth.

First aid   *Vapour inhaled:* standard treatment (p 109).
Affected eyes: standard treatment (p 109).
Skin contact: standard treatment (p 108).
If swallowed: standard treatment (p 109).

Fire hazard   Flash point below −18 °C; explosive limits 3.5–14 %; ignition temp. 384°C. *Extinguish fire with* water spray or dry powder.

Spillage disposal   Shut off all possible sources of ignition. Instruct others to keep at a safe distance. Wear breathing apparatus and gloves. Mop up with plenty of water and run to waste, diluting greatly with running water. Ventilate area well to evaporate remaining liquid and dispel vapour.

N-Ethylaniline

Colourless to brown liquid; bp 204 °C; insoluble in water.

TOXIC IN CONTACT WITH SKIN
GIVES OFF POISONOUS VAPOUR

Avoid breathing vapour. Prevent contact with skin and eyes.

Toxic effects   Excessive breathing of the vapour or absorption of the liquid through the skin can cause headache, drowsiness, cyanosis, mental confusion and, in severe cases, convulsions. The liquid damages the eyes, and the above effects can be expected if it is swallowed. **Chronic effects** Continued exposure to the vapour, or slight skin exposure to the liquid, over a period may affect the nervous system and the blood, causing fatigue, loss of appetite, headache and dizziness.

First aid   *Vapour inhaled:* standard treatment (p 109).
Affected eyes: standard treatment (p 109).
Skin contact: standard treatment (p 108).
If swallowed: standard treatment (p 109).

**Spillage
disposal**   Wear face-shield or goggles, and gloves. Mix with sand and shovel mixture into glass, enamel or polythene vessel for dispersion in an excess of dilute hydrochloric acid (1 volume of concentrated acid diluted with 2 volumes of water). Allow to stand, with occasional stirring, for 24 hours and then run acid extract to waste, diluting greatly with running water and washing the sand. The sand can be disposed of as normal refuse.

Ethyl azide

**Hazardous
reaction**   May detonate on rapid heating (B316).

Ethyl azidoformate

**Hazardous
reaction**   Liable to explode at 114 °C (B359).

Ethylbenzene

Colourless liquid with aromatic odour; bp 136 °C; immiscible with water.

HIGHLY FLAMMABLE
HARMFUL VAPOUR
IRRITATING TO SKIN, EYES AND RESPIRATORY SYSTEM

Avoid breathing vapour. Avoid contact with skin and eyes. TLV 100 ppm (435 mg m^{-3}).

Toxic effects The vapour irritates the eyes and respiratory system and may cause dizziness. The liquid irritates the skin and eyes. Assumed to be poisonous if taken by mouth.

First aid *Vapour inhaled:* standard treatment (p 109).
Affected eyes: standard treatment (p 109).
Skin contact: standard treatment (p 108).
If swallowed: standard treatment (p 109).

Fire hazard Flash point 15 °C; explosive limits 1–6.7%; ignition temp. 432 °C; *Extinguish fire with* foam, dry powder, carbon dioxide or vaporising liquid.

Spillage disposal Shut off all possible sources of ignition. Instruct others to keep at a safe distance. Wear breathing apparatus and gloves. Apply non-flammable dispersing agent if available and work to an emulsion with brush and water — run this to waste, diluting greatly with running water. If dispersant not available, absorb on sand, shovel into bucket(s) and transport to safe, open area for burial. Site of spillage should be washed thoroughly with water and soap or detergent.

Ethyl bromide – *see* Bromoethane

Ethyl butyrate

Colourless liquid with pineapple-like odour; bp 121 °C; sparingly soluble in water.

FLAMMABLE

Avoid contact with skin and eyes.

Toxic effects These are not marked. In high concentrations the vapour irritates the eyes and respiratory system, and is narcotic.

First aid *Vapour inhaled in high concentrations:* standard treatment (p 109).
Affected eyes: standard treatment (p 109).
Skin contact: standard treatment (p 108).
If swallowed: standard treatment (p 109).

Fire hazard Flash point 26 °C; ignition temp. 463 °C; *Extinguish fire with* water spray, foam, dry powder, carbon dioxide or vaporising liquid.

Spillage disposal Shut off all possible sources of ignition. Wear face-shield or goggles, and gloves. Apply non-flammable dispersing agent if available and work to an emulsion with brush and water — run this to waste, diluting greatly with running water. If dispersant not available, absorb on sand, shovel into bucket(s) and transport to safe, open area for atmospheric evaporation or burial. Ventilate site of spillage well to evaporate remaining liquid and dispel vapour.

Ethyl carbonate – *see* Diethyl carbonate

Ethyl chloroacetate

Colourless liquid with fruity odour; bp 144 °C: immiscible with water.

> FLAMMABLE
> IRRITATING TO SKIN, EYES AND RESPIRATORY SYSTEM

Avoid breathing vapour. Avoid contact with skin and eyes.

Toxic effects   The vapour irritates the respiratory system. The vapour and liquid irritate the eyes severely. Repeated or prolonged contact between the liquid and skin causes irritation. Assumed to be irritant if taken by mouth.

First aid   *Vapour inhaled:* standard treatment (p 109).
Affected eyes: standard treatment (p 109).
Skin contact: standard treatment (p 108).
If swallowed: standard treatment (p 109).

Fire hazard   Flash point 54 °C. *Extinguish fire with* water spray, foam, dry powder, carbon dioxide or vaporising liquid.

Spillage disposal   Shut off all possible sources of ignition. Instruct others to keep at a safe distance. Wear breathing apparatus and gloves. Apply non-flammable dispersing agent if available and work to an emulsion with brush and water — run this to waste, diluting greatly with running water. If dispersant not available absorb on sand, shovel into bucket(s) and transport to safe, open area for burial. Site of spillage should be washed thoroughly with water and soap or detergent.

Ethyl chloroformate (ethyl chlorocarbonate)

Colourless liquid with pungent odour; bp 95 °C; immiscible with water.

> HIGHLY FLAMMABLE
> GIVES OFF POISONOUS VAPOUR
> IRRITATING TO SKIN, EYES AND RESPIRATORY SYSTEM

Prevent inhalation of vapour. Prevent contact with skin and eyes.

Toxic effects   The vapour irritates the respiratory system. The vapour and liquid irritate the eyes severely. The liquid irritates the skin. Assumed to be irritant and poisonous if taken by mouth.

First aid   *Vapour inhaled:* standard treatment (p 109).
Affected eyes: standard treatment (p 109).
Skin contact: standard treatment (p 108).
If swallowed: standard treatment (p 109).

Fire hazard   Flash point 16 °C; *Extinguish fire with* water spray, foam, dry powder, carbon dioxide or vaporising liquid.

Spillage disposal   Shut off all possible sources of ignition. Instruct others to keep at a safe distance. Wear breathing apparatus and gloves. Apply non-flammable dispersing agent if available and work to an emulsion with brush and water — run this to waste, diluting greatly with running water. If dispersant not available, absorb on sand, shovel into bucket(s) and transport to safe, open area for burial. Site of spillage should be washed thoroughly with water and soap or detergent.

Ethyl diazoacetate

Hazardous reaction   Explosive, and distillation may be dangerous (B392).

Ethylene (ethene)

Colourless gas with sweetish smell; bp −103.7 °C; 1 volume dissolves in about 9 volumes of water at 25 °C.

EXTREMELY FLAMMABLE

Avoid breathing gas.

Toxic effects   Considered to be a simple asphyxiant gas which can act as an anaesthetic at high concentrations.

Hazardous reactions   Mixtures of ethylene and $AlCl_3$ are liable to explode in presence of nickel and other catalysts; mixtures with CCl_4 are liable to explode as are mixtures with chlorine; violent explosion with tetrafluoroethylene (B301–302).

First aid   *Gas inhaled in quantity:* standard treatment (p 109).
Fire hazard   Explosive limits 3.1–32%; ignition temp. 450 °C. Since the gas is supplied in a cylinder, turning off the valve will reduce any fire involving it; if possible, cylinders should be removed quickly from an area in which a fire has developed.
Disposal   Surplus gas or leaking cylinder can be vented slowly to air in a safe, open area or gas burnt off through a suitable burner.

Ethylene chlorohydrin − see 2-Chloroethanol

Ethylenediamine − see 1,2-Diaminoethane

Ethylene dibromide − see 1,2-Dibromoethane

Ethylene dichloride − see 1,2-Dichloroethane

Ethylene diperchlorate

Hazardous reaction   Highly sensitive, violently explosive material (B304).

Ethylene glycol − see Ethane-1,2-diol

Ethylene glycol monoethyl ether
– see **2-Ethoxyethanol**

Ethylene glycol monomethyl ether
– see **2-Methoxyethanol**

Ethylene glycol monomethyl ether acetate
–see **2-Methoxy ethyl acetate**

Ethyleneimine *– see* Aziridine

Ethylene oxide

Colourless liquid or gas with no very distinctive smell; bp 11 °C; soluble in water.

EXTREMELY FLAMMABLE
TOXIC BY INHALATION
IRRITATING TO EYES

Avoid breathing vapour. Avoid contact with eyes and skin. TLV 50 ppm (90 mg m^{-3}).

Toxic effects   If inhaled, the vapour irritates the respiratory tract and may give rise to bronchitis or pulmonary oedema; other possible effects include nausea, vomiting, convulsions and coma. The vapour irritates the eyes and may cause conjunctivitis and corneal damage. Excessive contact with liquid or solution can cause delayed burns and blistering.

Hazardous reactions   Vapour readily initiated into explosive decomposition (detailed recommendations on usage); liable to explode in autoclave reactions with some alkanethiols or alcohols; ammonia and other contaminants may cause explosive polymerisation (B308–310).

First aid   *Vapour inhaled:* standard treatment (p 109).
Affected eyes: standard treatment (p 109).
Skin contact: standard treatment (p 108).
If swallowed: standard treatment (p 109).

Disposal   The contents of a broken ampoule of ethylene oxide are best dispersed by thorough ventilation of the area concerned; care must be taken to shut off all possible sources of ignition and breathing apparatus must be worn. A leaking cylinder of the substance can be vented slowly to air in a safe, open area.

Ethylene ozonide

Hazardous reaction   Explodes very violently on heating, friction or shock (B311).

Ethyl ether – *see* Diethyl ether

Ethyl formate (formic ether)

Colourless, volatile liquid with a pleasant odour; bp 54 °C; somewhat soluble in water.

EXTREMELY FLAMMABLE
IRRITATING TO EYES AND RESPIRATORY SYSTEM

Avoid breathing vapour. Avoid contact with skin and eyes. TLV 100 ppm (300 mg m^{-3}).

Toxic effects The vapour irritates the respiratory system. Large concentrations affect the nervous system and can cause unconsciousness. The vapour and liquid irritate the eyes severely. Assumed to be irritant and poisonous if taken by mouth.

First aid *Vapour inhaled:* standard treatment (p 109).
Affected eyes: standard treatment (p 109).
If swallowed: standard treatment (p 109).

Fire hazard Flash point −20 °C; explosive limits 2.7–13.5%; ignition temp. 455 °C. *Extinguish fire with* water spray, foam, dry powder, carbon dioxide or vaporising liquid.

Spillage disposal Shut off all possible sources of ignition. Instruct others to keep at a safe distance. Wear breathing apparatus and gloves. Apply non-flammable dispersing agent if available and work to an emulsion with brush and water — run this to waste, diluting greatly with running water. If dispersant not available, absorb on sand, shovel into bucket(s) and transport to safe open area for atmospheric evaporation. Site of spillage should be washed thoroughly with water and soap or detergent.

Ethyl hypochlorite

Hazardous reaction Ignition or rapid heating of vapour causes explosion (B312).

Ethyl iodide – *see* Iodoethane

Ethyl isocyanide

Hazardous reaction Liable to explode (B358).

Ethyl lactate

Colourless liquid; bp 154 °C; soluble in water with partial hydrolysis.

FLAMMABLE

Toxic effects No evidence was found that ethyl lactate has significant toxic properties.

Fire hazard Flash point 46 °C; explosive limits 1.5–30%; ignition temp. 400 °C; *Extinguish fire with* water spray, foam, dry powder, carbon dioxide or vaporising liquid.

Spillage disposal Shut off all possible sources of ignition. Mop up with plenty of water and run to waste, diluting greatly with running water.

Ethyl mercaptan – *see* Ethanethiol

Ethyl methacrylate

Clear, colourless liquid; bp 119 °C; insoluble in water.

HIGHLY FLAMMABLE
HARMFUL SUBSTANCE IF TAKEN INTERNALLY
AND IF IN CONTACT WITH SKIN OR EYES
HARMFUL VAPOUR

Avoid contact with skin and eyes. Avoid breathing vapour.

Toxic effects The vapour irritates the eyes and respiratory system. The liquid irritates the skin and eyes. Less toxic when taken by mouth than ethyl acrylate.

First aid *Vapour inhaled:* standard treatment (p 109).
Affected eyes: standard treatment (p 109).
Skin contact: standard treatment (p 108).
If swallowed: standard treatment (p 109).

Fire hazard Flash point 20 °C. *Extinguish fire with* water spray, foam, dry powder, carbon dioxide or vaporising liquid.

Spillage disposal Shut off all possible sources of ignition. Instruct others to keep at a safe distance. Wear breathing apparatus and gloves. Apply non-flammable dispersing agent if available and work to an emulsion with brush and water. If dispersant not available, absorb on sand, shovel into bucket(s) and transport to safe open area for burial. Site of spillage should be washed thoroughly with water and detergent.

Ethyl methyl ketone – *see* Butanone

Ethyl methyl peroxide

Hazardous reactions Shock-sensitive as liquid or vapour; explodes violently when heated strongly (B372).

N-Ethylmorpholine

Colourless liquid; bp 138 °C; miscible with water.

FLAMMABLE
IRRITATING TO EYES

Avoid breathing vapour. Avoid contact with skin, eyes and clothing. TLV (skin) 20 ppm (94 mg m^{-3}).

Toxic effects	These are not well documented, but it is considered prudent to treat it with the same caution as its parent, morpholine, regarding it as hazardous by skin absorption, inhalation and ingestion.
First aid	*Vapour inhaled:* standard treatment (p 109). *Affected eyes:* standard treatment (p 109). *Skin contact:* standard treatment (p 109).
Fire hazard	Flash point 32 °C. *Extinguish fire with* water spray, foam, dry powder, carbon dioxide or vaporising liquid.
Spillage disposal	Shut off all possible sources of ignition. Wear face-shield or goggles, and gloves. Mop up with plenty of water and run to waste diluting greatly with running water. Ventilate area well to evaporate remaining liquid and dispel vapour.

Ethyl nitrate

Colourless liquid with pleasant odour; bp 89 °C; insoluble in water.

HIGHLY FLAMMABLE
RISK OF EXPLOSION BY SHOCK, FRICTION, FIRE OR OTHER SOURCES OF IGNITION

Avoid breathing vapour. Avoid contact with skin, eyes or clothing.

Toxic effects	Acute local effects are not documented but it is stated to be moderately toxic systemically by inhalation and ingestion.
Hazardous reaction	Explodes at 85 °C (B316).
First aid	*Vapour inhaled:* standard treatment (p 109). *Affected eyes:* standard treatment (p 109). *Skin contact:* standard treatment (p 108). *If swallowed:* standard treatment (p 109).
Fire hazard	Flash point 10 °C; in view of the probability of explosion shortly after ignition, the area involving an ethyl nitrate fire (or the risk of it occurring) should be evacuated and the fire brigade informed at once of the situation.
Spillage disposal	Wear face-shield or goggles, and gloves. Mix with sand and transport to a safe, open area for burial. Site of spillage should be washed thoroughly with water and soap or detergent.

Ethyl nitrite

Hazardous
reaction
Explodes above 90 °C (B315).

Ethyl perchlorate

Hazardous
reaction
Reputedly most explosive substance known—sensitive to impact, friction and heat (B313).

Ethylphosphine

Hazardous
reactions
Ignites in air; explodes on contact with Cl_2, Br_2, or fuming nitric acid (B329).

N-Ethylpiperidine

Colourless liquid; bp 131 °C; miscible with water.

HIGHLY FLAMMABLE
IRRITATING TO EYES

Avoid breathing vapour. Avoid contact with skin, eyes and clothing.

Toxic effects
These are not well documented, but the substance is considered to be harmful by vapour inhalation, skin absorption and ingestion.

First aid
Vapour inhaled: standard treatment (p 109).
Affected eyes: standard treatment (p 109).
Skin contact: standard treatment (p 108).
If swallowed: standard treatment (p 109).

Fire hazard
Flash point 19 °C. *Extinguish fire with* water spray, foam, dry powder, carbon dioxide or vaporising liquid.

Spillage
disposal
Shut off all possible sources of ignition. Wear face-shield or goggles, and gloves. Mop up with plenty of water and run to waste diluting greatly with running water. Ventilate area well to evaporate remaining liquid and dispel vapour.

Ethyl sulphate - *see* Diethyl sulphate

Ethyl vinyl ether

Hazardous
reaction
Methanesulphonic acid catalysed explosive polymerisation of the ether (B408).

Ethynediylbis(triethyltin) – *see* Bis(triethyltin) acetylene

2-Ethynylfuran

Hazardous reaction
Explodes on heating, or on contact with conc. nitric acid (B475).

Ethynyl vinyl selenide

Hazardous reaction
Explodes on heating (B387).

Ferric chloride anhydrous – *see* Iron(III) chloride

Fluorides (water soluble)

Normally colourless crystals or powders; soluble in water.

TOXIC IF TAKEN INTERNALLY

Avoid inhalation of dust. Prevent contact with eyes and skin. TLV (as F) (2.5 mg m^{-3}).

Toxic effects
The dust irritates all parts of the respiratory tract. The dust irritates the eyes and skin. Nausea, vomiting, diarrhoea and abdominal pains follow ingestion. **Chronic effect** Shortness of breath, cough, elevated temperature and cyanosis. (*See* note in Ch 6, p 84.)

First aid
Dust inhaled: standard treatment (p 109).
Affected eyes: standard treatment (p 109).
Skin contact: standard treatment (p 108).
If swallowed: standard treatment (p 109).

Spillage disposal
Wear face-shield or goggles, and gloves. Mop up with plenty of water and run to waste, diluting greatly with running water.

Fluorine

Pale yellow gas with sharp penetrating odour; bp −188 °C; reacts vigorously with water and may cause organic materials to inflame

MAY CAUSE FIRE
TOXIC BY INHALATION
IRRITATING TO SKIN, EYES AND RESPIRATORY SYSTEM

Prevent inhalation of vapour. Prevent contact with skin and eyes. TLV 0.1 ppm (0.2 mg m^{-3}).

Toxic effects	The gas is highly irritant to the eyes and respiratory system; high concentrations may produce thermal-type burns on the skin and chemical-type burns may result from lower concentrations.
Hazardous reactions	Anyone working with fluorine should study the whole section devoted to its dangerous reactions in Bretherick's *Handbook of reactive chemical hazards* (B728–735).
First aid	*Gas inhaled :* standard treatment (p. 109). *Affected eyes :* standard treatment (p 109). *Skin contact :* standard treatment (p 108).
Disposal	Instruct others to keep at a safe distance. Wear breathing apparatus and gloves. Organise ventilation of the area to dispel the gas completely. Leaking cylinder should be vented slowly in to a water-fed scrubbing tower or column in a fume cupboard, or into a fume cupboard served by such a tower.

Fluorine azide

Hazardous reaction	Unstable — explodes on vaporisation (at $-82\ °C$) (B727).

Fluorine nitrate

Hazardous reactions	Ignition in gas phase with NH_3, N_2O or H_2S; it may explode on contact with alcohol, ether, aniline or grease (B727).

Fluorine perchlorate

Hazardous reactions	Liquid explodes on freezing at $-167\,°C$; explosion of gas initiated by sparks, flame or contact with grease, dust or rubber tube; ignition occurs in excess of hydrogen (B634).

Fluoroacetylene

Hazardous reactions	Liquid explodes close to its bp, $-80\ °C$; ignition occurs in contact with solution of Br_2 in CCl_4; mercury salt decomposes but silver salt explodes on heating (B282).

Fluoroamine

Hazardous reaction	Impure material is very explosive (B725).

Fluoroboric acid and salts
– see Tetrafluoroboric acid and salts

1-Fluoro-2,4-dinitrobenzene

Hazardous preparation   Explosions with distillation residues (B465).

Fluorodinitromethyl compounds   Review of group (B55).

Fluoroethylene – see Vinyl fluoride

Fluorosilicic acid and salts
– see Hexafluorosilicic acid and salts

Fluorosulphuric acid (fluorosulphonic acid)

Colourless or yellow fuming liquid; soluble in water.

> CAUSES SEVERE BURNS
> IRRITATING TO SKIN, EYES AND RESPIRATORY SYSTEM

Avoid inhaling vapour. Prevent contact with eyes and skin.

Toxic effects   The vapour irritates all parts of the respiratory system. The vapour irritates and the liquid burns the eyes severely. The liquid and vapour burn the skin. If taken by mouth there would be severe internal irritation and damage.

First aid   *Vapour inhaled:* standard treatment (p 109).
Affected eyes: standard treatment (p 109).
Skin contact: standard treatment (p 108).
If swallowed: standard treatment (p 109).

Spillage disposal   Instruct others to keep at a safe distance. Wear breathing apparatus and gloves. Spread soda ash liberally over the spillage and mop up cautiously with water — run this to waste, diluting greatly with running water.

Formaldehyde solution (formalin)

Colourless, sometimes milky solution with pungent odour; bp 96 °C; miscible with water; the solution generally contains 37–41% formaldehyde and 11–14% methanol.

> FLAMMABLE
> SERIOUS RISK OF POISONING BY INHALATION OR SWALLOWING
> CAUSES BURNS
> IRRITATING TO SKIN, EYES AND RESPIRATORY SYSTEM

Avoid breathing vapour. Avoid contact with eyes and skin. TLV 2 ppm (3 mg m^{-3}).

Toxic effects   The vapour irritates all parts of the respiratory system. The liquid and vapour irritate the eyes severely. The liquid in contact with the skin has a hardening or tanning effect and causes irritation. Severe abdominal pains with nausea and vomiting and possibly loss of consciousness follow ingestion. **Chronic effects** High concentration of vapour inhaled for long periods can cause laryngitis, bronchitis or bronchial pneumonia; prolonged exposure may cause conjunctivitis; in contact with skin for long periods will cause cracking of skin and ulceration, particularly around fingernails.

First aid   *Vapour inhaled:* standard treatment (p 109).
Affected eyes: standard treatment (p 109).
Skin contact: standard treatment (p 108).
If swallowed: standard treatment (p 109).

Fire hazard   Flash point (37%) 50 °C. *Extinguish fire with* water spray, dry powder, carbon dioxide or vaporising liquid.

Spillage disposal   Shut off all possible sources of ignition. Wear face-shield or goggles, and gloves. Mop up with plenty of water and run to waste, diluting greatly with running water. Ventilate area well to evaporate remaining liquid and dispel vapour.

Formdimethylamide
– *see* **Dimethylformamide**

Formic acid

Colourless liquid which fumes in the higher concentrations; 98/100% acid boils at 100.5 °C; miscible with water.

CAUSES BURNS
IRRITATING TO SKIN, EYES AND RESPIRATORY SYSTEM

Avoid breathing vapour. Avoid contact with eyes and skin. TLV 5 ppm (9 mg m^{-3}).

Toxic effects   The vapour irritates all parts of the respiratory system. The vapour irritates the eyes and the liquid causes painful eye burns. The liquid burns the skin. If taken by mouth there is severe internal irritation and damage.

First aid   *Vapour inhaled:* standard treatment (p 109).
Affected eyes: standard treatment (p 109).
Skin contact: standard treatment (p 108).
If swallowed: standard treatment (p 109).

Spillage disposal   Instruct others to keep at a safe distance. Wear breathing apparatus and gloves. Spread soda ash liberally over the spillage and mop up cautiously with water — run this to waste, diluting greatly with running water.

2-Furaldehyde (furfuraldehyde, furfural)

Colourless to yellow liquid with distinctive odour; bp 162 °C; immiscible with water.

HARMFUL VAPOUR
IRRITATING TO SKIN, EYES AND RESPIRATORY SYSTEM

Avoid inhalation of vapour. Avoid contact with skin and eyes. TLV (skin) 5 ppm (20 mg m^{-3}).

Toxic effects   The vapour irritates the eyes and respiratory system. The liquid irritates the eyes. Assumed to be poisonous if taken by mouth. **Chronic effects** Exposure over prolonged periods may result in nervous disturbance, eye inflammation and disturbance of vision.

First aid   *Vapour inhaled:* standard treatment (p 109).
Affected eyes: standard treatment (p 109).
Skin contact: standard treatment (p 108).
If swallowed: standard treatment (p 109).

Spillage disposal   Instruct others to keep at a safe distance. Wear breathing apparatus and gloves. Apply nonflammable dispersing agent if available and work to an emulsion with brush and water—run this to waste, diluting greatly with running water. If dispersant not available, absorb on sand, shovel into bucket(s) and transport to safe, open area for burial. Site of spillage should be washed thoroughly with water and soap or detergent.

Furfuryl alcohol

Hazardous reactions   Reacts violently with formic acid; mixture with cyanoacetic acid exploded on heating; may react violently or explosively on contact with acids or acidic materials (B441).

Gallium (metal)

Hazardous reactions   Reacts violently with chlorine and bromine (B752).

Gallium triperchlorate

Hazardous preparation   Moist product may react violently with any organic material (B690).

Germane (germanium tetrahydride)

Hazardous reactions   Liable to ignite in air; contact with bromine may cause ignition at −112 °C (B753).

Germanium

Hazardous reactions   Powdered metal ignites in chlorine, lumps ignite on heating in chlorine or bromine; powdered metal reacts violently with nitric acid, mixtures with potassium chlorate and nitrate explode on heating (B752)

Germanium monohydride

Hazardous reaction   May decompose explosively in air (B753).

Germanium tetrachloride

Hazardous reaction   Reacts violently with water (B698).

Germanium tetrahydride – *see* Germane

Glutaryl diazide

Hazardous reaction   May explode on heating (B440).

Glycerol

Hazardous reactions   Reacts violently or explosively with many oxidising agents (B372).

Glycolonitrile

Storage hazard   Liable to polymerise explosively on storage (B298).

Glyoxal

Hazardous preparation   Oxidation of paraldehyde with nitric acid may become violent (B292).

Gold compounds Review of group (B56).

Guanidinium nitrate

Hazardous preparation   Explosions have occurred in preparation from dicyanodiamide and ammonium nitrate (B255).

Haloacetylene derivatives Review of class (B57).

Haloalkenes Review of group (B57).

Halocarbons Explanatory review (B59).

Halogen azides Review of group (B61).

N-Halogen compounds Review of class (B61–63).

Halogen oxides Review of group (B63).

N-Haloimides Review of group (B64).

Halosilanes Review of group (B64).

Heptane (and heptane fraction from petroleum)

Colourless liquid; bp 98 °C; insoluble in water.

EXTREMELY FLAMMABLE

Avoid breathing vapour. TLV 400 ppm (1600 mg m^{-3}).

Toxic effects	May irritate the respiratory system as vapour which, in high concentrations, is narcotic.
First aid	*Vapour inhaled in high concentrations:* standard treatment (p 109). *Affected eyes:* standard treatment (p 109). *If swallowed:* standard treatment (p 109).
Fire hazard	Flash point −4°C; explosive limits 1.2–6.7%; ignition temp. 223 °C. *Extinguish fire with* foam, dry powder, carbon dioxide or vaporising liquid.
Spillage disposal	Shut off all possible sources of ignition. Wear face-shield or goggles, and gloves. Apply non-flammable dispersing agent if available and work to an emulsion with brush and water — run this to waste, diluting greatly with running water. If dispersant not available, absorb on sand, shovel into bucket(s) and transport to safe, open area for atmospheric evaporation. Ventilate site of spillage well to evaporate remaining liquid and dispel vapour.

Hept-2-yn-1-ol

Hazardous preparation	Risk of explosion of distillation residue (B537).

Hexacarbonylchromium

Hazardous reaction	Explodes at 210 °C (B460).

Hexacarbonylmolybdenum

Storage hazard   Solutions liable to explode on long storage (B518).

Hexacarbonyltungsten

Hazardous preparation   Preparation from WCl_6, Al powder and CO in autoclave can be dangerous (B519).

1,2,3,4,5,6-Hexachlorocyclohexane

Hazardous reaction   Reaction with dimethylformamide in presence of iron may become violent (B486).

Hexadienynes

Hazardous reaction   Risks of explosions referred to (B484–485).

Hexafluorosilicic acid and salts
(fluorosilicic acid and salts)

Fluorosilicic acid is a colourless fuming liquid miscible with water. The salts are generally colourless, crystalline and soluble in water.

> CAUSES BURNS (acid)
> HARMFUL IF TAKEN INTERNALLY
> DANGER OF CUMULATIVE EFFECTS

Avoid inhaling vapour or dust. Prevent contact with eyes and skin.

Toxic effects   The vapour and dust irritate all parts of the respiratory system. The acid and salts burn the eyes severely. The acid burns the skin. If taken by mouth there would be severe internal irritation and damage.

First aid   *Vapour or dust inhaled:* standard treatment (p 109).
Affected eyes: standard treatment (p 109).
Skin contact: standard treatment (p 108).
If swallowed: standard treatment (p 109).

Spillage disposal   Wear face-shield or goggles, and gloves. Mop up with plenty of water and run to waste, diluting greatly with running water.

Hexalithium disilicide

Hazardous reactions   Reacts violently with water; explodes with nitric acid; ignites on warming in fluorine; incandescent reactions with P, Se or Te (B858).

Hexamethyldiplatinum

Hazardous reaction	Explodes on heating (B517).

Hexane (and hexane fraction from petroleum)

Colourless liquid; bp about 69 °C; insoluble in water.

EXTREMELY FLAMMABLE

Avoid breathing vapour. TLV 100 ppm (360 mg m^{-3}).

Toxic effects	The vapour may irritate the respiratory system and, in high concentrations, have narcotic action.
First aid	*Vapour inhaled in high concentrations:* standard treatment (p 109). *Affected eyes:* standard treatment (p 109). *If swallowed:* standard treatment (p 109).
Fire hazard	Flash point −23 °C; explosive limits 1.2–7.5%; ignition temp. 260 °C. *Extinguish fire with* foam, dry powder, carbon dioxide or vaporising liquid.
Spillage disposal	Shut off all possible sources of ignition. Wear face-shield or goggles, and gloves. Apply non-flammable dispersing agent if available and work to an emulsion with brush and water — run this to waste, diluting greatly with running water. If dispersant not available, absorb on sand, shovel into bucket(s) and transport to safe, open area for atmospheric evaporation. Ventilate site of spillage well to evaporate remaining liquid and dispel vapour.

Hexanoic acid (caproic acid)

Colourless liquid with unpleasant, sweat-like odour, bp 205 °C; immiscible with water.

CAUSES BURNS

Avoid contact with skin and eyes.

Toxic effects	The liquid burns the skin and eyes. Corrosive if taken by mouth.
First aid	*Affected eyes:* standard treatment (p 109). *Skin contact:* standard treatment (p 108). *If swallowed:* standard treatment (p 109).
Spillage disposal	Wear face-shield or goggles, and gloves. Spread soda ash liberally over the spillage and mop up cautiously with water — run this to waste, diluting greatly with running water.

Hexanolactam – see ε-Caprolactam

Hex-3-enedinitrile (1,4-dicyano-2-butene)

Hazardous reaction	Accelerating polymerisation — decomposition due to overheating in vacuum evaporator (B486).

Hydrazine solutions (hydrazine hydrate)

Colourless liquid; bp 113 °C; miscible with water.

CAUSES BURNS

Avoid contact with eyes and skin. TLV (skin) 1 ppm (1.3 mg m^{-3}).

Toxic effects	The liquid burns the eyes severely. The liquid burns the skin. If taken by mouth there is severe internal irritation and damage.
First aid	*Affected eyes:* standard treatment (p 109). *Skin contact:* standard treatment (p 108). *If swallowed:* standard treatment (p 109).
Spillage disposal	Wear face-shield or goggles, and gloves. Mop up with plenty of water and run to waste, diluting greatly with running water.

Hydrazinium chlorate

Hazardous reaction	Explodes violently at its mp, 80 °C (B651).

Hydrazinium chlorite

Hazardous reaction	Spontaneously flammable when dry (B650).

Hydrazinium nitrate

Hazardous reaction	Explosive properties have been studied in detail (B817).

Hydrazinium perchlorate

Hazardous reactions	Deflagration and thermal decomposition have been studied (B651).

Hydrazine salts

Colourless crystals; soluble in water.

CAUSE BURNS
IRRITATING TO SKIN AND EYES

Avoid contact with eyes and skin.

Toxic effects The crystals and solutions irritate or burn the eyes severely. The crystals and solutions burn the skin. If taken by mouth there would be severe internal irritation and damage.

First aid *Affected eyes:* standard treatment (p 109).
Skin contact: standard treatment (p 108).
If swallowed: standard treatment (p 109).

Spillage disposal Wear face-shield or goggles, and gloves. Mop up with plenty of water and run to waste, diluting greatly with running water.

Hydrazoic acid – *see* Hydrogen azide

Hydrazones Review of group (B67).

Hydriodic acid

Yellow to˙brown fuming liquid commonly available in concentrations of 55% and 66% HI; miscible with water.

HARMFUL VAPOUR
CAUSES BURNS
IRRITATING TO SKIN, EYES AND RESPIRATORY SYSTEM

Avoid breathing vapour. Prevent contact with eyes and skin.

Toxic effects The vapour irritates the respiratory system. The vapour irritates and the liquid burns the eyes severely. The vapour and liquid burn the skin. If taken by mouth there would be severe internal irritation and damage.

Hazardous preparation Blockage of condenser in preparation from iodine and phosphorus caused explosion (B755).

First aid *Vapour inhaled:* standard treatment (p 109).
Affected eyes: standard treatment (p 109).
Skin contact: standard treatment (p 108).
If swallowed: standard treatment (p 109).

Spillage disposal Instruct others to keep at a safe distance. Wear breathing apparatus and gloves. Spread soda ash liberally over the spillage and mop up cautiously with water — run this to waste, dilluting greatly with running water.

Hydrobromic acid

Colourless, fuming liquid with acrid smell commonly available in concentrations of 47%, 50% and 60%; miscible with water.

> HARMFUL VAPOUR
> CAUSES BURNS
> IRRITATING TO SKIN, EYES AND RESPIRATORY SYSTEM

Avoid breathing vapour. Prevent contact with eyes and skin. TLV (as HBr) 3 ppm (10 mg m^{-3}).

Toxic effects The vapour irritates all parts of the respiratory system. The vapour irritates, and the liquid burns, the eyes severely. The vapour and liquid irritate and may burn the skin. If taken by mouth there would be severe internal irritation and damage.

First aid *Vapour inhaled:* standard treatment (p 109).
Affected eyes: standard treatment (p 109).
Skin contact: standard treatment (p 108).
If swallowed: standard treatment (p 109).

Spillage disposal Instruct others to keep at a safe distance. Wear breathing apparatus and gloves. Spread soda ash liberally over the spillage and mop up cautiously with water — run this to waste, diluting greatly with running water.

Hydrochloric acid

Colourless, fuming liquid with pungent smell, commonly available in concentrations of 32% and 36% HCl; miscible with water.

> HARMFUL VAPOUR
> CAUSES BURNS

Avoid breathing vapour. Prevent contact with eyes and skin. TLV (as HCl) 5 ppm (7 mg m^{-3}).

Toxic effects The vapour irritates all parts of the respiratory system. The vapour irritates, and the liquid burns, the eyes severely. The vapour and liquid irritate and may burn the skin. If taken internally there would be severe irritation and damage.

First aid *Vapour inhaled:* standard treatment (p 109).
Affected eyes: standard treatment (p 109).
Skin contact: standard treatment (p 108).
If swallowed: standard treatment (p 109).

Spillage disposal Instruct others to keep at a safe distance. Wear breathing apparatus and gloves. Spread soda ash liberally over the spillage and mop up cautiously with water — run this to waste, diluting greatly with running water.

Hydrofluoric acid

Colourless, fuming liquid with pungent smell, commonly available in concentrations of 40%, 42% and 48% HF; miscible with water.

SERIOUS RISK OF POISONING BY INHALATION, SWALLOWING
OR SKIN CONTACT
CAUSES SEVERE BURNS

Prevent inhalation of fumes. Prevent contact with skin, eyes and clothing. TLV (as HF) 3 ppm
(2 mg m^{-3}).

Toxic effects The fume irritates severely all parts of the respiratory system. The fume or acid rapidly causes severe irritation of the eyes and eyelids; burns may develop. Skin burns do not usually cause pain until several hours have elapsed since contact. If taken by mouth there is immediate and severe internal irritation and damage.

First aid Special treatment (p 106).
Spillage Instruct others to keep at a safe distance. Wear breathing apparatus and
  disposal gloves. Spread soda ash liberally over the spillage and mop up cautiously with water — run this to waste, diluting greatly with running water.

Hydrogen

Colourless, odourless gas; bp −253 °C; sparingly soluble in water.

EXTREMELY FLAMMABLE

Toxic effects Classed as a simple asphyxiant gas.

Hazardous Catalytic Pt and similar metals containing adsorbed O_2 will heat and cause
  reactions ignition in contact with hydrogen; reaction hazards with such oxidants as chlorine dioxide, copper oxide, nitryl fluoride, NO, N_2O_4, PdO, O_2 are indicated — also those with BrF_3, Br_2, ClF_3, Cl_2, F_2; references given to literature on safe handling of liquid hydrogen (B780–781).

Fire hazard Explosive limits 4–75%; ignition temperature 585 °C. Extinguish small hydrogen fires with dry powder, carbon dioxide or vaporising liquid. Where the gas is supplied in a cylinder, turning off the valve will reduce fire.
Disposal Surplus gas or leaking cylinder can be vented slowly to air in a safe open space, or gas burnt off through a suitable burner.

Hydrogen azide (hydrazoic acid)

Hazardous Safe in dilute solutions, violently explosive in concentrated or pure states,
  reactions its properties are reviewed; readily forms explosive heavy metal azides (B775–776).

Hydrogen bromide

Colourless gas, fuming in moist air, with pungent acrid smell; dissolves readily in water forming hydrobromic acid.

> TOXIC BY INHALATION
> IRRITATING TO SKIN, EYES AND RESPIRATORY SYSTEM

Prevent inhalation of gas. Prevent contact with skin and eyes. TLV 3 ppm (10 mg m^{-3}).

Toxic effects   The gas irritates the respiratory system severely; it irritates and may burn the skin and eyes.

Hazardous reactions   Ignites on contact with fluorine; explodes with ozone (B204).

First aid   *Gas inhaled:* standard treatment (p 109).
Affected eyes: standard treatment (p 109).
Skin contact: standard treatment (p 108).

Disposal   Surplus gas or leaking cylinder can be vented slowly into a water-fed scrubbing tower or column in a fume cupboard, or into a fume cupboard served by such a tower.

Hydrogen chloride

Colourless gas, fuming in moist air, with pungent suffocating smell; bp −85 °C; dissolves readily in water forming hydrochloric acid.

> CORROSIVE, IRRITANT GAS
> TOXIC BY INHALATION
> IRRITATING TO SKIN, EYES AND RESPIRATORY SYSTEM

Prevent inhalation of gas. Prevent contact with skin and eyes. TLV 5 ppm (7 mg m^{-3}).

Toxic effects   The gas irritates the eyes and respiratory system severely; it irritates the skin and may cause severe burns to both eyes and skin.

Hazardous reactions   Violent reaction with aluminium; ignites on contact with fluorine; dangerous reactions with hexalithium disilicide, some acetylides, uranium dicarbide and tetraselenium tetranitride are recorded (B638–639).

First aid   *Gas inhaled:* standard treatment (p 109).
Affected eyes: standard treatment (p 109).
Skin contact: standard treatment (p 108).

Disposal   Surplus gas or leaking cylinder can be vented slowly into a water-fed scrubbing tower or column in a fume cupboard, or into a fume cupboard served by such a tower.

Hydrogen cyanide (hydrocyanic acid gas)

Colourless liquid or gas with faint odour of bitter almonds; bp 26 °C; very soluble in water, the solution being only weakly acidic.

EXTREMELY POISONOUS GAS AND LIQUID
POISONOUS BY SKIN ABSORPTION
HIGHLY FLAMMABLE

Prevent inhalation of gas. Prevent contact with skin and eyes. TLV (skin) 10 ppm (11 mg m^{-3}).

Toxic effects   Inhalation of high concentrations leads to shortness of breath, paralysis, unconsciousness, convulsions and death by respiratory failure. With lethal concentrations, death is extremely rapid although breathing may continue for some time. With low concentrations the effects are likely to be headache, vertigo, nausea and vomiting. Chronic exposure over long periods may induce fatigue and weakness. The average fatal dose is 55 mg which can also be assimilated by skin contact with the liquid. (*See* Ch 6, 7.)

Hazardous reaction   In absence of inhibitor, exothermic polymerisation occurs, at 184 °C this is explosive (B228).

First aid   Special treatment for hydrogen cyanide (p 104).

Fire hazard   Flash point −18 °C; explosive limits 6–14%; ignition temp. 538 °C. Since the liquid is supplied in a cylinder, turning off the valve will reduce any fire involving it; if possible, cylinders should be removed quickly from an area in which a fire has developed. Breathing apparatus must be worn during these operations.

Disposal   Breathing apparatus must be worn. Surplus gas or a leaking cylinder can be vented slowly into a water-fed scrubbing tower or column in a fume cupboard.

Hydrogen fluoride

Colourless, fuming gas or liquid; bp 19.5 °C; dissolves in water readily forming hydrofluoric acid.

GIVES OFF VERY POISONOUS VAPOUR
CAUSES SEVERE BURNS

Prevent inhalation of gas. Prevent contact with skin and eyes. TLV 3 ppm (2 mg m^{-3}).

Toxic effects   The gas irritates severely the eyes and respiratory system and may cause burns to the eyes. It irritates the skin and painful burns may develop after an interval. The liquid causes severe, painful burns on contact with all body tissues.

Hazardous reactions   Reference to handling precautions given; risk of violent reaction in fluorination of organic substances by passing HF into stirred suspension of mercury(II) oxide; arsenic trioxide and calcium oxide incandesce in contact with the liquid; violent evolution of SiF_4 when potassium tetrafluorosilicate comes in contact with the liquid (B724).

First aid   *Gas inhaled:* standard treatment (p 109).
Affected eyes: special treatment as for hydrofluoric acid (p 109).
Skin contact: special treatment as for hydrofluoric acid (p 108).

Disposal   Surplus gas or leaking cylinder can be vented slowly into a water-fed scrubbing tower or column in a fume cupboard, or into a fume cupboard served by such a tower.

Hydrogen iodide

Colourless gas which dissolves readily in water to form hydriodic acid.

CORROSIVE, IRRITANT GAS
TOXIC BY INHALATION
IRRITATING TO SKIN, EYES AND RESPIRATORY SYSTEM

Avoid breathing gas. Avoid contact with skin and eyes.

Toxic effects   The gas irritates the respiratory system; it irritates and may burn the eyes and skin.

Hazardous reactions   Causes momentary ignition of magnesium metal; mixture of the anhydrous liquid and potassium metal explodes very violently; the gas is ignited by molten potassium chlorate (B755).

First aid   *Gas inhaled:* standard treatment (p 109).
Affected eyes: standard treatment (p 109).
Skin contact: standard treatment (p 108).

Disposal   Surplus gas or leaking cylinder can be vented slowly into a water-fed scrubbing tower or column in a fume cupboard, or into a fume cupboard served by such a tower.

Hydrogen peroxide

Colourless liquid; miscible with water.

CAUSES BURNS
CONTACT WITH COMBUSTIBLE MATERIAL MAY CAUSE FIRE

Avoid contact with eyes and skin. TLV 1 ppm (1.4 mg m^{-3}).

Toxic effects   Hydrogen peroxide, especially in higher concentrations, is irritant and caustic to the mucous membranes, eyes and skin. If swallowed, the sudden evolution of oxygen may cause injury by acute distension of the stomach and may cause nausea, vomiting and internal bleeding.

Hazardous reactions   Several reviews of the hazards of using strong hydrogen peroxide solutions are referred to; hazardous reactions ranging from ignition to explosion are recorded with acetic acid, acetone, alcohols, alcohols/sulphuric acid, charcoal, carboxylic acids, ketones/nitric acid, mercury(II) oxide/nitric acid, metals, metal oxides, nitric acid/thiourea, nitrogenous bases, phosphorus, P_2O_5, $SnCl_2$, vinyl acetate (B785–791).

First aid   *Affected eyes:* standard treatment (p 109).
Skin contact: standard treatment (p 108).
If swallowed: standard treatment (p 109).

Spillage disposal   Wear face-shield or goggles, and gloves. Mop up with plenty of water and run to waste, diluting greatly with running water.

Hydrogen phosphide – *see* Phosphine

Hydrogen sulphide (sulphuretted hydrogen)

Colourless gas with an offensive odour; soluble in water.

EXTREMELY FLAMMABLE
TOXIC BY INHALATION

Avoid inhaling gas. TLV 10 ppm (15 mg m^{-3}).

Toxic effects In high concentrations may cause immediate unconsciousness followed by respiratory paralysis; at lower concentrations causes irritation of all parts of the respiratory system and eyes, headache, dizziness and weakness. Irritates the eyes and may cause conjunctivitis. (*See* Ch 6, p 88.)

Hazardous reactions May ignite on contact with a wide range of metal oxides; finely divided tungsten glows red-hot in stream of the gas; dangerous reactions with a variety of oxidants including CrO_3, Cl_2O, F_2, HNO_3, Na_2O_2, PbO_2; have been recorded; exothermic reaction with soda lime, which may become explosive in presence of oxygen (B798–799).

First aid *Gas inhaled:* standard treatment (p 109).
Affected eyes: standard treatment (p 109).

Fire hazard Explosive limits 4.3–46%; ignition temp. 260 °C. Since the gas is supplied in a cylinder, turning off the valve will reduce any fire involving it; if possible, cylinders should be removed quickly from an area in which a fire has developed.

Disposal Surplus gas or leaking cylinder can be vented slowly into a water-fed scrubbing tower or column in a fume cupboard, or into a fume cupboard served by such a tower.

Hydrogen trisulphide

Hazardous reactions Some metal oxides cause violent decomposition and ignition; reactions with NCl_3 and pentyl alcohol are explosive (B799).

2-Hydroxybut-3-enonitrile (1-cyano-2-propen-1-ol)

Hazardous reaction Liable to polymerise explosively (B388).

2-Hydroxyethylamine – *see* 2-Aminoethanol

2-Hydroxyethyl methacrylate

Pale yellow, mobile liquid; bp 84 °C at 5 mmHg; miscible with water.

CAUSES SEVERE BURNS

Prevent contact with skin, eyes and clothing.

Toxic effects	Splashes on skin and mucous surfaces can cause severe blistering.
First aid	*Affected eyes:* standard treatment (p 109).
	Skin contact: standard treatment (p 108).
	If swallowed: standard treatment (p 109).
Spillage disposal	Wear face-shield or goggles, and gloves. Mop up with plenty of water and run to waste, diluting greatly with running water. Ventilate area of spillage until dry.

1-Hydroxyethyl peracetate

Hazardous reaction	The low-melting solid is explosive (B413).

Hydroxylamine

Hazardous preparation and reactions	Risk of explosion in preparation from hydroxylamine hydrochloride and sodium hydroxide in methanol; ignites on contact with anhydrous copper-(II) sulphate; dangerous reactions with Na, Ca and finely divided zinc; it reacts violently or explosively with a large number of oxidants (B803–804).

Hydroxylammonium salts (hydroxylamine salts)

White crystals; soluble in water.

CORROSIVE
IRRITATING TO SKIN AND EYES

Avoid contact with eyes and skin.

Toxic effects	The salts and solutions irritate and burn the eyes severely. The salts and solutions burn the skin. If taken by mouth there is severe internal irritation and damage. **Chronic effects** Continued skin contact can cause dermatitis.
First aid	*Affected eyes:* standard treatment (p 109).
	Skin contact: standard treatment (p 108).
	If swallowed: standard treatment (p 109).
Spillage disposal	Wear face-shield or goggles, and gloves. Mop up with plenty of water and run to waste, diluting greatly with running water.

4-Hydroxy-4-methylpentan-2-one (diacetone alcohol)

Colourless liquid with faint, pleasant odour; bp 168 °C; miscible with water.

HARMFUL VAPOUR
IRRITATING TO SKIN, EYES AND RESPIRATORY SYSTEM

Avoid breathing vapour. Avoid contact with skin, eyes and clothing. TLV 50 ppm (240 mg m^{-3}).

Toxic effects The vapour irritates the eyes and respiratory system. The liquid irritates the eyes and mucous membranes and is absorbed by the skin, possibly with harmful effects. Taken by mouth, it has a narcotic effect; kidney and liver injury and anaemia have been produced in experimental animals.

First aid *Vapour inhaled:* standard treatment (p 109).
Affected eyes: standard treatment (p 109).
Skin contact: standard treatment (p 108).
If swallowed: standard treatment (p 109).

Spillage disposal Shut off all possible sources of ignition. Wear face-shield or goggles, and gloves. Mop up with plenty of water and run to waste, diluting greatly with running water. Ventilate area well to evaporate remaining liquid and dispel vapour.

2-Hydroxy-2-methylpropiononitrile

(acetone cyanohydrin; 2-cyanopropan-2-ol)

Colourless liquid; bp 82 °C at 23 mmHg; miscible with water.

SERIOUS RISK OF POISONING
BY INHALATION OR SWALLOWING
GIVES OFF POISONOUS VAPOUR

Prevent inhalation of vapour. Prevent contact with skin and eyes.

Toxic effects High vapour concentrations, if inhaled, rapidly cause giddiness, headache, unconsciousness, convulsions; in severe cases, breathing may cease due to hydrogen cyanide poisoning. It must be assumed that skin absorption and taking by mouth will have similar effects.

First aid Special treatment for hydrogen cyanide (p 104).

Spillage disposal Wear breathing apparatus and gloves. Instruct others to keep at a safe distance. Bleaching powder should be scattered liberally over the spillage, or an excess of sodium hypochlorite solution added. The treated spillage should then be mopped up into a bucket and allowed to stand for 24 hours before running to waste, diluting greatly with running water.

3-Hydroxytriazenes Review of group (B69).

Hypochlorous acid

Hazardous reactions Contact with alcohols forms unstable alkyl hypochlorites; explodes violently on contact with ammonia gas; ignites on contact with arsenic (B639).

Hypohalites Review of class (B69).

Hyponitrous acid

Hazardous   Extremely explosive solid (B783).
reaction

3,3'-Iminodipropiononitrile
(bis(2-cyanoethyl)amine)

Storage   Bottles in store for 18 months exploded (B494).
hazard

Indium compounds

The chloride, nitrate and sulphate of indium are soluble in water. The antimonide, arsenide, selenide and telluride are insoluble compounds used in semiconductor research and attention to toxicity limits has arisen from the risk of inhaling this type of dust during working.

> TOXIC BY INHALATION
> HARMFUL IF TAKEN INTERNALLY

TLV 0.1 mg m^{-3} (as In).

Toxic effects   Evidence on human toxicity is scarce, but solutions of indium have been shown to be poisonous when injected into animals: the blood, heart, liver, and kidneys being injured.

First aid   *If soluble salts swallowed:* standard treatment (p 109).
Spillage   Soluble salts should be mopped up with plenty of water and the solution
disposal   run to waste, diluting greatly with running water.

Iodic acid

Colourless crystals or powder; soluble in water.

> CONTACT WITH COMBUSTIBLE MATERIAL MAY CAUSE FIRE
> TOXIC BY INHALATION
> CAUSE BURNS
> POWERFUL OXIDISING AGENTS—ASSIST FIRE

Avoid breathing dust. Avoid contact with eyes and skin. Do not mix with combustible material.

Toxic effects   The dust irritates all parts of the respiratory system. The acid and solutions burn the eyes and skin. If taken by mouth is assumed to cause severe internal irritation and damage.

Hazardous   Reacts with boron below 40 °C and incandesces; deflagrates with charcoal,
reactions   phosphorus and sulphur on heating (B755).

First aid   *Dust inhaled:* standard treatment (p 109).
       Affected eyes: standard treatment (p 109).
       Skin contact: standard treatment (p 108).
       If swallowed: standard treatment (p 109).

Spillage Shovel into bucket of water and run solution or suspension to waste
disposal diluting greatly with running water. Site of spillage should be washed
       thoroughly to remove all oxidant, which is liable to render any organic
       matter (particularly wood, paper and textiles) with which it comes into
       contact, dangerously combustible when dry. Clothing wetted with the
       solution should be washed thoroughly.

Iodine

Bluish-black crystalline scales with a characteristic odour; almost insoluble in water.

     HARMFUL VAPOUR
     CAUSES BURNS

Avoid breathing vapour. Avoid contact with eyes and skin. TLV 0.1 ppm (1 mg m^{-3}).

Toxic effects The vapour irritates all parts of the respiratory system. The vapour and
       solid irritate the eyes. The solid burns the skin. If taken by mouth there is
       severe internal irritation and damage.

Hazardous Forms highly explosive addition compounds with ammonia; reaction with
reactions ethanol and phosphorus considered dangerous as school experiment;
       incandescent reaction with BrF_3; violent reaction with BrF_5; ignites with
       ClF_3; ignites with F_2; several metal acetylides and carbides react very
       exothermally with iodine; mixture with potassium explodes weakly on
       impact (B832–833).

First aid   *Vapour inhaled:* standard treatment (p 109).
       Affected eyes: standard treatment (p 109).
       Skin contact: standard treatment (p 108).
       If swallowed: standard treatment (p 109).

Spillage If the spillage is large and in a confined area, breathing apparatus should
disposal be worn. Large quantities are best disposed of by sweeping up with sand
       and burying in open waste land. Small quantities can be dealt with by
       dissolving in sodium thiosulphate or sodium metabisulphate solution and
       running the resulting solution to waste, diluting greatly with running water.
       Iodine stains on flooring can be cleared by mopping with thiosulphate or
       metabisulphite solution.

Iodine azide

Hazardous Shock- and friction-sensitive explosive (B831).
reaction

Iodine chloride

Reddish-brown liquid with pungent odour; soluble in water.

> HARMFUL VAPOUR
> CAUSES SEVERE BURNS

Avoid breathing vapour. Prevent contact with skin and eyes.

Toxic effects
The vapour irritates all parts of the respiratory system. The vapour and liquid burn the eyes severely. The liquid burns the skin. Assumed to cause severe internal irritation and damage if taken by mouth.

Hazardous reactions
Mixtures with sodium explode on impact, with potassium on contact. Al foil may ignite on long contact (B651).

First aid
Vapour inhaled: standard treatment (p 109).
Affected eyes: standard treatment (p 109).
Skin contact: standard treatment (p 108).
If swallowed: standard treatment (p 109).

Spillage disposal
Wear face-shield or goggles, and gloves; spread soda ash liberally over the spillage and mop up cautiously with plenty of water — run to waste, diluting greatly with running water.

Iodine heptafluoride

Hazardous reactions
Activated carbon ignites in gas; mixtures with methane ignite, those with hydrogen explode on heating or sparking; violent reactions with Ba, K and Na on contact but Al, Mg and Sn require heating to react; benzene, light petroleum, ethanol and ether ignite on contact; vigorous reactions with acetic acid, acetone or ethyl acetate (B746–747).

Iodine isocyanate

Storage hazard
Solutions gradually deposit touch-sensitive explosive (B257).

Iodine(V) oxide (iodine pentaoxide)

Hazardous reactions
Reacts violently with BrF_5; reacts explosively with warm carbon, sulphur, resin, sugar or powdered easily oxidisable elements (B833).

Iodine pentafluoride

Hazardous reactions
Reacts violently with benzene above 50 °C; reactions with water and dimethyl sulphoxide are violent; incandescence may occur with B, Si, red P, S, Sb, Bi, Mo, W; K or molten Na explodes on contact; calcium carbide or potassium hydride incandesce on contact (B744).

Iodine perchlorate

Hazardous reaction Exploded on laser irradiation at low temperature (B691).

Iodine triacetate

Hazardous reaction Explodes at 140 °C (B493).

Iodine trichloride

Fuming, orange-red crystalline masses with pungent odour; soluble in water.

HARMFUL VAPOUR
CAUSES SEVERE BURNS

Avoid breathing vapour. Prevent contact with eyes and skin.

Toxic effects The vapour irritates all parts of the respiratory system. The vapour and solid irritate and burn the eyes severely. The solid burns the skin. Assumed to cause severe internal irritation and damage if taken by mouth.

First aid *If inhaled:* standard treatment (p 109).
Affected eyes: standard treatment (p 109).
Skin contact: standard treatment (p 108).
If swallowed: standard treatment (p 109).

Spillage disposal Wear face-shield or goggles, and gloves; spread soda ash liberally over the spillage and mop up cautiously with plenty of water — run to waste, diluting greatly with running water.

Iodoacetic acid

Colourless crystals; mp 82 °C; soluble in water.

CAUSES SEVERE BURNS

Prevent contact with skin, eyes and clothing.

Toxic effects Causes severe irritation and burns to all tissues with which it comes in contact.

First aid *Affected eyes:* standard treatment (p 109).
Skin contact: standard treatment (p 108).
If swallowed: standard treatment (p 109).

Spillage disposal Wear face-shield or goggles, and gloves. Spread soda ash liberally over the spillage and mop up cautiously with plenty of water — run to waste, diluting greatly with running water.

1-Iodobuta-1,3-diyne

Hazardous reaction Crude material exploded during vacuum distillation; the pure compound exploded on scratching under illumination (B382).

Iododimethylarsine

Hazardous reaction Ignites when heated in air (B318).

Iodoethane (ethyl iodide)

Colourless to brown liquid, sensitive to air and light; bp 72 °C; 4g dissolve in 100g water at 20 °C.

HARMFUL VAPOUR

Avoid contact with skin, eyes and clothing.

Toxic effects These are not well documented though it is considered to be moderately toxic by inhalation and narcotic in high concentrations.

Hazardous preparation Reaction of ethanol, phosphorus and iodine considered too dangerous for school work (B313).

First aid *Vapour inhaled:* standard treatment (p 109).
Affected eyes: standard treatment (p 109).
Skin contact: standard treatment (p 108).
If swallowed: standard treatment (p 109).

Spillage disposal Wear face-shield or goggles, and gloves. Apply dispersing agent if available and work to an emulsion with brush and water — run this to waste, diluting greatly with running water. If dispersant not available, absorb on sand, shovel into bucket(s) and transport to safe, open area for atmospheric evaporation. Site of spillage should be washed thoroughly with water and soap or detergent.

Iodomethane (methyl iodide)

Heavy, colourless to yellow-brown liquid with a sweetish smell, bp 43 °C; sparingly soluble in water.

TOXIC IN CONTACT WITH SKIN
GIVES OFF POISONOUS VAPOUR
CAUSES BURNS

Avoid breathing vapour. Prevent contact with skin and eyes. TLV (skin) 5 ppm (28 mg m^{-3}).

Toxic effects Inhalation of vapour may cause dizziness, drowsiness, mental confusion, muscular twitching and delirium. The vapour and liquid irritate the eyes and distort the vision. The liquid irritates the skin and may cause blistering. The liquid must be assumed to be irritant and poisonous if taken by mouth.

**Hazardous
reaction**   Vigorously reactive with sodium dispersed in toluene (B242).

First aid   *Vapour inhaled:* standard treatment (p 109).
Affected eyes: standard treatment (p 109).
Skin contact: standard treatment (p 108).
If swallowed: standard treatment (p 109).

**Spillage
disposal**   Instruct others to keep at a safe distance. Wear breathing apparatus and gloves. Apply dispersing agent if available and work to an emulsion with brush and water — run this to waste, diluting greatly with running water. If dispersant not available, absorb on sand, shovel into bucket(s) and transport to safe, open area for atmospheric evaporation. Site of spillage should be washed thoroughly with water and soap or detergent.

3-Iodo-1-phenylpropyne (1-iodo-3-phenyl-2-propyne)

**Hazardous
reaction**   Exploded on distillation at about 180 °C (B566).

Iodosylbenzene

**Hazardous
reaction**   Explodes at 210 °C (B478).

Iodylbenzene

**Hazardous
reaction**   Explodes at 230 °C; extreme care required in heating, compressing or grinding iodyl compounds (B478).

Iodylbenzene perchlorate

**Hazardous
reaction**   Small sample exploded violently while still damp (B485).

Ion exchange resins

**Hazardous
reaction**   Expansion on moistening may fracture container (B72).

Iron(III) chloride anhydrous (ferric chloride anhydrous)

Black-brown crystals and aggregated masses; violently decomposed by water with the formation of hydrogen chloride.

CAUSES BURNS

Avoid contact with eyes and skin.

Toxic effects	Inhalation of fine crystals produces irritation or burns of the mucous membranes. Will cause painful eye burns. When moisture is present on skin, heat is produced on contact resulting in thermal and acid burns. If taken by mouth the immediate local reaction would cause severe burns.
First aid	*Dust inhaled:* standard treatment (p 109). *Affected eyes:* standard treatment (p 109). *Skin contact:* standard treatment (p 108). *If swallowed:* standard treatment (p 109).
Spillage disposal	Wear face-shield or goggles, and gloves. Mop up with plenty of water and run to waste, diluting greatly with running water.

Iron(II) maleate

Hazardous reaction	The finely divided compound (a product of phthalic anhydride manufacture) is rapidly oxidised in air above 150 °C and has been involved in plant fires (B383).

Iron(II) perchlorate

Hazardous preparation	Violent explosion when mixture of iron(II) sulphate and perchloric acid was being heated (B675).

Isobutane (2-methylpropane; isobutylene)

Colourless gas; bp −12 °C; somewhat soluble in water.

EXTREMELY FLAMMABLE

Avoid breathing vapour.

Toxic effects	The gas has an anaesthetic effect but is not toxic.
First aid	*Gas inhaled in quantity:* standard treatment (p 109).
Fire hazard	Explosive limits 1.8–8.4 %; ignition temp. 462 °C. Since the gas is supplied in a cylinder, turning off the valve will reduce any fire involving it; if possible cylinders should be removed quickly from an area in which a fire has developed.
Disposal	Surplus gas or leaking cylinder can be vented slowly to air in a safe, open area or gas burnt off through a suitable burner.

Isobutene –*see* 2-Methylpropene

Isobutyl alcohol (iso-butyl alcohol; 2-methylpropan-1-ol)

Colourless liquid with characteristic odour; bp 106 °C; sparingly soluble in water (1 part dissolves in about 20 parts water at 25 °C).

FLAMMABLE
HARMFUL VAPOUR
IRRITATING TO EYES AND RESPIRATORY SYSTEM

Avoid breathing vapour. Avoid contact with eyes. TLV 50 ppm (150 mg m^{-3}).

Toxic effects The vapour irritates the respiratory system and, in high concentrations, is narcotic. The liquid irritates the eyes and is harmful if taken internally.

First aid *Vapour inhaled in high concentrations:* standard treatment (p 109).
Affected eyes: standard treatment (p 109).
If swallowed: standard treatment (p 109).

Fire hazard Flash point 28 °C; explosive limits at 100 °C 1.7–10.9%; ignition temp. 427 °C. *Extinguish fire with* water spray, dry powder, carbon dioxide or vaporising liquid.

Spillage disposal Shut off all possible sources of ignition. Wear face-shield or goggles, and gloves. Apply non-flammable dispersing agent if available and work to an emulsion with brush and water — run this to waste, diluting greatly with running water. If dispersant not available, absorb on sand, shovel into bucket(s) and transport to safe, open area for atmospheric evaporation. Ventilate site of spillage well to evaporate remaining liquid and dispel vapour.

Isobutylene – *see* 2-Methylpropene

Isobutyl methyl ketone – *see* 4-Methylpentan-2-one

Isobutyric acid

Colourless, oily liquid with rancid odour; bp 154 °C; one part is soluble in five parts of water at 20 °C.

CAUSES BURNS
IRRITATING TO SKIN AND EYES

Avoid contact with skin, eyes and clothing.

Toxic effects The liquid burns the eyes and irritates or burns the skin. The liquid is irritant or corrosive if taken by mouth.

First aid *Affected eyes:* standard treatment (p 109).
Skin contact: standard treatment (p 108).
If swallowed: standard treatment (p 109).

Spillage disposal Wear face-shield or goggles, and gloves. Spread soda ash liberally over the spillage and mop up with plenty of water — run to waste, diluting greatly with running water.

Isobutyric anhydride *– see* Butyric and Isobutyric anhydrides

Isocyanatomethane (methyl isocyanate)

Colourless lachrymatory liquid; bp 44 °C; sparingly soluble in water.

> EXTREMELY FLAMMABLE
> SERIOUS RISK OF POISONING BY
> INHALATION, SWALLOWING OR SKIN CONTACT
> IRRITATING TO SKIN, EYES AND RESPIRATORY SYSTEM

Prevent inhalation of vapour. Prevent contact with skin, eyes and clothing. TLV (skin) 0.02 ppm (0.05 mg m^{-3}).

Toxic effects   The vapour irritates the respiratory system and eyes severely, causing lachrymation. The liquid irritates the eyes and skin severely and is assumed to be highly irritant and damaging if taken internally.

First aid   *Vapour inhaled:* standard treatment (p 109).
Affected eyes: standard treatment (p 109).
Skin contact: standard treatment (p 108).
If swallowed: standard treatment (p 109).

Fire hazard   Flash point below −7 °C. *Extinguish fire with* dry powder carbon dioxide or vaporising liquid.

Spillage disposal   Shut off all possible sources of ignition. Instruct others to keep at a safe distance. Wear breathing apparatus and gloves. Apply non-flammable dispersing agent and work to an emulsion with brush and water — run this to waste, diluting greatly with running water. If dispersant not available, absorb on sand, shovel into bucket(s) and transport to safe open place for atmospheric evaporation or burial. Site of spillage should be washed thoroughly with water and soap or detergent.

Isophorone (3,5,5-trimethylcyclohex-2-en-1-one)

Colourless liquid; bp 215 °C; slightly soluble in water.

> HARMFUL VAPOUR
> IRRITATING TO SKIN, EYES AND RESPIRATORY SYSTEM

Avoid breathing vapour. Avoid contact with skin, eyes and clothing. TLV 5 ppm (25 mg m^{-3}).

Toxic effects   The vapour irritates the eyes and respiratory system. The liquid irritates the eyes and may cause corneal damage; it may also irritate the skin because of its degreasing action. It is assumed to be toxic if taken by mouth.

First aid
Vapour inhaled: standard treatment (p 109).
Affected eyes: standard treatment (p 109).
Skin contact: standard treatment (p 108).
If swallowed: standard treatment (p 109).

Spillage disposal
Instruct others to keep at a safe distance. Wear breathing apparatus and gloves. Apply dispersing agent if available and work to an emulsion with brush and water—run this to waste, diluting greatly with running water. If dispersant not available, absorb on sand, shovel into bucket(s) and transport to safe, open area for burial. Site of spillage should be washed thoroughly with water and soap or detergent.

Isoprene (2-methylbuta-1,3-diene)

Colourless volatile liquid; bp 34 °C; insoluble in water.

EXTREMELY FLAMMABLE
IRRITATING TO SKIN, EYES AND RESPIRATORY SYSTEM

Avoid breathing vapour.

Toxic effects
The vapour is irritating to the respiratory system and is narcotic in high concentrations. The liquid irritates the skin and eyes, and is assumed to be irritant and harmful if taken internally.

Hazardous reactions
Absorbs oxygen from air forming explosive polymeric peroxide; explosion occurred during ozonisation in heptane (B442).

First aid
Vapour inhaled: standard treatment (p 109).
Affected eyes: standard treatment (p 109).
Skin contact: standard treatment (p 108).
If swallowed: standard treatment (p 109).

Fire hazard
Flash point −53 °C; ignition temp. 220 °C. *Extinguish fire with* foam, dry powder, carbon dioxide or vaporising liquid.

Spillage disposal
Shut off all possible sources of ignition. Instruct others to keep at a safe distance. Wear breathing apparatus and gloves. Apply non-flammable dispersing agent if available and work to an emulsion with brush and water—run this to waste, diluting greatly with running water. If dispersant not available, absorb on sand, shovel into bucket(s) and transport to safe, open area for atmospheric evaporation. Site of spillage should be washed thoroughly with water and soap or detergent.

Isopropyl acetate

Colourless liquid; bp 93 °C; 1 part is soluble in 23 parts water at 27 °C.

HIGHLY FLAMMABLE
IRRITATING TO EYES

Avoid breathing vapour. Avoid contact with eyes. TLV 250 ppm (950 mg m^{-3}).

Toxic effects
The vapour may irritate the eyes and respiratory system and is narcotic in high concentrations. The liquid irritates the eyes and will be irritant and narcotic if taken by mouth.

First aid	*Vapour inhaled in high concentrations:* standard treatment (p 109).
	Affected eyes: standard treatment (p 109).
	If swallowed: standard treatment (p 109).
Fire hazard	Flash point 4 °C; explosive limits 1.8–7.8%; ignition temp. 460 °C. *Extinguish fire with* dry powder, carbon dioxide or vaporising liquid.
Spillage disposal	Shut off all possible sources of ignition. Instruct others to keep at a safe distance. Wear breathing apparatus and gloves. Apply non-flammable dispersing agent and work to an emulsion with brush and water — run this to waste, diluting greatly with running water. If dispersant not available, absorb on sand, shovel into bucket and transport to safe, open area for atmospheric evaporation. Site of spillage should be washed thoroughly with water and soap or detergent.

Isopropyl alcohol – *see* Propan-2-ol

Isopropylbenzene – *see* Cumene

Isopropyl ether – *see* Di-isopropyl ether

Isopropyl hydroperoxide

Hazardous reaction   Explodes just above bp, 107–109 °C (B372).

Isopropyl hypochlorite

Hazardous reaction   Has extremely low stability — explosions have occurred during preparation (B367).

Isopropyl nitrate

Storage and handling   Technical Bulletin available for this rocket propellant (B370).

Ketone peroxides Review of group (B72).

Lactic acid

Colourless or slightly yellow, syrupy, hydroscopic liquid; mp 16.8 °C when anhydrous; miscible with water.

IRRITATING TO SKIN AND EYES

Avoid contact with skin, eyes and clothing.

Toxic effects   Irritates and may burn the eyes and skin. Irritant if taken by mouth.

First aid   *Affected eyes:* standard treatment (p 109).
Skin contact: standard treatment (p 108).
If swallowed: standard treatment (p 109).

Spillage disposal   Wear face-shield or goggles, and gloves. Spread soda ash liberally over the spillage and mop up cautiously with plenty of water — run to waste, diluting greatly with running water.

Lanthanum trihydride

Hazardous reaction   Ignites in air (B800).

Lauroyl peroxide – *see* Dodecanoyl peroxide

Lead(II) azide

Hazardous reaction   A detonator that has been studied in detail; in prolonged contact with copper or zinc may form extremely sensitive azides of these metals (B885)

Lead(IV) azide

Hazardous reaction   Liable to spontaneous decomposition which is sometimes explosive (B886)

Lead bromate

Hazardous reaction   An explosive salt (B211).

Lead dichlorite

Hazardous reactions   Explodes on heating above 100 °C or on rubbing with antimony sulphide or fine sulphur (B683).

Lead diperchlorate

Hazardous reaction   A saturated solution of the anhydrous salt in dry methanol exploded violently (B685).

Lead dithiocyanate

**Hazardous
reaction**   Explosive (B336).

Lead imide

**Hazardous
reactions**   Explodes on heating or in contact with water or dilute acids (B774).

Lead salts

White or coloured crystals or powders.

> **HARMFUL IF TAKEN INTERNALLY
> DANGER OF CUMULATIVE EFFECTS**

Avoid breathing dust. TLV (dust as Pb) (0.15 mg m^{-3}).

Toxic effects   Dust inhaled or material swallowed may cause severe internal injury with
vomiting, diarrhoea and collapse. **Chronic effects** Loss of appetite,
pallor, anaemia, constipation, colic, blue line on gums. (*See* Ch 6, 7.)

First aid   *If swallowed:* standard treatment (p 109).
Spillage   Small quantities of soluble lead salts can be dissolved in water and the
  disposal   solution run to waste, diluting greatly with running water. Insoluble compounds can be mixed with an excess of sand and disposed of as normal
refuse.

Lead tetrachloride

**Hazardous
reactions**   Liable to explode with potassium; explodes on warming with dilute sulphuric acid (B699).

Linseed oil

**Hazardous
reaction**   Cloths used to apply this to benches were dropped in waste bin — laboratory destroyed by fire some hours later (B74).

Liquid air

**Hazardous
reactions**   Review (B75).

Lithium (metal)

Silver-white metal, becoming yellowish on exposure to moist air; mp 180 °C; reacts with water.

CONTACT WITH WATER LIBERATES HIGHLY FLAMMABLE GASES

Avoid contact with skin, eyes and clothing.

Toxic effects   Lithium reacts with moisture forming lithium hydroxide which irritates the skin and mucous surfaces and may cause burns.

Hazardous reactions   Finely divided metal may ignite in air; will burn in nitrogen or carbon dioxide and is difficult to extinguish once alight; reacts violently with BrF_5 and explosively with diazomethane; mixtures with several halocarbons will explode on impact; formation of lithium amalgam may be explosive; ignites on contact with nitric acid; reacts violently or explosively with sulphur (B853–856).

First aid   *Affected eyes:* standard treatment (p 109).
Skin contact: standard treatment (p 108).
If swallowed: standard treatment (p 109).

Fire hazard   Extinguish fire with special extinguishant; powdered graphite, LiCl, KCl or zirconium silicate may also be used.

Disposal   May be allowed to react in large excess of cold water and solution run to waste.

Lithium aluminium hydride
– *see* Aluminium lithium hydride

Lithium azide

Hazardous reactions   Moist or dry salt decomposes explosively at 115–298 °C depending upon rate of heating (B857).

Lithium borohydride (lithium tetrahydroborate)

Hazardous reaction   Contact with limited amounts of water may cause ignition (B184).

Lithium hydride

Hazardous reactions   Ignites in warm air; mixtures with liquid oxygen are detonable explosives (B758).

Lithium hydroxide

White crystals; soluble in water.

CAUSES BURNS

Avoid contact with eyes and skin.

Toxic effects The solid and solution irritate the eyes severely and burn the skin. It is assumed that there is internal irritation and damage if taken by mouth.

First aid *Affected eyes:* standard treatment (p 109).
Skin contact: standard treatment (p 108).
If swallowed: standard treatment (p 109).

Spillage disposal Wear face-shield or goggles, and gloves. Mop up with plenty of water and run to waste, diluting greatly with running water.

Lithium tetrahydroborate
– *see* Lithium borohydride

Magnesium (metal)

Hazardous reactions As fine powder dispersed in air it is a serious explosion hazard; reaction with beryllium fluoride is violent; explosive acetylide may be formed from traces of acetylene in ethylene oxide if magnesium is contained in fittings used in ethylene oxide service; the powdered metal reacts and may explode on contact with chloromethane, chloroform or carbon tetrachloride and mixtures with carbon tetrachloride or trichloroethylene will flash on heavy impact; ignites, if moist, in fluorine or chlorine; may ignite, if finely divided, on heating in iodine vapour; reacts vigorously with certain cyanides; violently reduces some metal oxides; reaction with metal oxosalts may be explosive; reaction with methanol may become vigorous; heating with K_2CO_3, moist SiO_2, S, Te and numerous oxidants can be hazardous (B858–861).

Magnesium perchlorate

Hazardous reactions Explosions have followed the use of the anhydrous salt for drying organic solvents and solutions of organic compounds in such solvents; an explosion occurred in a drying tube containing the desiccant between cotton wool wads, which had been used for drying O_2/N_2O_4 mixture; may explode on contact with trimethyl phosphite (B677–679).

Maleic anhydride

White crystalline powder or lumps; mp 53 °C; dissolves in water forming maleic acid.

IRRITATING TO SKIN, EYES AND RESPIRATORY SYSTEM

Avoid breathing dust. Avoid contact with skin and eyes. TLV 0.25 ppm (1 mg m⁻³).

Toxic effects	The dust and vapour irritate the eyes, skin and respiratory system; prolonged contact with tissues may result in burns. Assumed to be irritant and harmful if taken by mouth.
Hazardous reactions	Decomposes exothermally in presence of alkali- or alkaline earth metal or ammonium ions, dimethylamine, triethylamine, pyridine or quinoline at temperatures above 150 °C; sodium ions and pyridine are particularly active (B383).
First aid	*Affected eyes:* standard treatment (p 109).
	Skin contact: standard treatment (p 108).
	If swallowed: standard treatment (p 109).
Spillage disposal	Sweep up and dissolve in water — run solution to waste, diluting greatly with running water.

Maleic anhydride ozonide

Hazardous reaction	Explodes on warming to −40 °C (B384).

Malononitrile (dicyanomethane, methylene cyanide)

Colourless crystals, mp 30.5 °C. Polymerises violently on heating to 120 °C and on contact with alkaline materials. Soluble in water.

SERIOUS RISK OF POISONING BY INHALATION OR SWALLOWING
GIVES OFF POISONOUS VAPOUR

Prevent contact with skin and eyes.

Toxic effects	It is said that the toxicity of malononitrile is of the same order as that of hydrogen cyanide though no cases of poisoning of humans have been traced. It is therefore considered wisest to treat the compound with the respect given to the alkali cyanides.
Hazardous reactions	May polymerise violently on heating at 130 °C or in contact with strong bases at lower temperatures (B346).
First aid	Special treatment for hydrogen cyanide (p 104).
Spillage disposal	Wear goggles or face-shield, and gloves. Sweep up, or absorb on sand, and place in a large volume of water to which is then added an excess of sodium hypochlorite solution. Allow to stand for 24 hours with occasional stirring and then run to waste, diluting greatly with running water. The site of the spillage should be mopped with sodium hypochlorite solution and rinsed with water.

Manganese diperchlorate

Hazardous reaction Explodes at 195 °C (B679).

Manganese trifluoride

Hazardous reaction Glass is attacked violently by heating in contact with it, SiF_4 being evolved (B740).

Mercaptoacetic acid (thioglycollic acid)

Colourless liquid with unpleasant smell; bp 123 °C at 29 mmHg; miscible with water. It is readily decomposed by mineral acids with liberation of poisonous hydrogen sulphide.

CAUSES BURNS
IRRITATING TO SKIN AND EYES

Avoid contact with skin and eyes.

Toxic effects The liquid irritates the eyes and skin and may cause burns. It must be assumed to be irritant and poisonous if taken by mouth.

First aid *Affected eyes:* standard treatment (p 109).
Skin contact: standard treatment (p 108).
If swallowed: standard treatment (p 109).

Spillage disposal Wear face-shield or goggles, and gloves. Spread soda ash liberally over the spillage and mop up cautiously with plenty of water — run to waste, diluting greatly with running water.

Mercury (quicksilver)

Heavy silvery liquid; insoluble in water.

GIVES OFF VERY POISONOUS VAPOUR
DANGER OF CUMULATIVE EFFECTS

Avoid breathing vapour. Avoid contact with eyes and skin. TLV 0.1 mg m^{-3} (skin).

Toxic effects High concentrations of vapour may cause metallic taste, nausea, abdominal pain, vomiting, diarrhoea and headache. **Chronic effects** Continued exposure to small concentrations of vapour may result in severe nervous disturbance, including tremor of the hands, insomnia, loss of memory, irritability and depression; other possible effects are loosening of teeth and excessive salivation. Continued skin contact with mercury may cause dermatitis and the above effects may be caused by absorption through the skin or following ingestion. Kidney damage may ensue. (*See* Ch 6, 7.)

Hazardous reactions	Prolonged contact between mercury and ammonia may result in formation of explosive solid; ease with which it forms amalgams with laboratory and electrical contact metals can cause severe corrosion problems; reacts violently with dry bromine; chlorine dioxide explodes when shaken with mercury (B824).
First aid	*Vapour inhaled:* standard treatment (p 109). *Skin contact:* standard treatment (p 108). *If swallowed:* standard treatment (p 109).
Spillage disposal	Because of the high toxicity of mercury vapour it is important to clean up mercury as thoroughly as possible, especially in confined areas. A small aspirator with a capillary tube and connected to a pump can be used for sucking up droplets. Mercury spilt into floor cracks can be made non-volatile by putting zinc dust down the cracks to form the amalgam.

Mercury(I) azide

Hazardous reaction	Explodes in air on heating to over 270 °C (B829).

Mercury(I) chlorite

Hazardous reaction	Explodes spontaneously when dry (B651).

Mercury compounds

Mercury compounds vary widely in appearance, solubility in water and toxicity. Mercuric compounds are generally more toxic than mercurous compounds. Some organic mercurial compounds are liquids with an extremely poisonous vapour, others are solids, the toxicity of which is not known with certainty. Thus phenylmercuric acetate appears relatively non-toxic but the alkyl mercurials are highly poisonous. Their toxicity by skin absorption is uncertain, but some, *eg* ethyl mercury phosphate and mercury fulminate, can cause dermatitis. The effects vary greatly according to the nature of the organic mercurial. Some compounds cause kidney damage while others can cause irreversible damage to the central nervous system.

> POISONOUS DUST
> SERIOUS RISK OF POISONING BY INHALATION, SWALLOWING OR SKIN CONTACT
> DANGER OF CUMULATIVE EFFECTS
> CAUSES IRRITATION OF SKIN AND EYES

Avoid breathing dusts. Avoid contact with skin and eyes. TLV (skin) (alkyl compounds) 0.01 mg m⁻³; (all forms except alkyl) as Hg 0.05 mg m⁻³.

Toxic effects	Inhalation of dust may cause nausea, abdominal pain, vomiting, diarrhoea and headache. Skin absorption may give rise to similar effects and the eyes may be damaged by direct contact with some salts. Abdominal pain, nausea, vomiting, diarrhoea and shock follow ingestion of the soluble mercuric salts. Kidney damage may ensue. **Chronic effects** The intake of small amounts of mercury compounds by inhalation, skin absorption

or ingestion over a long period may cause nervous disturbance, including tremor of the hands, insomnia, loss of memory, irritability and depression; other possible effects are loosening of the teeth and excessive salivation. (*See* numerous references to the toxicity of mercury compounds in Ch 6.)

First aid
Dusts inhaled: standard treatment (p 109).
Affected eyes: standard treatment (p 109).
Skin contact: standard treatment (p 108).
If swallowed: standard treatment (p 109).

Spillage disposal
Small quantities of soluble mercury compounds can be swept up, dissolved in water or acid and run to waste at very high dilution. Recovery may be warranted in the case of large amounts but the disposal of these must be considered carefully in the light of local conditions and regulations. If burial is carried out in an isolated area the solid compound should first be diluted 10–20 times by weight with sand.

Mercury(II) cyanide

Hazardous reactions
A friction- and impact-sensitive explosive which can detonate liquid HCN; reacts explosively with magnesium; explodes on heating with sodium nitrite (B332).

Mercury dichlorite

Hazardous reaction
Explodes spontaneously when dry (B677).

Mercury(II) fulminate

Hazardous reactions
A detonator initiated when dry by flame, heat, impact, friction or intense radiation; contact with sulphuric acid causes explosion (B333).

Mercury(II) nitrate

Hazardous reactions
Contact with acetylene in solution gives explosive mercury acetylide; with the aqueous solution ethanol gives mercury(II) fulminate; aqueous solution reacts with phosphine to give explosive complex; risk of violent reactions with petroleum hydrocarbons (B826).

Mercury(II) oxalate

Hazardous reactions
When dry, it explodes on percussion, grinding or heating to 105 °C; storage is inadvisable (B333).

Mercury(II) perbenzoate

Hazardous reaction   Explodes if heated above 110 °C (B600).

Mesityl oxide – *see* 4-Methylpent-3-en-2-one

Mesoxalonitrile

Hazardous reaction   Reacts explosively with water (B379).

Metal abietates – *see* Abietates, metal

Metal acetylides   Review of group (B76).

Metal azides and azide halides   Reviews of classes (B78).

Metal dusts   Review of class (B82).

Metal fulminates   Review of group (B83).

Metal hydrides   Review of group (B85).

Methacrylic acid

Colourless solid or liquid with unpleasant, acrid smell; mp 16 °C; bp 158 °C; soluble in water.

IRRITATING TO SKIN, EYES AND RESPIRATORY SYSTEM

Avoid breathing vapour. Avoid contact with skin, eyes and clothing.

Toxic effects   The vapour irritates the eyes and respiratory system. The liquid irritates the eyes and skin and is assumed to be very irritant and harmful if taken internally.

First aid   *Vapour inhaled:* standard treatment (p 109).
Affected eyes: standard treatment (p 109).
Skin contact: standard treatment (p 108).
If swallowed: standard treatment (p 109).

Spillage disposal	Wear goggles and gloves. Spread soda ash liberally over the spillage and mop up cautiously with plenty of water—run to waste, diluting greatly with running water.

Methane

Colourless gas, sparingly soluble in water.

> EXTREMELY FLAMMABLE

Toxic effects	Methane is non-toxic but can have narcotic effects in high concentrations in the absence of oxygen.
First aid	*Vapour inhaled in high concentrations:* standard treatment (p 109).
Fire hazard	Explosive limits 5–15%; ignition temp. 537 °C. Since the gas is supplied in cylinders, turning off the valve will reduce any fire involving it; if possible, cylinders should be removed quickly from an area in which a fire has developed.
Disposal	Surplus gas or leaking cylinder can be vented slowly to air in a safe, open area or gas burnt off through a suitable burner.

Methanethiol (methyl mercaptan)

Colourless gas with extremely disagreeable smell; bp 6 °C; sparingly soluble in water.

> EXTREMELY FLAMMABLE
> TOXIC BY INHALATION

Avoid breathing gas. Avoid contact with eyes. TLV 0.5 ppm (1 mg m^{-3}).

Toxic effects	The gas is nauseous and may be narcotic in high concentrations.
First aid	*Vapour inhaled:* standard treatment (p 109).
Fire hazard	Flash point below −18 °C; explosive limits 3.9–21.8%. Since the gas is supplied in a cylinder, turning off the valve will reduce any fire in which it is involved; if possible, cylinders should be removed quickly from an area in which a fire has developed.
Disposal	Surplus gas or gas from a leaking cylinder should be burnt through a suitable gas burner in a fume cupboard.

Methanol (methyl alcohol)

Colourless, volatile liquid, bp 65 °C; miscible with water.

> HIGHLY FLAMMABLE
> SERIOUS RISK OF POISONING BY INHALATION
> OR SWALLOWING

Avoid breathing vapour. Avoid contact with skin and eyes. TLV 200 ppm (260 mg m^{-3}.)

Toxic effects   Inhalation of high concentrations of vapour may cause dizziness, stupor, cramps and digestive disturbance. Lower concentrations may cause headache, nausea, vomiting and irritation of the mucous membranes. The vapour and liquid are very dangerous to the eyes, the effects sometimes being delayed for many hours. Ingestion damages the central nervous system, particularly the optic nerve (causing temporary or permanent blindness), and injures the kidneys, liver, heart and other organs; apart from the effects referred to above, unconsciousness may develop after some hours and this may be followed by death. **Chronic effects** Continued exposure to low concentrations of vapour may cause many of the above effects, continued skin contact may cause dermatitis. (Ch 6. p 84.)

Hazardous reactions   Violent explosion occurred when sodium was added to methanol/chloroform mixture; reaction with magnesium can be very vigorous; reaction with bromine can be violent, with sodium hypochlorite explosive; it is ignited by CrO_3; reactions with nitric acid or hydrogen peroxide may become explosive (B251).

First aid   *Vapour inhaled:* standard treatment (p 109).
Affected eyes: standard treatment (p 109).
Skin contact: standard treatment (p 108).
If swallowed: standard treatment (p 109).

Fire hazard   Flash point 10 °C; explosive limits 7.3–36.5%; ignition temp. 464 °C. *Extinguish fire with* water spray, dry powder, carbon dioxide or vaporising liquid.

Spillage disposal   Shut off all possible sources of ignition. Instruct others to keep at a safe distance. Wear breathing apparatus and gloves. Mop up with plenty of water and run to waste, diluting greatly with running water. Ventilate area well to evaporate remaining liquid and dispel vapour.

p-Methoxybenzyl chloride

Storage hazard   Bottles of this exploded on storage at room temperature (B548).

2-Methoxyethanol
(ethylene glycol monomethyl ether; methyl cellosolve)

Colourless volatile liquid with pleasant smell; bp 125 °C; miscible with water; liable to form explosive peroxides on exposure to light and air which should be decomposed before distilling the ether to small volume.

FLAMMABLE
HARMFUL VAPOUR
IRRITATING TO EYES AND RESPIRATORY SYSTEM

Avoid breathing the vapour. Avoid contact with skin and eyes. TLV 25 ppm (80 mg m^{-3}).

Toxic effects   The vapour irritates the respiratory system. The vapour and liquid irritate the eyes. Assumed to be poisonous if taken by mouth. **Chronic effects** Anaemia, blood abnormalities and symptoms of central nervous system damage have resulted from prolonged exposure.

First aid	*Vapour inhaled:* standard treatment (p 109). *Affected eyes:* standard treatment (p 109). *If swallowed:* standard treatment (p 109).
Fire hazard	Flash point 46 °C; explosive limits 2.5–14%; ignition temp. 288 °C. *Extinquish fire with* water spray, dry powder, carbon dioxide or vaporising liquid.
Spillage disposal	Shut off all possible sources of ignition. Instruct others to keep at a safe distance. Wear breathing apparatus and gloves. Mop up with plenty of water and run to waste, diluting greatly with running water. Ventilate area well to evaporate remaining liquid and dispel vapour.

2-Methoxyethyl acetate

(ethylene glycol monomethyl ether acetate)

Colourless liquid; bp 143 °C; miscible with water; liable to form explosive peroxides on exposure to light and air which must be decomposed before the ether ester is distilled to small volume.

FLAMMABLE
HARMFUL VAPOUR
IRRITATING TO EYES

Avoid contact with skin, eyes and clothing. TLV 25 ppm (120 mg m^{-3}) (skin).

Toxic effects	The liquid irritates the eyes and may irritate the skin; it may cause headache, dizziness, fatigue, nausea, vomiting and more serious disorders if taken by mouth or absorbed extensively through the skin.
First aid	*Affected eyes:* standard treatment (p 109). *Skin contact:* standard treatment (p 108). *If swallowed:* standard treatment (p 109).
Fire hazard	Flash point 54 °C. *Extinguish fire with* dry powder, carbon dioxide or vaporising liquid.
Spillage disposal	Shut off all possible sources of ignition. Wear face-shield or goggles, and gloves. Mop up with plenty of water and run to waste, diluting greatly with running water. Ventilate area of spillage well to evaporate remaining liquid and dispel vapour.

3-Methoxypropyne

Hazardous reaction	Explodes on distillation at 61 °C (B396).

Methyl acetate

Colourless, volatile liquid with pleasant odour; bp 58 °C; miscible with water.

EXTREMELY FLAMMABLE
IRRITATING TO EYES AND RESPIRATORY SYSTEM

Avoid breathing vapour. Avoid contact with skin and eyes. TLV 200 ppm (610 mg m^{-3}).

Toxic effects	The vapour irritates the eyes and respiratory system. High concentrations may cause dizziness and palpitations. Assumed to be poisonous if taken by mouth.
First aid	*Vapour inhaled:* standard treatment (p 109). *Affected eyes:* standard treatment (p 109). *Skin contact:* standard treatment (p 108). *If swallowed:* standard treatment (p 109).
Fire hazard	Flash point −9 °C; explosive limits 3.1–16%; ignition temp. 502 °C. *Extinguish fire with* water spray, foam, dry powder, carbon dioxide or vaporising liquid.
Spillage disposal	Shut off all possible sources of ignition. Wear face-shield or goggles, and gloves. Mop up with plenty of water and run to waste, diluting greatly with running water. Ventilate area of spillage well to evaporate remaining liquid and dispel vapour.

Methylacetylene – *see* Propyne

Methyl acrylate

Colourless liquid with acrid odour; bp 80 °C; immiscible with water.

EXTREMELY FLAMMABLE
HARMFUL VAPOUR
IRRITATING TO SKIN, EYES AND RESPIRATORY SYSTEM

Avoid breathing vapour. Avoid contact with skin and eyes. TLV (skin) 10 ppm (35 mg m⁻³).

Toxic effects	The vapour irritates the eyes and respiratory system; high concentrations may cause lethargy and lead to convulsions. The liquid irritates the skin and eyes. Assumed to be poisonous if taken by mouth.
First aid	*Vapour inhaled:* standard treatment (p 109). *Affected eyes:* standard treatment (p 109). *Skin contact:* standard treatment (p 108). *If swallowed:* standard treatment (p 109).
Fire hazard	Flash point −2.8 °C; explosive limits 2.8–25%. *Extinguish fire with* water spray, foam, dry powder, carbon dioxide or vaporising liquid.
Spillage disposal	Shut off all possible sources of ignition. Instruct others to keep at a safe distance. Wear breathing apparatus and gloves. Apply non-flammable dispersing agent if available and work to an emulsion with brush and water — run this to waste, diluting greatly with running water. If dispersant not available, absorb on sand, shovel into bucket(s) and transport to safe, open area for atmospheric evaporation or burial. Site of spillage should be washed thoroughly with water and soap or detergent.

Methylal – *see* Dimethoxymethane

Methyl alcohol – *see* Methanol

Methylamine

Colourless gas with pungent, fishy smell, bp −6.3 °C; very soluble in water (see below for solutions).

HIGHLY FLAMMABLE
IRRITATING TO SKIN, EYES AND RESPIRATORY SYSTEM

Avoid breathing gas. Avoid contact with skin and eyes. TLV 10 ppm (12 mg m⁻³).

Toxic effects	The gas irritates the skin, eyes and respiratory system and sustained contact may cause burns.
Hazardous reaction	Addition to nitromethane renders it susceptible to initiation by a detonator (B253).
First aid	*Gas inhaled:* standard treatment (p 109). *Affected eyes:* standard treatment (p 109). *Skin contact:* standard treatment (p 108).
Fire hazard	Flash point 0 °C; explosive limits 4.9–20.7%; ignition temp. 430 °C. Since the gas is supplied in a cylinder, turning off the valve will reduce any fire involving it; if possible, cylinders should be removed quickly from an area in which fire has developed.
Disposal	Surplus gas or leaking cylinder can be vented slowly into a water-fed scrubbing tower or column in a fume cupboard, or into a fume cupboard served by such a tower.

Methylamine solutions

Methylamine is commonly available in either aqueous or ethanolic solution in a concentration of 25–33%; these solutions are colourless and are miscible with water; they have a fishy smell.

HIGHLY FLAMMABLE
IRRITATING TO SKIN, EYES AND RESPIRATORY SYSTEM

Avoid breathing vapour. Avoid contact with skin and eyes. TLV (as $CH_3.NH_2$) 10 ppm (12 mg m⁻³).

Toxic effects	The vapour irritates the eyes and respiratory system. The solutions irritate the eyes and skin. The solution will cause irritation and damage if taken internally.
First aid	*Vapour inhaled:* standard treatment (p 109). *Affected eyes:* standard treatment (p 109). *Skin contact:* standard treatment (p 108). *If swallowed:* standard treatment (p 109).
Fire hazard	Flash point depends upon nature and strength of solution. *Extinguish fire with* water spray, foam, dry powder, carbon dioxide or vaporising liquid.
Spillage disposal	Shut off all possible sources of ignition. Instruct others to keep at a safe distance. Wear breathing apparatus and gloves. Mop up with plenty of water and run to waste diluting greatly with running water. Ventilate area well to evaporate remaining liquid and dispel vapour.

311

N-Methylaniline

Colourless to brown liquid; bp 196 °C; insoluble in water.

> TOXIC IN CONTACT WITH SKIN
> GIVES OFF POISONOUS VAPOUR

Avoid breathing vapour. Avoid contact with skin and eyes. TLV (skin) 2 ppm (9 mg m^{-3}).

Toxic effects   Excessive breathing of the vapour or absorption of the liquid through the skin can cause headache, drowsiness, cyanosis, mental confusion and, in severe cases, convulsions. The liquid is dangerous to the eyes, and the above effects can also be experienced if it is swallowed. **Chronic effects** Continued exposure to the vapour, or slight skin exposure to the liquid, over a period may affect the nervous system and the blood, causing fatigue, loss of appetite, headache and dizziness.

First aid   *Vapour inhaled:* standard treatment (p 109).
Affected eyes: standard treatment (p 109).
Skin contact: standard treatment (p 108).
If swallowed: standard treatment (p 109).

Spillage disposal   Wear face-shield or goggles, and gloves. Mix with sand and shovel mixture into glass, enamel or polythene vessel for dispersion in an excess of dilute hydrocholoric acid (1 volume of concentrated acid diluted with 2 volumes of water). Allow to stand, with occasional stirring, for 24 hours and then run acid extract to waste, diluting greatly with running water and washing the sand. The sand can be disposed of as normal refuse.

Methyl azide

Hazardous reactions   Stable at room temperature but may detonate on rapid heating; presence of mercury in the azide markedly reduces stability towards shock or electric discharge; a mixture with methanol and dimethyl malonate exploded while being sealed into a Carius tube (B246–247).

Methyl benzenediazoate

Hazardous reaction   Explodes on heating or after about an hour's storage in a sealed tube at ambient temperature (B535).

Methyl bromide – *see* Bromomethane

2-Methylbuta-1,3-diene – *see* Isoprene

2-Methylbutan-2-ol (t-amyl alcohol; t-pentyl alcohol)

Colourless, volatile liquid with characteristic odour and burning taste; bp 102 °C.

HIGHLY FLAMMABLE
HARMFUL VAPOUR

Avoid breathing vapour. Avoid contact with skin and eyes.

Toxic effects	Vapour may irritate the eyes and respiratory system. Liquid irritates the eyes severely and may irritate skin. If swallowed may cause headache, vertigo, nausea, vomiting, excitement and delirium followed by coma.
First aid	*Vapour inhaled:* standard treatment (p 109). *Affected eyes:* standard treatment (p 109). *Skin contact:* standard treatment (p 108). *If swallowed:* standard treatment (p 109).
Fire hazard	Flash point 19 °C; explosive limits 1.2–9%; *Extinguish fire with* water spray, dry powder, carbon dioxide or vaporising liquids.
Spillage disposal	Shut off all possible sources of ignition. Wear face-shield and gloves. Apply non-flammable dispersing agent if available and work to an emulsion with brush and water — run this to waste diluting greatly with running water. If dispersant not available, absorb on sand, shovel into bucket(s) and transport to safe open area for atmospheric evaporation or burial. Ventilate well to evaporate remaining liquid and dispel vapour.

Methyl carbonate – *see* Dimethyl carbonate

Methyl cellosolve – *see* 2-Methoxyethanol

Methyl chloride – *see* Chloromethane

Methylchloroform – *see* 1,1,1-Trichloroethane

Methyl cyanide – *see* Acetonitrile

Colourless liquid, bp 71 °C; immiscible with water.

HIGHLY FLAMMABLE
GIVES OFF POISONOUS VAPOUR
IRRITATING TO SKIN, EYES AND RESPIRATORY SYSTEM

Avoid breathing vapour. Avoid contact with skin and eyes.

Methyl chloroformate

Toxic effects The vapour irritates the respiratory system. The vapour irritates and the liquid burns the eyes. The liquid can irritate and burn the skin. Assumed to be irritant and poisonous if taken by mouth.

First aid *Vapour inhaled:* standard treatment (p 109).
Affected eyes: standard treatment (p 109).
Skin contact: standard treatment (p 108).
If swallowed: standard treatment (p 109).

Fire hazard Flash point 12 °C; ignition temp. 504 °C. *Extinguish fire with* water spray, foam, dry powder, carbon dioxide or vaporising liquid.

Spillage disposal Shut off all possible sources of ignition. Instruct others to keep at a safe distance. Wear breathing apparatus and gloves. Apply non-flammable dispersing agent and work to an emulsion with brush and water — run this to waste, diluting greatly with running water. If dispersant not available absorb on sand, shovel into bucket and transport to safe, open area for atmospheric evaporation. Site of spillage should be washed thoroughly with water and soap or detergent.

Methylcyclohexane

Colourless liquid; bp 100 °C; insoluble in water.

EXTREMELY FLAMMABLE

Avoid breathing vapour. Avoid contact with skin and eyes. TLV 400 ppm (1600 mg m^{-3}).

Toxic effects These are not well documented but animal experiments suggest that it is more toxic than cyclohexane. In high concentrations the vapour causes narcosis and anaesthesia.

First aid *Vapour inhaled in high concentrations:* standard treatment (p 109).
Affected eyes: standard treatment (p 109).
Skin contact: standard treatment (p 108).
If swallowed: standard treatment (p 109).

Fire hazard Flash point −4 °C; ignition temp. 285 °C. *Extinguish fire with* foam, dry powder, carbon dioxide or vaporising liquid.

Spillage disposal Shut off all possible sources of ignition. Wear face-shield or goggles, and gloves. Apply non-flammable dispersing agent if available and work to an emulsion with brush and water — run this to waste, diluting greatly with running water. If dispersant not available, absorb on sand, shovel into bucket(s) and transport to safe, open area for atmospheric evaporation. Ventilate site of spillage well to evaporate remaining liquid and dispel vapour.

Methylcyclohexanol (mixed isomers)

Colourless, viscous liquid with odour similar to that of menthol; bp of mixed isomers 155–180 °C; slightly soluble in water.

HARMFUL VAPOUR

Avoid breathing vapour. Avoid contact with skin, eyes and clothing. TLV 50 ppm (230 mg m^{-3}).

Toxic effects   The vapour irritates the eyes and respiratory system. The liquid irritates the eyes and may irritate the skin. Assumed to be irritant and harmful if taken by mouth.

First aid   *Vapour inhaled:* standard treatment (p 109).
Affected eyes: standard treatment (p 109).
Skin contact: standard treatment (p 108).
If swallowed: standard treatment (p 109).

Spillage disposal   Wear face-shield or goggles, and gloves. Apply dispersing agent if available and work to an emulsion with brush and water—run this to waste, diluting greatly with running water. If dispersant not available, absorb on sand, shovel into bucket(s) and transport to safe, open area for burial. Site of spillage should be washed thoroughly with water and soap or detergent.

Methylcyclohexanone (mixed isomers)

Colourless to pale yellow liquid with smell like that of acetone; bp of mixed isomers 160–170 °C; insoluble in water.
FLAMMABLE
HARMFUL VAPOUR

Avoid breathing vapour. Avoid contact with skin, eyes and clothing. TLV (skin) 50 ppm (230 mg m^{-3}).

Toxic effects   The vapour irritates the eyes and respiratory system. The liquid irritates the eyes and prolonged contact with the skin may result in kidney and liver damage. Assumed to cause irritation and damage if taken internally.

Hazardous reactions   Oxidation of the 4-isomer by addition to nitric acid at about 75 °C caused a detonation; reaction with mixtures of hydrogen peroxide and nitric acid caused the 3-isomer to form an oily explosive peroxide (B537).

First aid   *Vapour inhaled:* standard treatment (p 109).
Affected eyes: standard treatment (p 109).
Skin contact: standard treatment (p 108).
If swallowed: standard treatment (p 109).

Fire hazard   Flash point 48 °C. *Extinguish fire with* foam, dry powder, carbon dioxide or vaporising liquid.

Spillage disposal   Wear face-shield or goggles, and gloves. Apply dispersing agent if available and work to an emulsion with brush and water—run this to waste, diluting greatly with running water. If dispersant not available, absorb on sand, shovel into bucket(s) and transport to safe, open area for burial. Site of spillage should be washed thoroughly with water and soap or detergent.

3-Methyldiazirine

Hazardous reaction   The gas explodes on heating (B306).

Methyl diazoacetate

Hazardous reaction   Explodes with violence when heated (B352).

Methylene chloride – *see* Dichloromethane

Methylene chlorobromide – *see* Bromochloromethane

Methylene cyanide – *see* Malononitrile

4-Methyleneoxetan-2-one – *see* Diketen

Methyl ether – *see* Dimethyl ether

Methyl ethyl ketone – *see* Butanone

Methyl formate

Colourless liquid with pleasant odour; bp 32 °C; moderately soluble in water.

EXTREMELY FLAMMABLE
IRRITATING TO SKIN, EYES AND RESPIRATORY SYSTEM

Avoid breathing vapour. Avoid contact with skin, eyes and clothing. TLV 100 ppm (250 mg m⁻³).

Toxic effects The vapour irritates the eyes and respiratory system; in severe cases there may be retching, narcosis and pulmonary irritation that can result in death. Assumed to cause severe irritation and damage if taken internally.

First aid *Vapour inhaled:* standard treatment (p 109).
Affected eyes: standard treatment (p 109).
Skin contact: standard treatment (p 108).
If swallowed: standard treatment (p 109).

Fire hazard Flash point −19 °C; explosive limits 5.9–20%; ignition temp. 456 °C. *Extinguish fire with* dry powder, carbon dioxide or vaporising liquid.

Spillage disposal Shut off all possible sources of ignition. Instruct others to keep at a safe distance. Wear breathing apparatus and gloves. Mop up with plenty of water and run to waste, diluting greatly with running water. Ventilate area well to evaporate remaining liquid and dispel vapour.

Methylhydrazine

Hazardous reactions   A powerful reducing agent and fuel, hypergolic with many oxidants (B254).

Methyl hydroperoxide

Hazardous reactions   Violently explosive, shock-sensitive, especially on warming (B252).

Methyl hypochlorite

Hazardous reaction   Superheated vapour explodes as does liquid on ignition (B240).

Methyl iodide – *see* Iodomethane

Methyl isobutyl ketone – *see* 4-Methylpentan-2-one

Methyl isocyanate – *see* Isocyanatomethane

Methyl isocyanide

Hazardous reaction   Exploded when heated in a sealed ampoule (B298).

Methyllithium

Hazardous reaction   Ignites and burns in air (B243).

Methyl mercaptan – *see* Methanethiol

Methyl methacrylate

Colourless liquid; bp 101 °C; almost insoluble in water. This monomer normally contains a small quantity of stabiliser.

> HIGHLY FLAMMABLE
> HARMFUL VAPOUR
> IRRITATING TO SKIN, EYES AND RESPIRATORY SYSTEM

Avoid breathing vapour. Avoid contact with skin, eyes and clothing. TLV 100 ppm (410 mg m^{-3}).

Toxic effects	The vapour may irritate the eyes and respiratory system. The liquid will irritate the eyes and alimentary system if taken by mouth.
Hazardous reactions	Exposure to air at room temperature for two months generated an explosive ester/oxygen interpolymer; ignition occurred when dibenzoyl peroxide was added to a small amount of the ester (B445).
First aid	*Vapour inhaled:* standard treatment (p 109). *Affected eyes:* standard treatment (p 109). *Skin contact:* standard treatment (p 108). *If swallowed:* standard treatment (p 109).
Fire hazard	Flash point 10 °C; explosive limits 2.1–12.5%; ignition temp. 421 °C. *Extinguish fire with* water spray, foam, dry powder, carbon dioxide or vaporising liquid.
Spillage disposal	Shut off all possible sources of ignition. Wear face-shield or goggles, and gloves. Apply non-flammable dispersing agent if available and work to an emulsion with brush and water — run this to waste, diluting greatly with running water. If dispersant not available, absorb on sand, shovel into bucket(s) and transport to safe, open area for atmospheric evaporation. Ventilate site of spillage well to evaporate remaining liquid and dispel vapour.

Methyl nitrate

Hazardous reaction	Explodes at 65 °C and has high shock-sensitivity (B246).

Methyl-2-nitrobenzenediazoate

Hazardous reactions	Explodes on heating or on disturbing after 24 hours in a sealed tube at ambient temperature (B533).

1-Methyl-3-nitro-1-nitrosoguanidine

Hazardous reactions	A former diazomethane precursor, this compound detonates on high impact; sample exploded when heated in sealed capillary tube (B317).

2-Methyl-2-nitropropane (t-nitrobutane)

Hazardous reaction	Sample exploded during distillation (B418).

N-Methyl-*N*-nitrosotoluene-4-sulphonamide

Yellow crystals, insoluble in water; commonly used as a source of the highly reactive, toxic and explosive gas diazomethane which is generated when it is treated with alkalies; it may itself explode when heated to above 45 °C.

HEATING MAY CAUSE EXPLOSION

Avoid contact with skin and eyes.

Toxic effects No evidence has been found that this reagent is irritant or otherwise toxic, but its close association with the methylating technique involving highly toxic diazomethane suggests that protection against skin and eye contact should be used when it is being handled.

First aid *Affected eyes:* standard treatment (p 109).
Skin contact: standard treatment (p 108).
If swallowed: standard treatment (p 109).

Spillage disposal Wear face-shield or goggles, and gloves. Mix with sand and transport to safe, open area for burial. Site of spillage should be washed thoroughly with water and soap or detergent.

N-Methyl-*N*-nitrosourea

Hazardous reaction	Material stored at 20 °C exploded after 6 months (B317).

2-Methyloxiran – *see* Propylene oxide

4-Methylpentan-2-one (isobutyl methyl ketone)

Colourless liquid with faint camphor-like smell; bp 126 °C; slightly soluble in water.

HIGHLY FLAMMABLE
HARMFUL VAPOUR
IRRITATING TO SKIN, EYES AND RESPIRATORY SYSTEM

Avoid breathing vapour. Avoid contact with eyes. TLV 100 ppm (410 mg m^{-3}).

Toxic effects The vapour is somewhat irritating to the eyes and respiratory system and narcotic in high concentrations. The liquid irritates the eyes and will cause irritation and damage if taken internally.

First aid   *Vapour inhaled:* standard treatment (p 109).
                     Affected eyes: standard treatment (p 109).
                     If swallowed: standard treatment (p 109).

Fire hazard   Flash point 17 °C; explosive limits 1.2–8%; ignition temp. 460 °C. *Extinguish fire with* water spray, dry powder, carbon dioxide or vaporising liquid.

Spillage   Shut off all possible sources of ignition. Wear face-shield or goggles, and
disposal   gloves. Apply non-flammable dispersing agent if available and work to an emulsion with brush and water — run this to waste, diluting greatly with running water. If dispersant not available, absorb on sand, shovel into bucket(s) and transport to safe, open area for atmospheric evaporation or burial. Ventilate site of spillage well to evaporate remaining liquid and dispel vapour.

4-Methylpent-3-en-2-one (mesityl oxide)

Colourless, oily liquid with smell somewhat like honey; bp 130 °C; sparingly soluble in water.

FLAMMABLE
HARMFUL VAPOUR
IRRITATING TO SKIN, EYES AND RESPIRATORY SYSTEM

Avoid breathing vapour. Avoid contact with skin, eyes and clothing. TLV 25 ppm (100 mg m^{-3}).

Toxic effects   The vapour irritates the eyes and respiratory system. The liquid is highly irritating to the eyes and skin and will cause internal irritation and damage if taken by mouth.

First aid   *Vapour inhaled:* standard treatment (p 109).
                     Affected eyes: standard treatment (p 109).
                     Skin contact: standard treatment (p 108).
                     If swallowed: standard treatment (p 109).

Fire hazard   Flash point 31 °C; ignition temp. 344 °C. *Extinguish fire with* foam, dry powder, carbon dioxide or vaporising liquid.

Spillage   Shut off all possible sources of ignition. Instruct others to keep at a safe
disposal   distance. Wear breathing apparatus and gloves. Apply non-flammable dispersing agent if available and work to an emulsion with brush and water — run this to waste, diluting greatly with running water. If dispersant not available, absorb on sand, shovel into bucket(s) and transport to safe, open area for burial. Ventilate site of spillage well to evaporate remaining liquid and dispel vapour.

Methyl perchlorate

Hazardous   Explosive (B240).
reaction

Methylphosphine

Hazardous   Readily ignites in air (B253).
reaction

Methylpotassium

Hazardous reaction   Dry material is highly pyrophoric (B243).

2-Methylpropane – *see* Isobutane

2-Methylpropan-1-ol – *see* Isobutyl alcohol

2-Methylpropan-2-ol (t-butyl alcohol)

Colourless crystalline solid or liquid with camphor-like odour; mp 25 °C; bp 83 °C; miscible with water.

HIGHLY FLAMMABLE
HARMFUL VAPOUR

Avoid breathing vapour. Avoid contact with skin and eyes. TLV 100 ppm (300 mg m^{-3}).

Toxic effects   Vapour may irritate eyes and respiratory system. The liquid irritates the eyes and may irritate the skin causing dermatitis. If taken by mouth may cause headache, dizziness, drowsiness and narcosis.

First aid   *Vapour inhaled:* standard treatment (p 109).
Affected eyes: standard treatment (p 109).
Skin contact: standard treatment (p 108).
If swallowed: standard treatment (p 109).

Fire hazard   Flash point 10 °C; explosive limits 2.4–8%; ignition temp. 478 °C. *Extinguish fire with* dry powder, carbon dioxide or vaporising liquid.

Spillage disposal   Shut off all possible sources of ignition. Wear face-shield or goggles, and gloves. Mop up with plenty of water and run to waste, diluting greatly with running water. Ventilate area of spillage well to evaporate remaining liquid and dispel vapour.

2-Methylpropene (isobutylene; isobutene)

Colourless gas with smell like that of coal gas; bp −7 °C; practically insoluble in water.

EXTREMELY FLAMMABLE

Avoid breathing gas.

Toxic effects   The gas has an anaesthetic effect but is not toxic.

First aid   *Gas inhaled in quantity:* standard treatment (p 109).

Fire hazard   Flash point below −7 °C; explosive limits 1.8–8.8%; ignition temp. 465 °C. Since the gas is supplied in a cylinder, turning off the valve will reduce any fire involving it; if possible, cylinders should be removed quickly from an area in which a fire has developed.

Disposal   Surplus gas or leaking cylinder can be vented slowly to air in a safe, open area or gas burnt off through a suitable burner.

Methyl propyl ketone – *see* Pentan-2-one

Methylpyridines (picolines)

2-, 3- and 4-picolines are colourless liquids, boiling at 129 °C, 144 °C and 143 °C respectively; the 2- and 4-isomers have unpleasant smells. All are very soluble in water.

FLAMMABLE
IRRITATING TO EYES AND RESPIRATORY SYSTEM

Avoid breathing vapour. Avoid contact with skin and eyes.

Toxic effects The vapours irritate the respiratory tract to some extent. The liquids irritate the eyes and may be assumed to cause irritation and damage if taken internally.

First aid *Vapour inhaled:* standard treatment (p 109).
Affected eyes: standard treatment (p 109).
Skin contact: standard treatment (p 108).
If swallowed: standard treatment (p 109).

Fire hazard Flash points 28 °C, 40 °C and 57 °C respectively; 2-picoline has an ignition temperature of 538 °C. *Extinguish fire with* water spray, dry powder, carbon dioxide or vaporising liquid.

Spillage disposal Shut off all possible sources of ignition. Wear face-shield or goggles, and gloves. Mop up with plenty of water and run to waste, diluting greatly with running water. Ventilate area well to evaporate remaining liquid and dispel vapour.

Methylsodium

Hazardous reaction Ignites immediately in air (B248).

α-Methylstyrene (2-phenylpropylene)

Colourless liquid; bp 167 °C; insoluble in water.

FLAMMABLE
HARMFUL VAPOUR
IRRITATING TO SKIN, EYES AND RESPIRATORY SYSTEM

Avoid breathing vapour. Avoid contact with skin and eyes. TLV 100 ppm (480 mg m^{-3}).

Toxic effects The vapour irritates the eyes and respiratory system. The liquid irritates the eyes and may cause conjunctivitis. The liquid irritates the skin and may cause dermatitis. It is assumed to be irritant and harmful if taken by mouth.

First aid *Vapour inhaled:* standard treatment (p 109).
Affected eyes: standard treatment (p 109).
Skin contact: standard treatment (p 108).
If swallowed: standard treatment (p 109).

Fire hazard Flash point 54 °C; explosive limits 1.9–6.1%; ignition temp. 494 °C. *Extinguish fire with* foam, dry powder, carbon dioxide or vaporising liquid.

Spillage disposal Shut off all possible sources of ignition. Wear face-shield or goggles, and gloves. Apply non-flammable dispersing agent if available and work to an emulsion with brush and water — run this to waste, diluting greatly with running water. If dispersant not available, absorb on sand, shovel into bucket(s) and transport to safe, open area for burial. Ventilate site of spillage well to evaporate remaining liquid and dispel vapour.

Methyl sulphate – *see* Dimethyl sulphate

Methyltrichlorosilane (trichloro(methyl)silane)

Colourless to pale yellow, volatile liquid with pungent smell, which fumes strongly in moist air; bp 65·5 °C; it reacts vigorously with water forming hydrochloric acid and polymeric gels.

CAUSES BURNS
HIGHLY FLAMMABLE
IRRITATING TO SKIN, EYES AND RESPIRATORY SYSTEM

Prevent inhalation of vapour. Prevent contact with skin, eyes and clothing.

Toxic effects The vapour irritates the eyes and respiratory system severely. The liquid burns the skin and eyes and will cause severe internal damage if taken by mouth.

First aid *Vapour inhaled :* standard treatment (p 109).
Affected eyes : standard treatment (p 109).
Skin contact : standard treatment (p 108).
If swallowed : standard treatment (p 109).

Fire hazard Flash point 8.3 °C; explosive limits 8.5% to over 17%; ignition temp. 580 °C. *Extinguish fire with* dry powder, carbon dioxide or vaporising liquid.

Spillage disposal Shut off all possible sources of ignition. Instruct others to keep at safe distance. Wear breathing apparatus and gloves. Absorb on sand, shovel into bucket(s), transport to safe, open area and tip into large volume of water; leave to decompose before decanting the water to waste, diluting greatly with running water. Site of spillage should be ventilated after washing thoroughly with water and soap or detergent.

Methyl vinyl ether (vinyl methyl ether)

Colourless liquid or gas with sweetish odour; bp 8 °C; slightly soluble in water.

EXTREMELY FLAMMABLE

Avoid breathing gas.

Toxic effects   These have not been fully investigated. It has narcotic properties.

First aid   *Vapour inhaled :* standard treatment (p 109).
Fire hazard   Flash point −51 °C. Since the liquid is supplied in a cylinder, turning off the valve will reduce any fire involving it; if possible, cylinders should be removed from an area in which a fire has developed.
Disposal   Surplus gas or leaking cylinder can be vented slowly to air in a safe, open area or gas burnt off through a suitable burner.

Morpholine (diethylene oximide)

Colourless, mobile liquid with amine-like odour; bp 128 °C; miscible with water.

FLAMMABLE
IRRITATING TO SKIN, EYES AND RESPIRATORY SYSTEM

Avoid breathing vapour. Avoid contact with skin, eyes and clothing. TLV (skin) 20 ppm (70 mg m^{-3}) (skin).

Toxic effects   The vapour irritates the eyes and respiratory system. The liquid irritates the eyes and skin; it is also irritant when taken internally and may cause kidney and liver injury.

Hazardous reaction   Its addition to nitromethane makes it susceptible to initiation by a detonator (B417–418).

First aid   *Vapour inhaled :* standard treatment (p 109).
Affected eyes : standard treatment (p 109).
Skin contact : standard treatment (p 108).
If swallowed : standard treatment (p 109).
Fire hazard   Flash point 38 °C (open cup); ignition temp. 310 °C. *Extinguish fire with* dry powder, carbon dioxide or vaporising liquid.
Spillage disposal   Shut off all possible sources of ignition. Wear face-shield or goggles, and gloves. Mop up with plenty of water and run to waste, diluting greatly with running water. Ventilate area of spillage well to evaporate remaining liquid and dispel vapour.

Naphthalene

Hazardous reaction   Reacts explosively with dinitrogen pentaoxide (B577).

Naphthalene-1- and -2-diazonium salts

Hazardous reactions   They react with ammonium sulphide or hydrogen sulphide to form explosive compounds (B576).

1-Naphthylamine and salts

Colourless crystals when pure, darkening on exposure to light and air; the base, which has an unpleasant smell, melts at 50 °C and is insoluble in water; the hydrochloride is soluble in water. The use of 1-naphthylamine and its salts is controlled in the United Kingdom by The Carcinogenic Substances Regulations 1967.

> POISONOUS DUST
> TOXIC IN CONTACT WITH SKIN
> DANGER OF CUMULATIVE EFFECTS

Prevent inhalation of dust. Prevent contact with skin, eyes and clothing.

Toxic effects   The salts or their solutions irritate the eyes. **Chronic effects** Exposure to the dust or absorption through the skin may cause bladder tumours.

First aid   *Affected eyes:* standard treatment (p 109).
Skin contact: standard treatment (p 108).
If swallowed: standard treatment (p 109).

Spillage disposal   Wear breathing apparatus and gloves. Mix with sand and shovel mixture into glass, enamel or polythene vessel for dispersion in an excess of dilute hydrochloric acid (1 volume of concentrated acid diluted with 2 volumes of water). Allow to stand, with occasional stirring, for 24 hours and then run acid extract to waste, diluting greatly with running water and washing the sand. The sand can be disposed of as normal refuse.

2-Naphthylamine and salts

The use of these compounds in the United Kingdom is now prohibited under The Carcinogenic Substances Regulations 1968. Inhalation or absorption through the skin of the dust has been recognised as a cause of bladder tumours. It is not therefore considered appropriate to deal with their hazards more fully in this book.

Neopentane – *see* 2,2-Dimethylpropane

Nickel carbonyl – *see* Tetracarbonylnickel

Nickel salts

Green crystals or powder; mostly soluble in water.

> HARMFUL DUST
> IRRITATING TO SKIN AND EYES

Avoid breathing dust. Avoid contact with eyes and skin. TLV (metal and soluble compounds) 0.1 mg m^{-3} (as Ni).

Toxic effects　　The salts and their solutions will irritate the eyes. Assumed to be poisonous if taken by mouth. **Chronic effects** Continued skin contact can cause dermatitis.

First aid　　*Inhaled dust:* standard treatment (p 109).
Affected eyes: standard treatment (p 109).
Skin contact: standard treatment (p 108).
If swallowed: standard treatment (p 109).

Spillage disposal　　Soluble nickel salts should be dissolved in water and the solution run to waste diluting greatly with running water. Insoluble compounds can be mixed with sand and put out as ordinary refuse.

Nitric acid

Colourless or pale yellow fuming liquid; miscible with water.

CONTACT WITH COMBUSTIBLE MATERIALS MAY CAUSE FIRES
CAUSES SEVERE BURNS

Avoid breathing vapour. Prevent contact with eyes and skin. TLV 2 ppm (5 mg m^{-3}).

Toxic effects　　The vapour irritates all parts of the respiratory system. The vapour irritates and the liquid burns the eyes severely. The vapour and liquid burn the skin. If taken by mouth there is severe internal irritation and damage.

Hazardous reactions　　The range of vigorous, violent and explosive reactions, in which the stronger forms of nitric acid participate, is very wide. Several pages in *Handbook of reactive chemical hazards* are devoted to them, the other reactive participants including acetic acid, acetic anhydride, acetone, acetonitrile, acrylonitrile, alcohols, ammonia, aromatic amines, BrF_5, butanethiol, cellulose, crotonaldehyde, copper nitride, cyclohexylamine, dichloromethane, diethyl ether, 1,1-dimethylhydrazine, divinyl ether, fluorine, hydrazine, hydrocarbons, hydrogen iodide, hydrogen peroxide, ion exchange resins, iron(II) oxide, lactic acid, metal acetylides, metals, metal salicylates, 4-methylcyclohexanone, nitroaromatics, nitrobenzene, nitromethane, non-metal hydrides, non-metals, organic matter, phenylacetylene, phosphine derivatives, phosphorus halides, phthalic anhydride/sulphuric acid, polyalkenes, sulphur dioxide, sulphur halides, thioaldehydes, thioketones, thiophene, 2,4,6-trimethyltrioxane (B759–774).

First aid　　*Vapour inhaled:* standard treatment (p 109).
Affected eyes: standard treatment (p 109).
Skin contact: standard treatment (p 108).
If swallowed: standard treatment (p 109).

Spillage disposal　　Instruct others to keep at a safe distance. Wear breathing apparatus and gloves. Spread soda ash liberally over the spillage and mop up cautiously with water — run this to waste, diluting greatly with running water.

Nitric amide (nitramide, nitroamine)

Hazardous reactions　　Various preparations have been violent or explosive; drop of concentrated alkali added to solid causes a flame and explosive decomposition; it explodes on contact with concentrated sulphuric acid (B783).

Nitric oxide – *see* Nitrogen oxide

Nitrites of nitrogenous bases Review of group (B100).

Nitro-acyl halides Review of group (B101).

Nitroalkanes Review of group (B101).

Nitroanilines

The nitroanilines (*o-*, *m-* and *p-*) are yellow to orange-red crystals or powders; slightly soluble in water.

TOXIC IN CONTACT WITH SKIN
GIVES OFF POISONOUS VAPOUR

Avoid breathing dusts. Prevent contact with skin and eyes. TLV (skin) (*p*-isomer) 1 ppm (6 mg m^{-3}).

Toxic effects   Inhalation of dusts or excessive skin absorption of the solids may result in headache, flushing of the face, difficulty in breathing, nausea and vomiting; weakness, drowsiness, irritability and cyanosis may follow. Dermatitis may follow skin contact. The dusts will damage the eyes and effects similar to the above may be expected if the substances are taken by mouth.

First aid   *Dust inhaled:* standard treatment (p 109).
Affected eyes: standard treatment (p 109).
Skin contact: standard treatment (p 108).
If swallowed: standard treatment (p 109).

Spillage disposal   Wear face-shield or goggles, and gloves. Mix with sand and shovel mixture into glass, enamel or polythene vessel for dispersion in an excess of dilute hydrochloric acid (1 volume of concentrated acid diluted with 2 volumes of water). Allow to stand, with occasional stirring, for 24 hours and then run acid extract to waste, diluting greatly with running water and washing the sand. The sand can be disposed of as normal refuse.

Nitrobenzene

Pale yellow or yellow liquid with odour of bitter almonds; bp 211 °C; immiscible with water.

SERIOUS RISK OF POISONING BY INHALATION,
SWALLOWING OR SKIN CONTACT
DANGER OF CUMULATIVE EFFECTS

Avoid breathing vapour. Prevent contact with skin and eyes. TLV (skin) 1 ppm (5 mg m^{-3}).

Toxic effects   Inhalation of the vapour may cause a burning sensation in the chest, difficulty in breathing, cyanosis and, in severe cases, unconsciousness. The liquid injures the eyes and if it is absorbed excessively through the skin may give rise to the above symptoms. Drowsiness, vomiting, cyanosis and unconsciousness may follow ingestion.

Hazardous reactions   Addition of $AlCl_3$ to large volume of nitrobenzene containing 5% of phenol caused violent explosion; mixture with sodium chlorate is highly explosive; plant explosion occurred when it was being reacted with nitric acid and water; mixtures with dinitrogen tetraoxide once used as liquid explosives; (B479).

First aid   *Vapour inhaled:* standard treatment (p 109).
Affected eyes: standard treatment (p 109).
Skin contact: standard treatment (p 108).
If swallowed: standard treatment (p 109).

Spillage disposal   Instruct others to keep at a safe distance. Wear breathing apparatus and gloves. Apply dispersing agent if available and work to an emulsion with brush and water — run this to waste, diluting greatly with running water. If dispersant not available, absorb on sand, shovel into bucket(s) and transport to safe, open area for burial. Site of spillage should be washed thoroughly with water and soap or detergent.

m-Nitrobenzenediazonium perchlorate

Hazardous reaction   Explosive, very sensitive to heat and shock (B470).

2-Nitrobenzonitrile

Hazardous preparation   Explosion occurred (B523).

2-Nitrobenzoyl chloride

Hazardous preparation   Explosion of distillation residue (B521).

4-Nitrobenzoyl chloride

Yellow crystals with pungent odour; mp 75 °C; reacts with water forming benzoic and hydrochloric acids.

CAUSES BURNS
IRRITATING TO SKIN AND EYES

Avoid breathing vapour. Avoid contact with skin and eyes.

Toxic effects   These result mainly from its reaction with moisture on the tissues to form hydrochloric acid, which is the primary irritant. Thus irritation or burns may be caused at the point of contact.

First aid   *Vapour inhaled:* standard treatment (p 109).
Affected eyes: standard treatment (p 109).
Skin contact: standard treatment (p 108).
If swallowed: standard treatment (p 109).

Spillage   Wear face-shield or goggles, and gloves. Spread soda ash liberally over
disposal   the spillage and mop up cautiously with plenty of water — run to waste,
diluting greatly with running water.

3-Nitrobenzoyl nitrate

Hazardous   Explodes if heated rapidly (B523).
reaction

t-Nitrobutane – *see* 2-Methyl-2-nitropropane

Nitrocellulose – *see* Cellulose nitrate

Nitroethane

Colourless, oily liquid with pleasant odour; bp 114 °C; immiscible with water; may form explosive compounds with certain amines and alkalies.

FLAMMABLE
HARMFUL VAPOUR
IRRITATING TO SKIN, EYES AND RESPIRATORY SYSTEM

Avoid breathing vapour. Avoid contact with skin and eyes. TLV 100 ppm (310 mg m^{-3}).

Toxic effects   The vapour irritates the eyes and respiratory system. The liquid irritates
the eyes. Absorption by skin contact or ingestion may give rise to liver and
kidney damage.

First aid   *Vapour inhaled:* standard treatment (p 109).
Affected eyes: standard treatment (p 109).
Skin contact: standard treatment (p 108).
If swallowed: standard treatment (p 109).

Fire hazard   Flash point 28 °C; ignition temp. 415 °C. *Extinguish fire with* dry powder,
carbon dioxide or vaporising liquid.

Spillage   Shut off all possible sources of ignition. Instruct others to keep at a safe
disposal   distance. Wear breathing apparatus and gloves. Apply non-flammable
dispersing agent and work to an emulsion with brush and water – run this
to waste, diluting greatly with running water. If dispersant not available,
absorb on sand, shovel into bucket and transport to safe, open area for
atmospheric evaporation or burial. Site of spillage should be washed
thoroughly with water and soap or detergent.

2-Nitroethanol

Hazardous reaction   Explosion towards end of vacuum distillation (B316).

Nitrogen dioxide – *see* Dinitrogen tetraoxide

Nitrogen oxide (nitric oxide)

Colourless gas which, on release to atmosphere, is rapidly oxidised to nitrogen dioxide, a red gas with a pungent odour; bp −152 °C; about 7 cm³ dissolves in 100 g water at 0° C.

VERY TOXIC BY INHALATION

Prevent inhalation of gas. TLV 25 ppm (30 mg m⁻³).

Toxic effects   The toxic effects must be assumed to be those of nitrogen dioxide which is rapidly formed when nitric oxide mixes with air; nitrogen dioxide is a particularly dangerous gas because of its insidious mode of attack — several hours may elapse before the person exposed to it develops lung irritation and great discomfort. If gassing has been extensive, pulmonary oedema (flooding of lungs) may develop and this can result, several days later, in death.

Hazardous reactions   The liquid is sensitive to detonation in the absence of fuel; when mixed with carbon disulphide an explosion occurred; it ignited hydrogen/oxygen mixtures; pyrophoric chromium incandesces in the gas while Ca, K or U need heating before ignition occurs; acts as initiator to explosion of NCl₃; reacts with boron at ambient temperature with brilliant flashes while charcoal and phosphorus burn more brilliantly than in air (B870–872).

First aid   *It is important to treat any case of considerable exposure as serious and to obtain medical attention even if symptoms of respiratory irritation have not shown themselves. If exposed to other than very low concentrations of the gas, the casualty should be made to rest and kept warm until medical attention is received.*

Disposal   Surplus gas or leaking cylinder can be vented slowly into a water-fed scrubbing tower or column in a fume cupboard, or into a fume cupboard served by such a tower.

Nitrogen trichloride

Hazardous reaction   Wide variety of solids, liquids and gases will initiate the violent and often explosive decomposition of NCl₃ (B691).

Nitrogen trifluoride

Colourless gas with pungent, 'mouldy' smell; bp −129 °C; insoluble in water. Shock exposure of the gas to heat, flame or electric spark, or active contact with organic material, may cause fire and possibly explosion.

> VERY TOXIC BY INHALATION
> IRRITATING TO SKIN, EYES AND RESPIRATORY SYSTEM

Prevent inhalation of gas. Prevent contact with skin, eyes and clothing. TLV 10 ppm (29 mg m^{-3}).

Toxic effects   The gas irritates eyes, skin and respiratory system severely. Prolonged exposure to low concentrations may cause mottling of the teeth and skeletal change.

First aid   *Vapour inhaled:* standard treatment (p 109).
Affected eyes: standard treatment (p 109).
Skin contact: standard treatment (p 109).

Disposal   If the cylinder develops a leak it should be vented slowly in a well-ventilated fume cupboard until discharged.

Nitroguanidine

Hazardous reaction   Explosive though difficult to detonate (B250).

Nitroindane

Hazardous reaction   Crude mixture of 4- and 5-isomers obtained by nitration of indane is explosive in final stages of preparation (B568).

Nitromethane

Colourless, oily liquid; bp 101 °C; slightly soluble in water; may form shock-sensitive, explosive compounds with certain amines and alkalies.

> FLAMMABLE

Avoid breathing vapour. TLV 100 ppm (250 mg m^{-3}).

Toxic effects   These arise out of ingestion, and inhalation of the vapour. The vapour irritates the respiratory system while there will be irritation and damage internally if the liquid is taken by mouth.

Hazardous reactions   May explode by detonation, heat or shock; addition of bases or acids renders it susceptible to initiation by detonator; risk of explosion in preparation of 2-nitroethanol after reaction with formaldehyde; risk of explosion on heating with hydrocarbons; explosions have occurred with lithium perchlorate; mixture with nitric acid is extremely explosive (B244–246).

First aid	*Vapour inhaled:* standard treatment (p 109).
	Affected eyes: standard treatment (p 109).
	Skin contact: standard treatment (p 108).
	If swallowed: standard treatment (p 109).
Fire hazard	Flash point 35 °C; ignition temp. 418 °C. *Extinguish fire with* dry powder, carbon dioxide or vaporising liquid. Shock or heat may cause nitromethane to explode.
Spillage disposal	Shut off all possible sources of ignition. Instruct others to keep at a safe distance. Wear breathing apparatus and gloves. Apply non-flammable dispersing agent and work to an emulsion with brush and water — run this to waste, diluting greatly with running water. If dispersant not available, absorb on sand, shovel into buckets and transport to safe, open area for atmospheric evaporation or burial. Site of spillage should be washed thoroughly with water and soap or detergent.

N-Nitromethylamine

Hazardous reaction	Decomposed explosively by concentrated sulphuric acid (B250).

Nitroperchlorylbenzene

Hazardous reactions	Explosive; shock-sensitive (B469).

Nitrophenols

o-Nitrophenol is more volatile than the *m*- and *p*- compounds, and has a phenolic (carbolic) odour; the nitrophenols are pale yellow or yellow crystals or powders; sparingly soluble in water.

HARMFUL SUBSTANCE IF TAKEN
INTERNALLY OR IF IN CONTACT WITH SKIN

Avoid breathing vapour or dust. Prevent contact with skin and eyes.

Toxic effects	Excessive intake by inhalation of the dust, absorption by the skin or ingestion may cause irritation, headache, drowsiness and cyanosis. Effects may be cumulative. Assumed to irritate and injure the eyes.
First aid	*Vapour or dust inhaled:* standard treatment (p 109).
	Affected eyes: standard treatment (p 109).
	Skin contact: standard treatment (p 107). But also see note on phenol (p 108).
	If swallowed: standard treatment (p 109).
Spillage disposal	Wear face-shield or goggles, and gloves. Mix with sand and transport to a safe, open area for burial. Site of spillage should be washed thoroughly with water and soap or detergent.

2-Nitrophenylacetyl chloride

Hazardous preparation	Distillation residue liable to explode (B546).

p-Nitrophenylhydrazine

Orange-yellow powder; sparingly soluble in water.

HARMFUL IN CONTACT WITH SKIN

Prevent contact with skin and eyes.

Toxic effects	These have not been recorded but it must be assumed that intake, whether by inhalation of the dust, absorption through the skin or ingestion, will result in irritation or poisoning. Irritation of the eyes by the dust must also be assumed.
First aid	*Dust inhaled:* standard treatment (p 109). *Affected eyes:* standard treatment (p 109). *Skin contact:* standard treatment (p 108). *If swallowed:* standard treatment (p 109).
Spillage disposal	Wear face-shield or goggles, and gloves. Mix with sand and shovel mixture into glass, enamel or polythene vessel for dispersion in an excess of dilute hydrochloric acid (1 volume of concentrated acid diluted with 2 volumes of water). Allow to stand, with occasional stirring, for 24 hours and then run acid extract to waste, diluting greatly with running water and washing the sand. The sand can be disposed of as normal refuse.

3-Nitrophthalic acid

Hazardous preparation	Eruptive decomposition in nitration of phthalic anhydride (B545).

Nitroso compounds Review of group (B104).

Nitrosodimethylaniline
– *see* *N,N*-Dimethyl-*p*-nitrosoaniline

1-Nitroso-2-naphthol

Yellow-brown powder; insoluble in water.

IRRITATING TO SKIN, EYES AND RESPIRATORY SYSTEM

Avoid breathing dust. Avoid contact with skin and eyes.

Toxic effects	The dust irritates the respiratory system. The dust irritates the eyes and skin and may cause dermatitis. Assumed to be irritant and poisonous if taken by mouth.

First aid	*Dust inhaled:* standard treatment (p 109).
	Affected eyes: standard treatment (p 109).
	Skin contact: standard treatment (p 108).
	If swallowed: standard treatment (p 109).
Spillage disposal	Wear face-shield or goggles, and gloves. Mix with sand and transport to a safe, open area for burial. Site of spillage should be washed thoroughly with water and soap or detergent.

Nitrosophenols

Hazardous reactions	The *o-* isomer explodes on heating or in contact with concentrated acids. Barrels of the *p-* isomer heated spontaneously and caused a fire (B480–481).

3-Nitrosotriazenes Review of group (B105).

Nitrosyl chloride

Reddish brown gas with irritant, penetrating smell; bp −5.8 °C; decomposes on contact with water or moisture.

TOXIC BY INHALATION
IRRITATING TO SKIN, EYES AND RESPIRATORY SYSTEM

Prevent inhalation of gas. Prevent contact with skin.

Toxic effects	These have not been fully investigated but may be expected to be intermediate between those of chlorine and nitrogen oxides. The gas is intensely irritating but provides good warning of its dangers because of its penetrating odour.
Hazardous reaction	Cold sealed tube containing NOCl, Pt wire and traces of acetone exploded on warming up (B657).
First aid	*Vapour inhaled:* standard treatment (p 109).
	Affected eyes: standard treatment (p 109).
	Skin contact: standard treatment (p 109).
Disposal	Surplus gas or leaking cylinder can be vented slowly into a water-fed scrubbing tower or column in a fume cupboard, or into a fume cupboard served by such a tower.

Nitrosyl fluoride

Hazardous reactions	Reaction of mixture with unspecified haloalkene in pressure vessel at −78 °C caused it to rupture when moved; incandescent reactions with Sb, Bi, As, B, red P and Si; explodes on mixing with oxygen difluoride (B726).

Nitrosyl perchlorate

Hazardous reactions
Explodes on contact with pinene; it ignites acetone and ethanol and then explodes; with ether there is gassing followed by explosion; small amounts of primary aromatic amines are ignited on contact while explosions result with larger amounts; urea ignites on stirring with it (B658).

Nitrosylsulphuric acid

Hazardous reactions
In preparation from SO_2 and nitric acid, absence of N_2O_4 may result in explosion; explosion occurred during plant-scale diazotisation of a di-nitroaniline hydrochloride (B774).

Nitrosyl tribromide

Hazardous reaction
Powdered sodium antimonide ignites when dropped into the vapour (B213).

5-Nitrotetrazole

Hazardous reaction
An evaporated solution of the sodium salt exploded after 2 weeks (B230).

Nitrotoluenes

o-Nitrotoluene is a yellow liquid (bp 220 °C; *m*-nitrotoluene a yellow or brown-yellow liquid or crystalline solid (mp 15 °C) and *p*-nitrotoluene is a pale yellow or yellow crystalline solid (mp 52 °C). All are insoluble in water.

SERIOUS RISK OF POISONING BY INHALATION, SWALLOWING OR SKIN CONTACT
DANGER OF CUMULATIVE EFFECTS

Avoid breathing vapour or dust. Avoid contact with skin and eyes. TLV (skin) (*p*-) 5 ppm (30 mg m^{-3}).

Toxic effects
Inhalation of vapour or dust or skin absorption may cause difficulty in breathing, cyanosis and, in severe cases, unconsciousness. Assumed to injure the eyes and cause the above effects if taken by mouth.

First aid
Vapour or dust inhaled: standard treatment (p 109).
Affected eyes: standard treatment (p 109).
Skin contact: standard treatment (p 108).
If swallowed: standard treatment (p 109).

Spillage disposal
Wear face-shield or goggles, and gloves. Mix with sand and transport to a safe, open area for burial. Site of spillage should be washed thoroughly with water and soap or detergent.

Nitrourea

Hazardous reaction Unstable explosive (B247).

Nitrous acid

Hazardous reactions Causes phosphorus trichloride to explode (B759).

Nitrous fumes – *see* Dinitrogen tetraoxide

Nitrous oxide – *see* Dinitrogen oxide

Nitryl chloride

Hazardous reactions Reacts violently with ammonia or sulphur trioxide even at −75 °C; it attacks organic matter rapidly, sometimes explosively (B657).

Nitryl hypofluorite

Hazardous reaction Dangerously explosive as solid, liquid or gas (B727).

Nonacarbonyldiiron

Hazardous reaction Commercial material (flash point 35 °C) has autoignition temperature of 93 °C. (B565).

Oleum (fuming sulphuric acid, Nordhausen acid)

Colourless to yellow viscous fuming liquid; reacts violently with water.

CAUSES SEVERE BURNS
IRRITATING TO SKIN, EYES AND RESPIRATORY SYSTEM

Avoid inhaling vapour. Prevent contact with eyes and skin.

Toxic effects   The vapour irritates all parts of the respiratory system. The vapour irritates and the liquid burns the eyes severely. The vapour and liquid burn the skin. If swallowed there would be most severe internal irritation and damage.

First aid   *Vapour inhaled:* standard treatment (p 109).
Affected eyes: standard treatment (p 109).
Skin contact: standard treatment (p 108).
If swallowed: standard treatment (p 109).

Spillage disposal   Wear face-shield or goggles, and gloves. Spread soda ash liberally over the spillage and mop up cautiously with plenty of water — run to waste, diluting greatly with running water.

Organic peroxides Review of group (B110).

Organolithium reagents Review of group (B111).

Orthophosphoric acid – *see* Phosphoric acid

Osmic acid (osmium tetroxide)

Colourless or pale yellow crystals with pungent odour; mp 40 °C; soluble in water.

HARMFUL VAPOUR
CAUSES BURNS

Prevent inhalation of vapour. Prevent contact with eyes and skin. TLV as Os, 0.0002 ppm (0.002 mg m^{-3}).

Toxic effects   The vapour irritates all parts of the respiratory system. The vapour, solid and solution irritate and burn the eyes severely. The acid and its solution burn the skin. If taken by mouth there would be severe internal irritation and damage. **Chronic effects** Continued exposure to vapour causes disturbances of the vision; continued skin contact results in dermatitis and ulceration.

First aid   *Vapour inhaled:* standard treatment (p 109).
Affected eyes: standard treatment (p 109).
Skin contact: standard treatment (p 108).
If swallowed: standard treatment (p 109).

Spillage disposal   Wear face-shield or goggles, and gloves. Mop up with plenty of water and run to waste, diluting greatly with running water.

Oxalates

Most oxalates are colourless; ammonium, sodium and potassium oxalates are soluble in water.

HARMFUL IF TAKEN INTERNALLY

Avoid contact with eyes and skin.

Toxic effects	If swallowed there would be severe internal pain followed by collapse.
First aid	*Affected eyes:* standard treatment (p 109). *Skin contact:* standard treatment (p 108). *If swallowed:* standard treatment (p 109).
Spillage disposal	Wear face-shield or goggles, and gloves. Mop up with plenty of water and run to waste, diluting greatly with running water.

Oxalic acid

Colourless crystals; soluble in water.

HARMFUL IF TAKEN INTERNALLY

Avoid contact with eyes and skin. TLV 1 mg m^{-3}.

Toxic effects	The dust irritates the respiratory system. The dust and solutions irritate the eyes. If swallowed there would be severe internal pain followed by collapse.
First aid	*Dust inhaled:* standard treatment (p 109). *Affected eyes:* standard treatment (p 109). *Skin contact:* standard treatment (p 108). *If swallowed:* standard treatment (p 109).
Spillage disposal	Wear face-shield or goggles, and gloves. Mop up with plenty of water and run to waste, diluting greatly with running water.

Oxalyl chloride

Colourless fuming liquid with acrid odour; bp 64 °C; reacts vigorously with water forming hydrochloric and oxalic acids.

TOXIC BY INHALATION
CAUSES BURNS

Avoid breathing vapour. Prevent contact with skin and eyes.

Toxic effects	The vapour irritates the respiratory system severely. The vapour and liquid irritate the eyes. The liquid irritates the skin and may cause burns; must be assumed to be very irritant and poisonous if taken by mouth.
First aid	*Vapour inhaled:* standard treatment (p 109). *Affected eyes:* standard treatment (p 109). *Skin contact:* standard treatment (p 108). *If swallowed:* standard treatment (p 109).
Spillage disposal	Wear face-shield or goggles, and gloves. Spread soda ash liberally over the spillage and mop up cautiously with plenty of water — run to waste, diluting greatly with running water.

Oxodisilane

Hazardous reaction	Ignites in air (B815).

2-Oxohexamethyleneimine
– *see* ε-**Caprolactam**

Oxosilane

Hazardous   Ignites in air (B785).
reaction

Oxygen

Handbook of reactive chemical hazards (pp 904–912) covers many incidents with liquid and gaseous oxygen.

Oxygen fluorides Review of group (B119).

1-Oxyperoxy compounds Review of group (B119).

Ozone

Colourless gas or dark blue liquid (bp −111°C) with characteristic odour.

TOXIC BY INHALATION
IRRITATING TO RESPIRATORY SYSTEM

Avoid inhalation. TLV 0.1 ppm (0.2 mg m^{-3}).

Toxic effects   The gas irritates the upper respiratory system strongly and may cause headache. High concentrations have caused death by lung congestion in animals.

Hazardous   Solid and liquid ozone are highly explosive; reacts with alkenes to form
reactions   peroxides which are often explosive; gelatinous explosive ozonides formed with benzene, aniline and other aromatic compounds; reactions with bromine, N_2O_4 and HBr are usually explosive (B917–920).

First aid   *Gas inhaled:* standard treatment (p 109).

Ozonides Review of group (B120).

Paraformaldehyde

White crystalline powder; depolymerised on heating to form gaseous formaldehyde; slowly soluble in cold water, more readily in hot.

HARMFUL IF TAKEN INTERNALLY

Avoid breathing dust. Avoid contact with skin and eyes.

Toxic effects   The dust irritates the eyes, skin and respiratory system and is irritant and damaging if taken internally.

First aid   *Dust inhaled:* standard treatment (p 109).
Affected eyes: standard treatment (p 109).
If swallowed: standard treatment (p 109).

Spillage disposal   Wear face-shield or goggles, and gloves. Mop up with plenty of water and run to waste, diluting greatly with running water.

Paraldehyde (paracetaldehyde; 2,4,6-trimethyl-1,3,5-trioxan)

Colourless liquid with characteristic, aromatic odour; bp 128 °C; 1 part dissolves in about 8 parts of water at 25 °C.

FLAMMABLE

Avoid breathing vapour. Avoid contact with skin, eyes and clothing.

Toxic effects   The vapour has narcotic effects, and large doses taken internally cause prolonged unconsciousness, respiratory difficulty and pulmonary oedema.

Hazardous reaction   Reaction with nitric acid to form glyoxal is liable to become violent (B507).

First aid   *Vapour inhaled:* standard treatment (p 109).
Affected eyes: standard treatment (p 109).
Skin contact: standard treatment (p 108).
If swallowed: standard treatment (p 109).

Fire hazard   Flash point 36 °C; ignition temp. 238 °C. *Extinguish fire with* water spray, dry powder, carbon dioxide or vaporising liquid.

Spillage disposal   Shut off all possible sources of ignition. Instruct others to keep at a safe distance. Wear breathing apparatus and gloves. Apply non-flammable dispersing agent and work to an emulsion with brush and water — run this to waste, diluting greatly with running water. If dispersant not available, absorb on sand, shovel into bucket(s) and transport to safe, open area for burial. Site of spillage should be washed thoroughly with water and soap or detergent.

Pentaboranes

Hazardous reactions B_5H_9 ignites spontaneously if impure; forms shock-sensitive solution in solvents containing carbonyl, ether or ester functional groups and/or halogen substituents. B_5H_{11} ignites in air (B192).

Pentachloroethane

Colourless, heavy liquid of relatively low volatility with chloroform-like odour; bp 162 °C; immiscible with water.

GIVES OFF POISONOUS VAPOUR

Avoid breathing vapour. Avoid contact with skin, eyes and clothing.

Toxic effects   The vapour irritates the eyes, nose and lungs and may cause drowsiness, giddiness, headache, and in high concentrations, unconsciousness. Assumed to be poisonous if taken by mouth. **Chronic effects** Continuous breathing of low concentrations of vapour over a long period may cause jaundice by action on the liver. It may also affect the nervous system and the blood.

First aid   *Vapour inhaled:* standard treatment (p 109).
Affected eyes: standard treatment (p 109).
Skin contact: standard treatment (p 108).
If swallowed: standard treatment (p 109).

Spillage disposal   Instruct others to keep at a safe distance. Wear breathing apparatus and gloves. Apply dispersing agent if available and work to an emulsion with brush and water — run this to waste, diluting greatly with running water. If dispersant not available, absorb on sand, shovel into bucket(s) and transport to safe, open area for burial. Site of spillage should be washed thoroughly with water and soap or detergent.

Pentachlorophenol and sodium pentachlorophenate

Pentachlorophenol is a colourless to yellow crystalline solid with a phenolic (carbolic) odour; mp 190 °C; insoluble in water. The sodium salt is a buff powder or flaked solid, soluble in water.

SERIOUS RISK OF POISONING BY INHALATION, SWALLOWING OR SKIN CONTACT
CAUSES IRRITATION OF SKIN AND EYES

Avoid breathing dust. Prevent contact with skin and eyes. TLV (skin) 0.5 mg m⁻³.

Toxic effects   The dusts irritate the nose and eyes. Absorption through the skin or ingestion may cause accelerated breathing, feverishness and muscular weakness; in severe cases, convulsions and unconsciousness may follow. Skin contact may cause dermatitis.

First aid   *Dust inhaled:* standard treatment (p 109).
Affected eyes: standard treatment (p 109).
Skin contact: standard treatment (p 108).
If swallowed: standard treatment (p 109).

Spillage disposal   Wear face-shield or goggles, and gloves. Mix with sand and transport to a safe, open area for burial. Site of spillage should be washed thoroughly with water and soap or detergent.

Penta-1,3-diyne

**Hazardous
reaction**　Explodes on distillation (B435).

Pentafluoroguanidine

**Hazardous
reaction**　Extremely explosive (B224).

Pentane

Colourless liquid; bp 36 °C; almost insoluble in water.

EXTREMELY FLAMMABLE

Avoid breathing vapour. TLV 600 ppm (1800 mg m^{-3}).

Toxic effects　The vapour is narcotic in high concentrations.

First aid　*Vapour inhaled in high concentrations:* standard treatment (p 109).
Affected eyes: standard treatment (p 109).
If swallowed: standard treatment (p 109).

Fire hazard　Flash point −49°C; explosive limits 1.4–8%; ignition temp. 309 °C. *Extinguish fire with* foam, dry powder, carbon dioxide or vaporising liquid.

**Spillage
disposal**　Shut off all possible sources of ignition. Wear face-shield or goggles, and gloves. Apply non-flammable dispersing agent if available and work to an emulsion with brush and water — run this to waste, diluting greatly with running water. If dispersant not available, absorb on sand, shovel into bucket(s) and transport to safe, open area for atmospheric evaporation. Ventilate site of spillage well to evaporate remaining liquid and dispel vapour.

Pentane-2,4-dione (acetylacetone)

Colourless or slightly yellow liquid with pleasant smell; bp 140 °C. one part dissolves in about eight parts of water at 25 °C.

FLAMMABLE

Avoid breathing vapour. Avoid contact with skin and eyes.

Toxic effects　The vapour may irritate the respiratory system. The liquid irritates the eyes. The liquid may irritate the skin. If swallowed, may cause internal irritation and more severe damage.

First aid *Vapour inhaled:* standard treatment (p 109).
Affected eyes: standard treatment (p 109).
Skin contact: standard treatment (p 108).
If swallowed: standard treatment (p 109).

Fire hazard Flash point 34 °C. *Extinguish fire with* water spray, dry powder, carbon dioxide or vaporising liquids.

Spillage disposal Shut off all possible sources of ignition. Wear face-shield and gloves. Apply non-flammable dispersing agent and work to an emulsion with brush and water — run this to waste diluting greatly with running water. If dispersant not available, absorb on sand, shovel into bucket(s) and transport to safe open area for burial. Ventilate area well to evaporate remaining liquid and dispel vapour.

Pentan-1-ol and pentan-2-ol (n- and s-amyl alcohols)

Colourless liquids; bp 138 °C and 119 °C respectively; both are slightly soluble in water.

FLAMMABLE

Avoid breathing vapour. Avoid contact with skin and eyes.

Toxic effects The vapours may irritate the eyes and respiratory system. The alcohols are not readily absorbed by the skin but, as liquids, will irritate the eyes. They are both harmful when taken by mouth and may cause giddiness, headache coughing, deafness and delirium.

First aid *Vapour inhaled in high concentrations:* standard treatment (p 109).
Affected eyes: standard treatment (p 109).
If swallowed: standard treatment (p 109).

Fire hazard Pentan-1-ol: flash point 33 °C; explosive limits 1–10%; ignition temp. 300 °C. Pentan-2-ol: flash point 40 °C. *Extinguish fire with* water spray, foam, dry powder, carbon dioxide or vaporising liquid.

Spillage disposal Shut off all possible sources of ignition. Wear face-shield or goggles, and gloves. Apply non-flammable dispersing agent if available and work to an emulsion with brush and water — run this to waste, diluting greatly with running water. If dispersant not available, absorb on sand, shovel into bucket(s) and transport to safe, open area for atmospheric evaporation or burial. Ventilate site of spillage well to evaporate remaining liquid and dispel vapour.

Pentan-2-one (methyl propyl ketone)

Colourless liquid with acetone/ether-like odour; bp 102 °C; about 5.5 g dissolves in 100 g water at 25 °C.

HIGHLY FLAMMABLE

Avoid breathing vapour. Avoid contact with skin and eyes. TLV 200 ppm (700 mg m^{-3}).

Toxic effects The vapour may irritate the eyes and respiratory system; it is narcotic in high concentrations. The liquid will irritate the eyes and is assumed to be irritant and narcotic if taken internally.

Hazardous reaction	May react explosively with BrF₃ (B451).
First aid	*Vapour inhaled:* standard treatment (p 109). *Affected eyes:* standard treatment (p 109). *Skin contact:* standard treatment (p 108). *If swallowed:* standard treatment (p 109).
Fire hazard	Flash point 7 °C; explosive limits 1.5–8%; ignition temp. 505 °C. *Extinguish fire with* water spray, dry powder, carbon dioxide or vaporising liquid.
Spillage disposal	Shut off all possible sources of ignition. Wear face-shield or goggles, and gloves. Apply non-flammable dispersing agent if available and work to an emulsion with brush and water—run this to waste, diluting greatly with running water. If dispersant not available, absorb on sand, shovel into bucket(s) and transport to safe, open area for atmospheric evaporation. Ventilate site of spillage well to evaporate remaining liquid and dispel vapour.

Pentan-3-one (diethyl ketone)

Hazardous reaction	Gives shock- and heat-sensitive oily peroxide with hydrogen peroxide and nitric acid (B451).

Pent-2-en-4-yn-3-ol

Hazardous reaction	Distillation residue exploded at over 90 °C (B441).

Pentyl acetate (n-amyl acetate and commercial amyl acetate)

(consisting mainly of 3-methylbutyl acetate). Colourless liquids with pear-like odour; bp 148 °C and 128–132 °C respectively; slightly soluble in water.

FLAMMABLE

Avoid breathing vapour. Avoid contact with eyes. TLV 100 ppm (525 mg m⁻³).

Toxic effects	High concentrations of the vapour irritate the eyes and may cause headache and fatigue. Must be considered harmful if taken by mouth.
First aid	*Vapour inhaled in high concentrations:* standard treatment (p 109). *Affected eyes:* standard treatment (p 109). *If swallowed:* standard treatment (p 109).
Fire hazard	Flash points 25 °C and 23 °C respectively; explosive limits 1–7.5% respectively; ignition temps. 379 °C and 380 °C. *Extinguish fire with* dry powder, carbon dioxide or vaporising liquid.
Spillage disposal	Shut off all possible sources of ignition. Wear face-shield or goggles, and gloves. Apply non-flammable dispersing agent if available and work to an emulsion with brush and water—run this to waste, diluting greatly with running water. If dispersant not available, absorb on sand, shovel into bucket(s) and transport to safe, open area for atmospheric evaporation or burial. Ventilate site of spillage well to evaporate remaining liquid and dispel vapour.

Perchlorates

The perchlorates of ammonium, magnesium, sodium and potassium are colourless crystalline solids, all except that of potassium being readily soluble in water. Ammonium perchlorate is explosive when dry.

EXPLOSIVE WHEN MIXED WITH COMBUSTIBLE MATERIAL

Avoid contact with combustible materials. Avoid contact with skin, eyes and clothing.

Toxic effects   The dust and strong solutions will irritate the skin, eyes and respiratory system. They are also irritant and harmful if taken internally.

Hazardous reactions   Mixtures of inorganic perchlorates with combustible materials are readily ignited; mixtures with finely divided combustible materials frequently react explosively. Organic perchlorates are self-contained explosives. Specific accidents with perchlorates of Ag, Be, Ga, K, Li, Mg, Mn, Na, Ni, Pb, Ti, U, V and Zr are recorded in *Handbook of reactive chemical hazards* (*see* Metal Oxohalogenates, p 91) and some of these appear in this text.

First aid   *Affected eyes:* standard treatment (p 109).
Skin contact: standard treatment (p 108).
If swallowed: standard treatment (p 109).

Spillage disposal   Shovel into bucket of water and run solution to waste, diluting greatly with running water. Site of spillage should be washed thoroughly to remove all oxidant, which is liable to render any organic matter (particularly paper, textiles and wood) with which it comes into contact dangerously combustible when dry. Clothing wetted with the solution should be washed thoroughly.

Perchloric acid

Colourless liquid; miscible with water; commonly sold commercially as 60–62% or 72% constant boiling acid.

CONTACT WITH COMBUSTIBLE MATERIAL MAY CAUSE FIRE
CAUSES BURNS

Prevent contact with eyes and skin.

Toxic effects   Burns the eyes and skin severely. Assumed to cause severe internal irritation and damage if taken by mouth.

Hazardous reactions   Everyone using perchloric acid or perchlorates should become familiar with the chapter 'Handling perchloric acid and perchlorates' by Everett and Graf in *Handbook of laboratory safety* edited by N. V. Steere (The Chemical Rubber Co., Cleveland). There is a very long history of accidents with perchloric acid in particular and a study of the summaries contained on pp 641–648 of *Handbook of reactive chemical hazards* shows that the partners of perchloric acid in explosions and violent reactions include acetic acid, acetic anhydride, alcohols, aniline/formaldehyde, antimony-(III) compounds, bismuth and alloys, carbon, cellulose and derivatives, dehydrating agents, diethyl ether, glycerol/lead oxide cement, glycols and their ethers, hydrogen, iodides, iron(II) sulphate, ketones, nitric acid/organic matter, oleic acid, phosphine, pyridine, sodium phosphinate, organic sulphoxides, trichloroethylene, trizinc diphosphide (B641–648).

The distillation of 70–72% commercial perchloric acid under vacuum concentrates the acid to 75% and over, and these stronger acids are liable to explode on heating; such distillation, when justified, should be carried out at atmospheric pressure so that the constant-boiling (and much safer) acid is obtained. The use of wooden bench tops for analytical and other work with perchloric acid and perchlorates is very undesirable. Such benches, used for these purposes regularly and subjected to occasional perchloric acid spills, have been known to explode on percussion.

First aid *Affected eyes:* standard treatment (p 109).
Skin contact: standard treatment (p 108).
If swallowed: standard treatment (p 109).

Spillage disposal Wear face-shield or goggles, and gloves. Spread soda ash liberally over the spillage and mop up cautiously with plenty of water — run to waste, diluting greatly with running water.

Perchloroethylene – *see* Tetrachloroethylene

Perchloromethyl mercaptan – *see* Trichloromethane sulphenyl chloride

Perchlorylbenzene

Hazardous reaction After an interval, a mixture with aluminium chloride suddenly exploded (B477).

Perchloryl compounds Review of group (B122).

Perchloryl fluoride

Hazardous reactions Reference to booklet on safe handling given; at temperatures over 100 °C, calcium acetylide, potassium cyanide, potassium thiocyanate may react explosively in the gas; may react with carbon powder, foamed elastomers, lampblack, sawdust very violently even at −78 °C; reactions with hydrocarbons, H_2S, NO, SCl_2 or vinylidene chloride may be explosive above 100 °C, or if mixtures are ignited; explosive compounds may be formed by reaction with nitrogenous bases (B632).

1-Perchlorylpiperidine

Hazardous reactions Has exploded on storage, on heating or in contact with piperidine (B449).

Periodic acid

Hazardous reaction   1.5M solutions in dimethyl sulphoxide explode after short interval (B755).

Permanganic acid

Hazardous reaction   Likely to explode at room temperature (B758).

Peroxides and peroxides in solvents

Reviews of groups (B123).

Peroxidisable compounds

A particularly useful review (B123–126).

Peroxoacids and salts Reviews of group (B126–127).

Peroxodisulphuric acid

Hazardous reactions   May cause aniline, benzene, ethanol, ether, nitrobenzene or phenol to explode on contact (B797).

Peroxomonophosphoric acid

Hazardous reactions   80% solution may ignite organic matter (B805).

Peroxomonosulphuric acid

Hazardous reactions   Explosions have occurred alone, or with acetone, alcohols (secondary and tertiary), aromatic compounds, catalysts; it rapidly carbonises wool or cellulose while cotton is ignited (B796).

Peroxonitric acid

Hazardous reaction   Decomposes explosively at −30 °C (B774).

Peroxyacetic acid

Hazardous reaction   Explodes violently at 110 °C (B311).

Peroxy acids Review of group (B128).

Peroxybenzoic acid

Hazardous reaction   Explodes weakly on heating (B528).

Peroxycarbonate esters Review of group (B128).

Peroxyesters Review of group (B129).

Peroxyhexanoic acid

Hazardous reaction   Explodes and ignites on rapid heating (B506).

Peroxytrifluoroacetic acid
– *see* Trifluoroperoxyacetic acid

Petroleum spirit (petroleum ether)

Petroleum spirits are supplied in a variety of fractions boiling between 30 °C and 160 °C; all are colourless liquids whose smell varies with the volatilities of the fractions.

EXTREMELY FLAMMABLE or
HIGHLY FLAMMABLE

Avoid breathing vapour. Avoid contact with skin and eyes. TLV 500 ppm (2000 mg m^{-3}).

Toxic effects   Inhalation of high concentrations of the vapour, particularly of the lower boiling fractions, can cause intoxication, headache, nausea and coma. The liquids irritate the eyes and skin contact results in defatting of the area of contact, increasing the risk of dermatitis from other agents. If taken by mouth they may cause burning sensation, vomiting, diarrhoea and drowsiness.

First aid	*Vapour inhaled in high concentrations:* standard treatment (p 109).
	Affected eyes: standard treatment (p 109).
	Skin contact: standard treatment (p 108).
	If swallowed: standard treatment (p 109).
Fire hazard	Flash point (lower fractions) below −17 °C; explosive limits approx. 1–6%; ignition temperatures range from about 290 °C. *Extinguish fire with* foam, dry powder, carbon dioxide or vaporising liquid.
Spillage disposal	Shut off all possible sources of ignition. Wear face-shield or goggles and gloves. Apply non-flammable dispersing agent if available and work to an emulsion with brush and water — run this to waste, diluting greatly with running water. If dispersant not available, absorb on sand, shovel into bucket(s) and transport to safe, open area for atmospheric evaporation. Ventilate site of spillage well to evaporate remaining liquid and dispel vapour.

Phenacyl bromide (ω-bromoacetophenone)

Colourless to greenish crystals; mp 50 °C; insoluble in water.

and

Phenacyl chloride (ω-chloroacetophenone)

Colourless to yellow crystals or powder; mp 54 °C; insoluble in water.

IRRITATING TO SKIN, EYES AND RESPIRATORY SYSTEM

Avoid breathing vapour. Avoid contact with skin and eyes.

Toxic effects	The vapour and dust irritate the respiratory system and skin; they irritate the eyes severely. Assumed to be very irritant if taken by mouth.
First aid	*Vapour inhaled:* standard treatment (p 109).
	Affected eyes: standard treatment (p 109).
	Skin contact: standard treatment (p 108).
	If swallowed: standard treatment (p 109).
Spillage disposal	Wear face-shield or goggles, and gloves. Mix with sand and transport to a safe, open area for burial. Site of spillage should be washed thoroughly with water and soap or detergent.

Phenetidines – *see* Ethoxyanilines

Phenol (carbolic acid)

Colourless to pink crystalline substance with distinctive odour; mp 43 °C; somewhat soluble in water (1 g dissolves in about 15 cm³ water at 25 °C).

TOXIC IF TAKEN INTERNALLY
CAUSES BURNS

Avoid breathing vapour. Avoid contact with skin and eyes. TLV (skin) 5 ppm (19 mg m⁻³).

Toxic effects The vapour irritates the respiratory system and eyes. Skin contact causes softening and whitening followed by the development of painful burns; its rapid absorption through the skin may cause headache, dizziness, rapid and difficult breathing, weakness and collapse. If taken by mouth it causes severe burns, abdominal pain, nausea, vomiting and internal damage. **Chronic effects** The inhalation of vapour over a long period may cause digestive disturbances, nervous disorders, skin eruptions and damage to the liver and kidneys; dermatitis may result from prolonged contact with weak solutions. (*See* notes in Ch 6 p 87 and Ch 7 p 91.)

First aid *Vapour inhaled:* standard treatment (p 109).
Affected eyes: standard treatment (p 109).
Skin contact: standard treatment (p 108). But see also note on phenol (p 107).
If swallowed: standard treatment (p 109).

Spillage disposal Wear face-shield or goggles, and gloves. Mix with sand and transport to a safe, open area for burial. Site of spillage should be washed thoroughly with water and soap or detergent.

Phenol-disulphonic and -sulphonic acids
- see **Sulphonic acids**

Phenoxyacetylene

Hazardous reaction Samples rapidly heated in sealed tubes to about 100 °C exploded (B546).

Phenylacetonitrile (benzyl cyanide)

Colourless to yellow liquid; bp 234 °C; insoluble in water.

SERIOUS RISK OF POISONING BY INHALATION, SWALLOWING OR SKIN CONTACT
IRRITATING TO EYES

Avoid breathing vapour. Avoid contact with skin and eyes.

Toxic effects Inhalation of the vapour may cause pallor, faintness, headache and possibly vomiting. The liquid irritates the eyes and may irritate the skin. If taken by mouth, internal irritation and poisoning must be assumed.

Hazardous reaction When sodium hypochlorite solution was used to destroy acidified residues, a violent explosion occurred (B547).

First aid *Vapour inhaled:* standard treatment (p 109).
Affected eyes: standard treatment (p 109).
Skin contact: standard treatment (p 108).
If swallowed: standard treatment (p 109).

Spillage disposal Wear face-shield or goggles, and gloves. Apply dispersing agent if available and work to an emulsion with brush and water—run this to waste, diluting greatly with running water. If dispersant not available, absorb on sand, shovel into bucket(s) and transport to safe, open area for burial. Site of spillage should be washed thoroughly with water and soap or detergent.

Phenylchlorodiazirine

Hazardous reaction Is about three times as shock-sensitive as glyceryl nitrate (B524).

Phenylenediamines

p-Phenylenediamine — pale mauve to mauve crystals or powder. *m*-Phenylenediamine — colourless to brown or black crystals or lumps. The *m*-isomer is moderately soluble in water, the *p*-isomer only sparingly so.

> HARMFUL SUBSTANCE IF TAKEN INTERNALLY
> OR IF IN CONTACT WITH SKIN
> IRRITATING TO THE EYES

Avoid contact with skin and eyes. TLV (skin) (*p*-) 0.1 mg m^{-3}.

Toxic effects Eye irritation and injury follow contact. May irritate the skin causing dermatitis; the skin becomes blackened. Assumed to be poisonous if taken by mouth.

First aid *Affected eyes:* standard treatment (p 109).
Skin contact: standard treatment (p 108).
If swallowed: standard treatment (p 109).

Spillage disposal Wear face-shield or goggles, and gloves. Shovel into a glass, enamel or polythene vessel for dispersion in an excess of dilute hydrochloric acid (one volume concentrated acid diluted with two volumes of water). Allow to stand, with occasional stirring, for 24 hours and then run the base hydrochloride solution to waste, diluting greatly with running water. Site of spillage should be washed with water and detergent.

Phenylethylene – *see* Styrene

Phenylhydrazine and hydrochloride

Phenylhydrazine is a yellow to red-brown liquid or solid (mp 19 °C); insoluble in water. The hydrochloride consists of colourless to brown crystals or powder, soluble in water.

> HARMFUL SUBSTANCE IF TAKEN INTERNALLY
> OR IF IN CONTACT WITH SKIN
> IRRITATING TO THE EYES

Avoid inhalation of vapour or dust. Prevent contact with skin and eyes. TLV (skin) 5 ppm (22 mg m^{-3}).

Toxic effects Absorption resulting from inhalation, skin contact or ingestion may result in blood and liver damage, giving rise to nausea, vomiting and jaundice. Dermatitis may follow skin contact. Must be considered injurious to the eyes. (*See* note in Ch 6 p 85)

First aid   *Vapour or dust inhaled:* standard treatment (p 109).
Affected eyes: standard treatment (p 109).
Skin contact: standard treatment (p 108).
If swallowed: standard treatment (p 109).

Spillage disposal   Wear face-shield or goggles, and gloves. Mix the base with sand and shovel mixture into glass, enamel or polythene vessel for dispersion in an excess of dilute hydrochloric acid (1 volume of concentrated acid diluted with 2 volumes of water). Allow to stand, with occasional stirring, for 24 hours and then run acid extract to waste, diluting greatly with running water and washing the sand. The sand can be disposed of as normal refuse. A spillage of the hydrochloride can be mopped up with water and the solution run to waste, diluting greatly with running water.

Phenylmercury salts
– see **Mercury compounds**

Phenylsilver

Hazardous reactions   Explodes on warming to room temperature or on light friction (B475).

5-Phenyltetrazole

Hazardous reaction   Explodes on attempted distillation (B527).

Phosgene (carbonyl chloride) and solutions

The gas is colourless and has a musty smell. As a pale yellow liquid it boils at 7.6 °C. Solution of the gas in toluene (HIGHLY INFLAMMABLE) is available commercially.

TOXIC BY INHALATION

Do not breath gas. Avoid contact of solutions with skin and eyes. TLV 0.1 ppm (0.4 mg m^{-3}).

Toxic effects   The gas produces delayed secretion of fluid into the lung (pulmonary oedema) when inhaled and there may be delay of several hours before effects develop. These include breathlessness, cyanosis and the coughing up of frothy fluid. (*See* note in Ch 6 p 83)

Hazardous reaction   Mixture with potassium is shock-sensitive (B218).

First aid   *Vapour inhaled:* standard treatment (p 109).
Affected eyes: standard treatment (p 109).
Skin contact with solution: standard treatment (p 108).
If solution swallowed: standard treatment (p 109).

Spillage disposal

(a) Solution in toluene: Shut off all possible sources of ignition. Instruct others to keep at a safe distance. Wear breathing apparatus and gloves. Apply non-flammable dispersing agent and work to an emulsion with brush and water — run this to waste, diluting greatly with running water. If dispersant not available, absorb on sand, shovel into bucket(s) and transport to safe, open area for burial. Site of spillage should be washed thoroughly with water and soap or detergent. Ventilate area of spillage well to dispel remaining vapour. (b) Gas: Surplus gas or leaking cylinder can be vented slowly into a water-fed scrubbing tower or column in a fume cupboard, or into a fume cupboard served by such a tower.

Phosphine (hydrogen phosphide)

Colourless gas with smell somewhat like that of rotting fish; bp $-88\,°C$; slightly soluble in water. With a trace of P_2H_4 in it, phosphine (PH_3) is spontaneously flammable in air, burning with a luminous flame.

EXTREMELY FLAMMABLE
TOXIC BY INHALATION

Prevent inhalation of gas.

Toxic effects

The sequence of effects of inhalation have been stated to be pain in chest, sensation of coldness, weakness, vertigo, shortness of breath, lung damage, convulsions, coma and death.

Hazardous reactions

The impure gas ignites spontaneously in air; ignition occurs on contact with chlorine or bromine or their aqueous solutions; oxidised explosively by fuming nitric acid (B806).

First aid

Gas inhaled: standard treatment (p 109).

Fire hazard

Impure phosphine must be regarded as spontaneously flammable. Since the gas is supplied in a cylinder, turning off the valve will reduce any fire involving it; if possible, cylinders should be removed quickly from an area in which a fire has developed.

Disposal

Surplus gas or leaking cylinder can be vented slowly to air in a safe, open area or gas burnt off through a suitable burner in a fume cupboard.

Phosphinic acid

Hazardous reaction

Redox reaction with mercury(II) oxide is explosive (B805).

Phosphonium perchlorate

Hazardous reaction

An explosive, sensitive to moist air, friction or heat (B650).

Phosphoric acid (orthophosphoric acid)

Colourless viscous liquid (88–93%) or moist white crystals (100%); miscible with water.

CAUSES BURNS

Avoid contact with eyes and skin. TLV 1 mg m^{-3}.

Toxic effects	The liquid burns the eyes severely. The liquid burns the skin. If taken by mouth there would be severe internal irritation and damage.
First aid	*Affected eyes:* standard treatment (p 109). *Skin contact:* standard treatment (p 108). *If swallowed:* standard treatment (p 109).
Spillage disposal	Wear face-shield or goggles, and gloves. Spread soda ash liberally over the spillage and mop up cautiously with plenty of water — run to waste, diluting greatly with running water.

Phosphoric oxide
– *see* Phosphorus pentaoxide

Phosphorus, red and white (yellow)

Red (amorphous) phosphorus sublimes at 416 °C; insoluble in water. White (yellow) phosphorus is a pale yellow, waxy, translucent solid; mp 44 °C; usually stored under water; if exposed to the air it rapidly ignites giving off fumes of the oxide.

RED	**WHITE (YELLOW)**
HIGHLY FLAMMABLE EXPLOSIVE WHEN MIXED WITH OXIDISING SUBSTANCES	SPONTANEOUSLY FLAMMABLE IN AIR SERIOUS RISK OF POISONING BY INHALATION OR SWALLOWING Prevent contact with skin and eyes. TLV 0.1 mg m^{-3}.

Toxic effects	Red phosphorus is relatively harmless physiologically unless it contains the white allotrope. The vapour from ignited phosphorus irritates the nose, throat, lungs and eyes. Solid white phosphorus burns the skin and eyes and causes severe internal damage if taken by mouth. **Chronic effects** Continued absorption of small amounts can cause anaemia, intestinal weakness, pallor, bone and liver damage.
Hazardous reactions	Reference to brochure on handling is given; contact of both allotropes with boiling alkalies evolves phosphine, which usually ignites in air; yellow phosphorus reacts vigorously with chlorosulphuric acid, accelerating to explosion — the same occurs at a higher temperature with the red allotrope; both forms ignite on contact with fluorine and chlorine, the red allotrope in liquid bromine; the white form explodes in liquid bromine or chlorine, and ignites in contact with bromine vapour or solid iodine; the white form reacts far more readily than the red with air or oxygen; both forms react violently with hydrogen peroxide if incompletely immersed;

both forms are liable to explode when mixed with perchlorates, chlorates, bromates and subjected to friction, impact or heat; explodes with chromyl chloride when moist and on impact or grinding with ammonium, mercury(I) or silver nitrates or potassium permanganate; incandescent or violent reactions occur with some metals and oxides (B927–932).

First aid	*Vapour of burning phosphorus inhaled:* standard treatment (p 109).
	Affected eyes: standard treatment (p 109).
	Skin contact: standard treatment (p 108).
	If swallowed: standard treatment (p 109).
Fire hazard	Water is the best medium for fighting a phosphorus fire caused by its spontaneous ignition.
Spillage disposal	*Red form:* Moisten with water, shovel into a bucket, transport to a safe, isolated area where the moisture can be allowed to dry off and the phosphorus burnt off. This applies to small spillages only.
	White form: Wear face-shield or goggles, and gloves; cover with wet sand, shovel into bucket and remove to safe, open, isolated area where the phosphorus can be burnt off under supervision after drying out. Small spillages of phosphorus can be burnt off in a fume cupboard.

Phosphorus(III) oxide

Hazardous reactions	Reacts violently with disulphur dichloride, liquid bromine, PCl_5, sulphur, sulphuric acid, hot water; ignition likely with air and oxygen at elevated temperatures (B920).

Phosphorus oxychloride
– *see* Phosphoryl chloride

Phosphorus pentachloride

White to pale yellow, fuming, crystalline masses with pungent, unpleasant odour. Violently decomposed by water with formation of hydrochloric acid and phosphoric acid.

IRRITANT VAPOUR AND DUST
CAUSES BURNS

Avoid breathing vapour and dust. Prevent contact with eyes and skin. TLV 1 mg m^{-3}.

Toxic effects	Vapour and dust severely irritate the mucous membranes and all parts of the respiratory system. The vapour severely irritates and the solid burns the eyes. The vapour and solid burn the skin. If taken by mouth there would be severe internal irritation and damage. **Chronic effects** Continued exposure to low concentrations of vapour may cause damage to lungs.
Hazardous reactions	Reacts violently with P_2O_5, nitrobenzene and water, explosively with sodium and urea; ignites with Al powder and fluorine (B701–702).

First aid	*Vapour inhaled:* standard treatment (p 109). *Affected eyes:* standard treatment (p 109). *Skin contact:* standard treatment (p 108). *If swallowed:* standard treatment (p 109).
Spillage disposal	Wear goggles and gloves. Mix with dry sand, shovel into an enamel or polythene bucket, transport to a safe, open area and add, a little at a time, to a large volume of water; after reaction is complete, run to waste, diluting greatly with running water. Dispose of sand as normal refuse. Wash spillage site with water.

Phosphorus pentafluoride

Colourless gas, fuming strongly in moist air, with a pungent smell; bp −85 °C; it is hydrolysed by water with formation of hydrogen fluoride and, ultimately, phosphoric acid.

**SERIOUS RISK OF POISONING BY INHALATION, SWALLOWING OR SKIN CONTACT
CAUSES SEVERE BURNS**

Prevent inhalation of gas. Prevent contact with skin and eyes.

Toxic effects	Because of its ready hydrolysis by moisture or water to hydrogen fluoride its toxic effects are similar to the latter. It irritates severely the eyes and respiratory system and may cause burns to the eyes. It irritates the skin and, if exposure has been considerable, painful burns may develop after an interval.
First aid	*Gas inhaled:* standard treatment (p 109). *Affected eyes:* standard treatment (p 109). *Skin contact:* standard treatment (p 108).
Disposal	Surplus gas or leaking cylinder can be vented slowly into a water-fed scrubbing tower or column in a fume cupboard, or into a fume cupboard served by such a tower.

Phosphorus pentaoxide

(phosphorus(V) oxide; phosphoric oxide)

White crystalline deliquescent powder; reacts violently with water.

CAUSES SEVERE BURNS

Prevent contact with eyes and skin.

Toxic effects	The dust irritates all parts of the respiratory system, burns the eyes severely and burns the skin. If taken by mouth there would be severe internal irritation and damage.
Hazardous reactions	Reference given to brochure on safe handling; attempted dehydration of 95% formic acid caused rapid evolution of CO; reacts vigorously with HF below 20 °C; dry mixtures with Na_2O or CaO react violently if warmed or moistened; reaction with warm Na or K is incandescent; reaction explosive when heated with Ca; reacts violently with concentrated hydrogen peroxide; ignition occurs on contact with F_2O; the reaction with water is highly exothermic and enough to ignite combustible materials in contact (B925).

First aid   *Dust inhaled:* standard treatment (p 109).
                  Affected eyes: standard treatment (p 109).
                  Skin contact: standard treatment (p 108).
                  If swallowed: standard treatment (p 109).

Spillage   Wear face-shield or goggles, and gloves. Mix with dry sand, shovel into
disposal   an enamel or polythene bucket, transport to a safe, open area and add, a
                  little at a time, to a large volume of water; after reaction is complete, run
                  to waste, diluting greatly with running water. Dispose of sand as normal
                  refuse. Wash site of spillage with water.

Phosphorus(III) and (V) sulphides

Yellow crystalline powders with a peculiar odour; react with water, forming phosphorus acids
and hydrogen sulphide.

HIGHLY FLAMMABLE
CORROSIVE

Avoid breathing dust. Prevent contact with eyes and skin. TLV (P_4S_{10}) 1 mg m^{-3}.

Toxic effects   The dust irritates the mucous membranes and all parts of the respiratory
                      system; it irritates the eyes and burns the skin. Assumed to cause severe
                      internal irritation and damage if taken by mouth.

Hazardous P_4S_{10} ignites by friction, sparks or flames; it heats and may ignite with
reactions   limited amounts of water (B932).

First aid   *Dust inhaled:* standard treatment (p 109).
                      Affected eyes: standard treatment (p 109).
                      Skin contact: standard treatment (p 108).
                      If swallowed: standard treatment (p 109).

Fire hazard   Ignition temp. (P_4S_3) 282 °C, (P_4S_{10}) 142 °C, the latter temperature being
                      readily obtained by friction. *Extinguish fire with* dry powder, carbon dioxide
                      or sand.

Spillage   Wear face-shield or goggles, and gloves. Spread soda ash liberally over
disposal   the spillage and mop up cautiously with plenty of water — run to waste,
                      diluting greatly with running water.

Phosphorus tribromide

Colourless fuming liquid; bp 175 °C; reacts violently with water, forming hydrobromic acid and
phosphorous acid.

HARMFUL VAPOUR
CAUSES BURNS
IRRITATING TO SKIN, EYES AND RESPIRATORY SYSTEM

Avoid breathing vapour. Prevent contact with eyes and skin.

Toxic effects   Vapour severely irritates the mucous membranes and all parts of the
                      respiratory system. The vapour irritates and the liquid burns the eyes.
                      The vapour and liquid burn the skin. Assumed to cause severe internal
                      irritation and damage if taken by mouth. **Chronic effects** Continued
                      exposure to low concentrations of vapour may cause damage to lungs.

Hazardous reactions	Reacts rapidly with warm water, violently with limited quantities; sodium floats on PBr_3 without reaction but the addition of a little water causes violent explosion; potassium ignites on contact with the liquid or vapour (B213).
First aid	*Vapour inhaled:* standard treatment (p 109). *Affected eyes:* standard treatment (p 109). *Skin contact:* standard treatment (p 108). *If swallowed:* standard treatment (p 109).
Spillage disposal	Instruct others to keep at a safe distance. Wear breathing apparatus and gloves. Spread soda ash liberally over the spillage and mop up cautiously with water — run this to waste, diluting greatly with running water.

Phosphorus trichloride

Colourless fuming liquid; bp 75 °C; violently decomposed by water, forming hydrochloric acid and phosphorous acid.

HARMFUL VAPOUR
CAUSES BURNS
IRRITATING TO SKIN, EYES AND RESPIRATORY SYSTEM

Avoid breathing vapour. Prevent contact with eyes and skin. TLV 0.5 ppm (3 mg m^{-3}).

Toxic effects	Vapour severely irritates the mucous membranes and all parts of the respiratory system. The vapour and liquid burn the eyes and skin. Assumed to cause severe internal irritation and damage if taken by mouth. **Chronic effects** Continued exposure to low concentrations of vapour may cause lung damage.
Hazardous reactions	Residue in preparation of acetyl chloride from PCl_3 and acetic acid may decompose violently with evolution of flammable phosphine; reacts violently or explosively with dimethyl sulphoxide; potassium ignites in PCl_3, molten sodium explodes on contact; may explode on contact with chromyl chloride; ignites on contact with fluorine; explodes with nitric acid; reacts violently or explosively with sodium peroxide; reacts violently with water with liberation of some diphosphane which ignites (B695).
First aid	*Vapour inhaled:* standard treatment (p 109). *Affected eyes:* standard treatment (p 109). *Skin contact:* standard treatment (p 108). *If swallowed:* standard treatment (p 109).
Spillage disposal	Instruct others to keep at a safe distance. Wear breathing apparatus and gloves. Mix with dry sand, shovel into enamel or polythene bucket(s) and add, a little at a time, to a large volume of water; when reaction is complete run solution to waste, diluting greatly with running water. Wash area of spillage thoroughly with water.

Phosphorus tricyanide

Hazardous reaction	Explosions occurred during vacuum sublimation (B379).

Phosphorus trifluoride

Hazardous reactions Ignites on contact with fluorine; explosion has occurred at low temperatures with dioxygen difluoride (B741).

Phosphorus trisulphide
– *see* Phosphorus(III) sulphide

Phosphoryl chloride (phosphorus oxychloride)

Colourless fuming liquid; bp 107 °C; violently decomposed by water, forming hydrochloric acid and phosphoric acid.

HARMFUL VAPOUR
CAUSES BURNS
IRRITATING TO SKIN, EYES AND RESPIRATORY SYSTEM

Avoid breathing vapour. Prevent contact with eyes and skin.

Toxic effects The vapour severely irritates the mucous membranes and all parts of the respiratory system; there may be sudden or delayed pulmonary oedema. The vapour severely irritates the eyes and skin. The liquid burns the eyes and skin. Assumed to cause severe internal irritation and damage if taken by mouth.

Hazardous reactions There is a considerable delay in its reaction with water which ultimately becomes violent — closed or only slightly vented vessels therefore represent a hazard; may react explosively with dimethyl sulphoxide; zinc dust ignites on contact with a little $POCl_3$ (B693–694).

First aid *Vapour inhaled:* standard treatment (p 109).
Affected eyes: standard treatment (p 109).
Skin contact: standard treatment (p 108).
If swallowed: standard treatment (p 109).

Spillage disposal Instruct others to keep at a safe distance. Wear breathing apparatus and gloves. Mix with dry sand, shovel into enamel or polythene bucket(s) and add, a little at a time, to a large volume of water; when reaction is complete run solution to waste, diluting greatly with running water. Wash area of spillage thoroughly with water.

Phthalic anhydride

White crystalline needles; mp 131 °C; dissolves in hot water forming phthalic acid.

IRRITANT DUST

Avoid contact with skin and eyes. TLV 2 ppm (12 mg m^{-3}).

Toxic effects The dust irritates the eyes, skin and respiratory system. It will cause internal irritation if taken by mouth.

Hazardous reactions	Mixture with anhydrous CuO exploded on heating; nitration with nitric acid/sulphuric acid presents dangers; mixture with sodium nitrite will explode on heating (B544).
First aid	*Affected eyes:* standard treatment (p 109).
	Skin contact: standard treatment (p 108).
	If swallowed: standard treatment (p 109).
Spillage disposal	Clear up with dust-pan and brush or bucket and shovel, placing in large volume of water. The acid solution produced on standing may be run to waste, diluting greatly with running water.

Phthaloyl diazide

Hazardous reaction	Extremely explosive (B544).

Phthaloyl peroxide

Hazardous reactions	Detonatable by impact or by melting at 123 °C (B545).

Picolines – *see* Methylpyridines

Picrates Review of group (B130).

Picric acid

Picric acid (yellow crystals) should be kept moist with not less than half its own weight of water. It is commonly used in alcoholic solutions in the laboratory.

> RISK OF EXPLOSION BY SHOCK, FRICTION, FIRE OR OTHER SOURCE OF IGNITION
> FORMS VERY SENSITIVE EXPLOSIVE METALLIC COMPOUNDS
> HARMFUL IF TAKEN INTERNALLY

Avoid contact with skin and eyes. TLV (skin) 0.1 mg m^{-3}.

Toxic effects	Skin contact may result in dermatitis. Poisonous if taken by mouth. **Chronic effects** Absorption through the skin or inhalation of dust over a long period may result in skin eruptions, headache, nausea, vomiting or diarrhoea; the skin may become yellow.
Hazardous reactions	Explosive which is usually stored as water-wet paste; forms salts with many metals some of which (Pb, Hg, Cu or Zn) are sensitive to heat friction or impact; contact of acid with concrete floors may form friction-sensitive calcium salt (B466).

First aid	*Affected eyes:* standard treatment (p 109). *Skin contact:* standard treatment (p 108). *If swallowed:* standard treatment (p 109).
Spillage disposal	Wear face-shield or goggles, and gloves. Moisten well with water and mix with sand. Transport to isolated area for burial. Site of spillage must be washed thoroughly with water and detergent.

Picryl azide

Hazardous reaction	Explodes weakly on impact (B464).

Piperazine and piperazine hydrate

Both take the form of colourless crystals; mp 106 °C and 44 °C respectively; both are very soluble in water, the solutions being strongly alkaline.

CAUSE BURNS
IRRITATING TO SKIN, EYES AND RESPIRATORY SYSTEM

Avoid breathing vapour. Avoid contact with skin, eyes and clothing.

Toxic effects	The vapour irritates the eyes and respiratory system. The solution irritates and may burn the eyes and skin. The solid and aqueous solutions will cause internal burning and damage if taken by mouth.
First aid	*Vapour inhaled:* standard treatment (p 109). *Affected eyes:* standard treatment (p 109). *Skin contact:* standard treatment (p 108). *If swallowed:* standard treatment (p 109).
Spillage disposal	Wear face-shield or goggles, and gloves. Mop up with plenty of water and run to waste, diluting greatly with running water.

Piperidine (hexahydropyridine)

Colourless liquid with amine-like odour, soapy to the touch; bp 106 °C; miscible with water.

HIGHLY FLAMMABLE
IRRITATING TO SKIN, EYES AND RESPIRATORY SYSTEM

Avoid breathing vapour. Avoid contact with skin, eyes and clothing.

Toxic effects	The vapour irritates the eyes and respiratory system. The liquid irritates the eyes and skin and is assumed to be irritant and harmful if taken internally.
First aid	*Vapour inhaled:* standard treatment (p 109). *Affected eyes:* standard treatment (p 109). *Skin contact:* standard treatment (p 108). *If swallowed:* standard treatment (p 109).
Fire hazard	Flash point 16 °C. *Extinguish fire with* dry powder, carbon dioxide or vaporising liquid.

Spillage disposal	Shut off all possible sources of ignition. Wear face-shield or goggles, and gloves. Mop up with plenty of water and run to waste, diluting greatly with running water. Ventilate area of spillage well to evaporate remaining liquid and dispel vapour.

Pivaloyl azide

Hazardous reaction	Explodes on warming (B447).

Platinum (metal)

Hazardous reactions	Used Pt catalysts present explosion risk; mixture of Pt and As in sealed tube at 270 °C exploded; addition of Pt-black catalyst to ethanol caused ignition; addition of Pt-black to concentrated hydrogen peroxide may cause explosion; reacts violently with Li at about 540 °C; platinum burns in phosphorus vapour below red-heat; finely divided Pt reacts incandescently when heated with selenium or tellurium (B934–936).

Platinum compounds Review of group (B130).

Polynitroalkyl and polynitroaryl compounds Reviews of classes (B131).

Polyperoxides Review of group (B132).

Potassium (metal)

Soft silvery white masses normally coated with a grey oxide or hydroxide skin; mp 64 °C; reacts violently with water with the formation of potassium hydroxide and hydrogen gas, which will ignite. Explosions have occurred when old heavily-crusted potassium metal has been cut with a knife. Such old stock should be disposed of by dissolving, uncut, in isopropyl alcohol.

CONTACT WITH WATER LIBERATES HIGHLY FLAMMABLE GASES CAUSES BURNS

Avoid contact with skin and eyes.

Toxic effects	The metal in contact with moisture on the skin can cause thermal and caustic burns. In the same way the potassium hydroxide resulting from the reaction of the metal with water can cause burning of the skin and eyes.

Hazardous reactions   Reference given to detailed procedures for safe handling; comparison of properties of K and Na reviewed showing former to be invariably more hazardous; K reacts readily with CO to form explosive carbonyl; K does not react with air or oxygen in complete absence of moisture, but in its presence oxidation becomes fast and melting and ignition take place; prolonged but restricted access to air results in the formation of coatings of yellow superoxide on top of the monoxide—percussion or dry cutting of the metal brings the metal underneath in contact with the superoxide and a very violent explosion occurs; the disposal of scrap potassium is described, the use of ethanol or methanol being excluded because of their violent reactions with the metal; the metal reacts vigorously with various forms of carbon and, if air is present, ignition and explosion are likely; mixtures of the metal with a wide range of halocarbons are shock sensitive and may explode with violence on slight impact; K ignites in fluorine and dry chlorine; it explodes violently in liquid bromine and incandesces in the vapour; mixture of metal with anhydrous HI explodes violently; reaction with mercury to form amalgams is vigorous or violent; reacts violently or explosively with many metal halides or oxides; reactions with non-metal halides and oxides are usually explosive as are those with oxidants; the reaction with sulphuric acid is explosive; with water, the heat evolved is enough to ignite the hydrogen evolved—large pieces of the metal explode on the surface of water scattering burning particles over a wide area (B835–841).

First aid   *Affected eyes:* standard treatment (p 109).
Skin contact: standard treatment (p 108).
Mouth contact: standard treatment (p 109).

Fire hazard   A fire resulting from the ignition of potassium metal is best extinguished by smothering it with dry soda ash.

Spillage disposal   Instruct others to keep at a safe distance. Wear face-shield or goggles, and gloves. Cover with dry soda ash, shovel into dry bucket, transport to safe, open area and add, a little at a time, to a large excess of dry propan-2-ol. Leave to stand for 24 hours and run solution to waste, diluting greatly with water.

Potassium amide

Hazardous reactions   More violently reactive than sodium amide; reaction with water is violent and ignition may occur; explodes when heated with potassium nitrite under vacuum (B781).

Potassium antimonate
– *see* Antimony compounds

Potassium arsenate and arsenite
– *see* Arsenic and compounds

Potassium azide

Hazardous reactions   Melts on heating then decomposes evolving nitrogen, the residue igniting with a feeble explosion; reacts violently with manganese dioxide on warming; explodes at 120 °C when heated in liquid sulphur dioxide (B847).

Potassium bichromate
- *see* Chromates and dichromates

Potassium bisulphate
- *see* Potassium hydrogensulphate

Potassium chlorate

Colourless crystals or crystalline powder; mp 368 °C; soluble in water.

EXPLOSIVE WHEN MIXED WITH COMBUSTIBLE MATERIAL
HARMFUL IF TAKEN INTERNALLY

Avoid contact with skin, eyes and clothing.

Toxic effects   When taken internally may irritate the intestinal tract and kidneys.

Hazardous reactions   Fabric gloves (wrongly used) became impregnated with the chlorate during handling operations and were subsequently ignited by cigarette ash; molten $KClO_3$ ignites HI gas; impact, *etc* — sensitive explosives are formed by mixtures of the chlorate with tricopper diphosphide, trimercury tetraphosphide, finely divided Al, Cu, Mg or Zn, many metal sulphides, or thiocyanates, arsenic, carbon, phosphorus, sulphur or other readily oxidised materials; addition of $KClO_3$ in portions to H_2SO_4 below 60 °C and above 200 °C causes brisk effervescence — between these temperatures, explosions are caused by the ClO_2 produced; a mixture with sodium amide exploded (B652–655).

First aid   *Affected eyes:* standard treatment (p 109).
If swallowed: standard treatment (p 109).

Spillage disposal   Shovel into bucket of water and run solution to waste, diluting greatly with running water. Site of spillage and clothing should be washed thoroughly to remove all oxidant.

Potassium chromate
- *see* Chromates and dichromates

Potassium cyanide
- see **Cyanides (water-soluble)**

Potassium dichromate
- see **Chromates and dichromates**

Potassium dioxide (potassium superoxide)

Hazardous reactions Stable when pure, but impact-sensitive when formed as a layer on metallic potassium; reacts violently with diselenium dichloride; explosions may occur with hydrocarbons; oxidises arsenic, antimony, copper, tin or zinc with incandescence (B848).

Potassium ferricyanide and ferrocyanide
- see **Potassium hexacyanoferrate (3— and 4—)**

Potassium fluoride *- see* **Fluorides** (water soluble)

Potassium fluorosilicate
- see **Hexafluorosilicic acid and salts**

Potassium hexacyanoferrate (3— and 4—)

Hazardous reactions Contact of the hexacyanoferrate(4—) with ammonia may be explosive; mixtures of the (4—) salt with CrO_3 explode on heating above 196 °C; both of the salts mixed with sodium nitrite explode on heating (B461).

Potassium hydride

Hazardous reactions Ignites on contact with fluorine; reacts slowly with oxygen in dry air, more rapidly in moist (B756).

Potassium hydrogen fluoride
- see **Fluorides** (water soluble)

Potassium hydrogensulphate (potassium bisulphate)

Colourless crystals or fused masses; soluble in water.

CAUSES BURNS
IRRITATING TO SKIN AND EYES

Avoid contact with skin and eyes.

Toxic effects	The solid and its solutions severely irritate and burn the eyes and skin If taken by mouth there is severe internal irritation and damage.
First aid	*Affected eyes:* standard treatment (p 109). *Skin contact:* standard treatment (p 108). *If swallowed:* standard treatment (p 109).
Spillage disposal	Wear face-shield or goggles, and gloves. Spread soda ash liberally over the spillage and mop up cautiously with plenty of water — run to waste, diluting greatly with running water.

Potassium hydroxide (caustic potash)

Colourless sticks, flakes, powder or pellets soluble in water.

CAUSES SEVERE BURNS

Prevent contact with eyes and skin.

Toxic effects	The solid and its solutions severely irritate and burn the eyes and skin. If taken by mouth there would be severe internal irritation and damage.
Hazardous reactions	Liquid or gaseous ClO_2 will explode on contact with solid KOH or its concentrated solution; germanium is oxidised by fused KOH with incandescence; as little as two flakes of moist KOH dropped on 2 kg potassium peroxodisulphate caused a vigorous self-sustained fire; may cause the explosion of peroxidised tetrahydrofuran and must not be used to dry this (B756).
First aid	*Affected eyes:* standard treatment (p 109). *Skin contact:* standard treatment (p 108). *If swallowed:* standard treatment (p 109).
Spillage disposal	Wear face-shield or goggles, and gloves. Shovel into large volume of water in an enamel or polythene vessel and stir to dissolve; run the solution to waste diluting greatly with running water.

Potassium methylamide

Hazardous reactions	Extremely hygroscopic and pyrophoric; may explode on contact with air (B249).

Potassium nitrate

Colourless crystals, soluble in water.

EXPLOSIVE WHEN MIXED WITH COMBUSTIBLE MATERIALS

Keep out of contact with all combustible material.

Hazardous reactions
Mixture of KNO_3 and calcium silicide is readily ignited, high temperature primer capable of initiating many high temperature reactions; mixtures with powdered Ti, Sb or Ge explode on heating; mixtures with antimony trisulphide, barium or calcium sulphides all explode on heating; mixtures with arsenic disulphide or molybdenum disulphide are detonatable; finely divided mixture with boron ignites and explodes on percussion; the oldest explosive, gunpowder, is a mixture of the nitrate with charcoal and sulphur; mixture with red phosphorus reacts vigorously on heating; mixtures of the nitrate with arsenic explode when ignited; in general, mixtures with combustible materials are readily ignited, if finely divided they can be explosive (B846–847).

Fire hazard
Mixtures of potassium nitrate and combustible materials are readily ignited; mixtures with finely divided combustible materials can react explosively.

Spillage disposal
Shovel into bucket of water and run solution to waste, diluting greatly with running water. Site of spillage should be washed thoroughly to remove all oxidant, which is liable to render any organic matter (particularly wood, paper and textiles) with which it comes into contact, dangerously combustible when dry. Clothing wetted with the solution should be washed thoroughly.

Potassium nitride

Hazardous reactions
Usually ignites in air; on heating with phosphorus a highly flammable mixture is formed which evolves ammonia and phosphine with water; with sulphur a similar mixture is formed which evolves ammonia and hydrogen sulphide on contact with water (B852).

Potassium nitrite

White or slightly yellow, deliquescent, crystalline solid; mp 441 °C; soluble in water.

CONTACT WITH COMBUSTIBLE MATERIAL MAY CAUSE FIRE
HARMFUL SUBSTANCE IF TAKEN INTERNALLY

Avoid contact with clothing and other absorbent fabrics.

Hazardous reactions
Addition of ammonium sulphate to the fused nitrite causes effervescence and ignition; addition of boron to fused nitrite causes violent decomposition; explosion occurs when mixture with potassium amide is heated under vacuum (B845).

Spillage disposal
Shovel into bucket of water and run solution to waste diluting greatly with running water. Site of spillage should be washed thoroughly to remove all oxidant. Clothing wetted by a solution of the nitrite should be washed thoroughly.

Potassium perchlorate – *see* Perchlorates

Potassium permanganate

Hazardous reactions

Risk of explosion with acetic acid or acetic anhydride; antimony ignites on grinding in a mortar with the solid permanganate, while arsenic explodes; contact of glycerol with solid $KMnO_4$ caused a vigorous fire; reaction with concentrated hydrochloric acid is occasionally explosive; hydrogen peroxide and permanganate (solid) react violently and cause a fire; mixtures with phosphorus or sulphur react explosively on grinding or heating respectively; explosion follows the addition of concentrated sulphuric acid to damp $KMnO_4$ (B843–845).

Potassium peroxodisulphate

Hazardous reactions

A vigorous self-sustaining fire resulted when as little as two flakes of moist KOH was dropped on the surface of 2 kg of the dry salt. Dry salt gives off oxygen rapidly at 100 °C, but at only 50 °C when wet (B851–852).

Potassium selenate and selenite – *see* Selenium and compounds

Potassium tellurite – *see* Tellurium and compounds

Propadiene (allene)

Hazardous reaction

May decompose explosively under a pressure of 2 bar (2×10^5 Pa) (B351).

Propanal – *see* Propionaldehyde

Propane

Colourless gas which burns with a smoky flame; bp −42 °C; 100 volumes of water dissolve 6.5 volumes of the gas at 17.8 °C and 753 mmHg.

EXTREMELY FLAMMABLE

Avoid breathing gas.

Toxic effects   The gas is anaesthetic in high concentrations but is not toxic.

First aid   *Gas inhaled in high concentrations:* standard treatment (p 109).
Fire hazard   Flash point −104 °C; explosive limits 2.2–9.5%; ignition temp. 468 °C. Since the gas is supplied in a cylinder, turning off the valve will reduce any fire involving it; if possible, cylinders should be removed quickly from an area in which a fire has developed.
Disposal   Surplus gas or leaking cylinder can be vented slowly to air in a safe, open area or gas burnt off through a suitable burner.

Propan-1-ol (propyl alcohol)

Colourless liquid with alcoholic odour; bp 97 °C; miscible with water.

FLAMMABLE

Avoid breathing vapour. TLV 200 ppm (500 mg m⁻³).

Toxic effects   The vapour may irritate the eyes and respiratory system and may be narcotic in high concentrations. The liquid irritates the eyes and is narcotic if taken internally.

First aid   *Vapour inhaled:* standard treatment (p 109).
Affected eyes: standard treatment (p 109).
If swallowed: standard treatment (p 109).
Fire hazard   Flash point 25 °C; explosive limits 2.1–13.5%; ignition temp. 433 °C. *Extinguish fire with* water spray, dry powder, carbon dioxide or vaporising liquid.
Spillage disposal   Shut off all possible sources of ignition. Wear face-shield or goggles, and gloves. Mop up with plenty of water and run to waste, diluting greatly with running water. Ventilate area well to evaporate remaining liquid and dispel vapour.

Propan-2-ol (isopropyl alcohol, 'IPA')

Colourless liquid with somewhat alcoholic odour; bp 82 °C; miscible with water.

HIGHLY FLAMMABLE

Avoid breathing vapour. TLV 400 ppm (980 mg m⁻³).

Toxic effects   Inhalation of the vapour in high concentrations and ingestion of the liquid may result in headache, dizziness, mental depression, nausea, vomiting, narcosis, anaesthesia and coma; the fatal dose is about 100 cm³. The liquid may damage the eyes severely.

Hazardous reactions   In the preparation of aluminium isopropoxide, the dissolution of Al in propan-2-ol becomes vigorously exothermic but its onset is frequently delayed—only small amounts of Al should be present until the reaction starts; in recovering propan-2-ol from the reduction of crotonaldehyde by aluminium isopropoxide, a violent explosion occurred which was attributed to either by-product peroxidised diisopropyl ether or peroxidised crotonaldehyde; an explosion occurred during the thawing of a frozen mixture of

369

the alcohol with trinitromethane; ignition occurs on grinding with CrO_3; it forms explosive mixtures with hydrogen peroxide; a bottle of the alcohol, exposed to sunlight for a long period, became 0.36M in peroxide and potentially explosive (B371).

First aid	*Vapour inhaled in high concentrations:* standard treatment (p 109).
	Affected eyes: standard treatment (p 109).
	If swallowed: standard treatment (p 109).
Fire hazard	Flash point 12 °C; explosive limits 2.3–12.7%; ignition temp. 399 °C. *Extinguish fire with* water spray, dry powder, carbon dioxide or vaporising liquid.
Spillage disposal	Shut off all possible sources of ignition. Wear face-shield or goggles, and gloves. Mop up with plenty of water and run to waste, diluting greatly with running water. Ventilate area of spillage well to evaporate remaining liquid and dispel vapour.

Propargyl bromide – *see* 3-Bromopropyne

Propene (propylene)

Colourless gas which burns with a sooty flame; bp −48 °C; slightly soluble in water.

EXTREMELY FLAMMABLE

Avoid breathing gas.

Toxic effects	The gas is a simple asphyxiant which is anaesthetic if inhaled in high concentrations; it has no significant toxic properties.
First aid	*Vapour inhaled in high concentrations:* standard treatment (p 109).
Fire hazard	Flash point −108 °C; explosive limits 2–11.1%; ignition temp. 460 °C. Since the gas is supplied in a cylinder, turning off the valve will reduce any fire involving it; if possible, cylinders should be removed quickly from an area in which a fire has developed.
Spillage disposal	Surplus gas or leaking cylinder can be vented slowly to air in a safe, open area or gas burnt off through a suitable burner.

Propene ozonide

Hazardous reaction	Liable to explode at ambient temperature (B366).

Prop-2-en-1-ol – *see* Allyl alcohol

β-Propiolactone (oxetan-2-one)

Colourless liquid; bp 155 °C with decomposition; slowly hydrolysed in water to hydracrylic acid.

TOXIC IN CONTACT WITH SKIN
IRRITATING TO SKIN, EYES AND RESPIRATORY SYSTEM

Prevent contact with skin, eyes and clothing. Prevent inhalation of vapour.

Toxic effects   It is highly irritant to the skin and has produced skin cancer in experimental animals. It is assumed to be similarly irritant and dangerous if taken by mouth. It must be regarded as a potential human carcinogen.

First aid   *Affected eyes:* standard treatment (p 109).
Skin contact: standard treatment (p 108).
If swallowed: standard treatment (p 109).

Spillage disposal   Wear face-shield or goggles, and gloves. Apply dispersing agent if available and work to an emulsion with brush and water—run this to waste, diluting greatly with running water. If dispersant not available, absorb on sand, shovel into bucket(s) and transport to safe, open area for burial. Site of spillage should be washed thoroughly with water and soap or detergent.

Propiolaldehyde

Hazardous reaction   Undergoes almost explosive polymerisation in the presence of alkalies or pyridine (B346).

Propionaldehyde (propanal)

Colourless liquid with suffocating odour; bp 49 °C; 1 part dissolves in about 5 parts water at 20 °C. It may form explosive peroxides.

EXTREMELY FLAMMABLE
IRRITATING TO EYES AND RESPIRATORY SYSTEM

Avoid breathing vapour.

Toxic effects   The vapour irritates the respiratory system. The liquid irritates the eyes and is assumed to be irritant and damaging if taken internally.

First aid   *Vapour inhaled:* standard treatment (p 109).
Affected eyes: standard treatment (p 109).
If swallowed: standard treatment (p 109).

Fire hazard   Flash point −9 °C: explosive limits 2.9–17%; ignition temp. 207 °C. *Extinguish fire with* water spray, dry powder, carbon dioxide or vaporising liquid.

Spillage disposal   Shut off all possible sources of ignition. Instruct others to keep at a safe distance. Wear breathing apparatus and gloves. Apply non-flammable dispersing agent and work to an emulsion with brush and water—run this to waste, diluting greatly with running water. If dispersant not available, absorb on sand, shovel into bucket(s) and transport to safe, open area for atmospheric evaporation. Site of spillage should be washed thoroughly with water and soap or detergent.

Propionic acid

Colourless, oily liquid with rancid odour; bp 141 °C; miscible with water.

> FLAMMABLE
> CAUSES BURNS
> IRRITATING TO SKIN, EYES AND RESPIRATORY SYSTEM

Avoid contact with skin and eyes.

Toxic effects	The vapour irritates the eyes and respiratory system. The liquid burns the skin and eyes. The acid is corrosive if taken by mouth.
First aid	*Affected eyes:* standard treatment (p 109). *Skin contact:* standard treatment (p 108). *If swallowed:* standard treatment (p 109).
Fire hazard	Flash point 54 °C; ignition temp. 513 °C. *Extinguish fire with* water spray, dry powder, carbon dioxide or vaporising liquid.
Spillage disposal	Shut off all possible sources of ignition; wear face-shield or goggles, and gloves. Spread soda ash liberally over the spillage and mop up cautiously with plenty of water — run to waste, diluting greatly with running water.

Propionyl chloride

Colourless liquid with acrid odour; bp 80 °C; reacts with water forming propionic and hydro-chloric acids.

> HIGHLY FLAMMABLE
> CAUSES BURNS
> IRRITATING TO SKIN, EYES AND RESPIRATORY SYSTEM

Avoid breathing vapour. Prevent contact with skin and eyes.

Toxic effects	The vapour irritates the eyes and respiratory system. The liquid burns the skin and eyes. Assumed to cause severe internal irritation and damage if taken by mouth.
First aid	*Vapour inhaled:* standard treatment (p 109). *Affected eyes:* standard treatment (p 109). *Skin contact:* standard treatment (p 108). *If swallowed:* standard treatment (p 109).
Fire hazard	Flash point 12 °C. *Extinguish fire with* dry powder, carbon dioxide or vaporising liquid.
Spillage disposal	Shut off all possible sources of ignition. Instruct others to keep at a safe distance. Wear breathing apparatus and gloves. Absorb on sand, shovel into bucket(s), transport to safe, open area and tip into large volume of water; leave to decompose before decanting the water to waste, diluting greatly with running water. Site of spillage should be ventilated after washing thoroughly with water and soap or detergent.

isoPropyl acetate – *see* Isopropyl acetate

Propyl alcohol – *see* Propan-1-ol

isoPropyl alcohol – *see* Propan-2-ol

Propylamines

The primary propylamines (n- and iso-) are colourless, volatile liquids with an ammoniacal odour; bp 49 °C and 32 °C respectively; miscible with water.

> EXTREMELY FLAMMABLE
> HARMFUL VAPOUR
> IRRITATING TO SKIN, EYES AND RESPIRATORY SYSTEM

Avoid breathing vapour. Avoid contact with skin and eyes. TLV (iso-) 5 ppm (12 mg m^{-3}).

Toxic effects The vapour irritates the respiratory system. The vapour and liquid irritate and may burn the eyes. The liquid may cause skin burns. Assumed to be very irritant and poisonous if taken by mouth.

First aid *Vapour inhaled:* standard treatment (p 109).
Affected eyes: standard treatment (p 109).
Skin contact: standard treatment (p 108).
If swallowed: standard treatment (p 109).

Fire hazard Flash point (n- and iso-) −37 °C; explosive limits (n- and iso-) about 2–10%; ignition temp. 318 °C (n-), 420 °C (iso-). *Extinguish fire with* water spray, dry powder, carbon dioxide or vaporising liquid.

Spillage disposal Shut off all possible sources of ignition. Instruct others to keep at a safe distance. Wear breathing apparatus and gloves. Mop up with plenty of water and run to waste, diluting greatly with running water. Ventilate area well to evaporate remaining liquid and dispel vapour.

Propyl chlorides – *see* Chloropropanes

Propyl cyanide – *see* Butyronitrile

3-Propyldiazirine

Hazardous reaction Exploded on attempted distillation at about 75 °C (B406).

Propylene oxide (1,2-epoxypropane; 2-methyloxiran)

Colourless, volatile liquid with ethereal odour; bp 35 °C; somewhat soluble in water.

> EXTREMELY FLAMMABLE
> CAUSES BURNS
> IRRITATING TO SKIN, EYES AND RESPIRATORY SYSTEM

Avoid breathing vapour. Avoid contact with skin and eyes. TLV 100 ppm (240 mg m^{-3}).

Toxic effects The vapour irritates the eyes and respiratory system. The liquid irritates the eyes and has a delayed blistering action upon the skin in which it is rapidly absorbed. Assumed to be irritant and poisonous when taken by mouth.

Hazardous reactions Use as a biological sterilant is hazardous because of ready formation of explosive mixtures with air; drum of crude oxide and sodium hydroxide catalyst exploded and ignited (B364).

First aid *Vapour inhaled:* standard treatment (p 109).
Affected eyes: standard treatment (p 109).
Skin contact: standard treatment (p 108).
If swallowed: standard treatment (p 109).

Fire hazard Flash point $-37\,°C$; explosive limits 2.1–28.5%. *Extinguish fire with* dry powder, carbon dioxide or vaporising liquid.

Spillage disposal Shut off all possible sources of ignition. Instruct others to keep at a safe distance. Wear breathing apparatus and gloves. Organise ventilation of the area to dispel vapour completely.

Propyne (allylene; methylacetylene)

Colourless gas; bp $-23\,°C$; slightly soluble in water.

EXTREMELY FLAMMABLE

Avoid breathing gas. TLV 1000 ppm (1650 mg m^{-3}).

Toxic effects The gas is anaesthetic in high concentrations but is not toxic.

First aid *Gas inhaled in high concentrations:* standard treatment (p 109).

Fire hazard Explosive limits about 2.4–11.7%. Since the gas is supplied in a cylinder, turning off the valve will reduce any fire involving it; if possible, cylinders should be removed quickly from an area in which a fire has developed.

Disposal Surplus gas or leaking cylinder can be vented slowly to air in a safe, open area or burnt off through a suitable burner.

Prop-2-yn-1-ol (propargyl alcohol)

Hazardous reactions If dried with alkali before distillation, the residue is liable to explode; violent exothermic eruption may occur in preparation of hydroxyacetone from mercury(II) sulphate, sulphuric acid and the alcohol (B354).

Prop-2-yn-1-thiol

Hazardous reaction Polymerised explosively when distilled at atmospheric pressure (B354).

Prop-2-ynyl vinyl sulphide

Hazardous reaction Decomposed explosively above $85\,°C$ (B441).

Pyridine

Colourless liquid with a sharp penetrating odour; bp 115 °C; miscible with water.

> HIGHLY FLAMMABLE
> HARMFUL VAPOUR
> IRRITATING TO EYES, SKIN AND RESPIRATORY SYSTEM

Avoid breathing vapour. Avoid contact with skin and eyes. TLV 5 ppm (15 mg m^{-3}).

Toxic effects	The vapour irritates the respiratory system and may cause headache, nausea, giddiness and vomiting. The vapour and liquid irritate the eyes and may cause conjunctivitis. The liquid may irritate the skin causing dermatitis. Affects the central nervous system if taken by mouth and large doses act as a heart poison.
Hazardous reactions	Maleic anhydride decomposes exothermally in the presence of pyridine; the solid obtained by reaction with BrF_3 ignites when dry; the pyridine complex with CrO_3 is unstable; it incandesces on contact with fluorine; reacts violently with N_2O_4 (B437).
First aid	*Vapour inhaled:* standard treatment (p 109). *Affected eyes:* standard treatment (p 109). *Skin contact:* standard treatment (p 108). *If swallowed:* standard treatment (p 109).
Fire hazard	Flash point 20 °C; explosive limits 1·8–12·4%; ignition temp. 482 °C. *Extinguish fire with* water spray, foam, dry powder, carbon dioxide or vaporising liquid.
Spillage disposal	Shut off all possible sources of ignition. Instruct others to keep at a safe distance. Wear breathing apparatus and gloves. Mop up with plenty of water and run to waste, diluting greatly with running water. Ventilate area well to evaporate remaining liquid and dispel vapour.

Pyridinium perchlorate

Hazardous reactions	Can be detonated by impact and has occasionally exploded when disturbed (B439).

Pyrocatechol – *see* Catechol

Pyrophoric metals Review (B134).

Pyrrolidine

Colourless mobile liquid; bp 89 °C; miscible in water.

> HIGHLY FLAMMABLE
> HARMFUL VAPOUR
> IRRITATING TO SKIN AND EYES

Avoid breathing vapour. Avoid contact with skin, eyes and clothing.

Toxic effects The vapour irritates the respiratory system; the liquid irritates the skin and mucous surfaces.

First aid *Vapour inhaled:* standard treatment (p 109).
Affected eyes: standard treatment (p 109).
Skin contact: standard treatment (p 108).
If swallowed: standard treatment (p 109).

Fire hazard Flash point 3 °C; *Extinguish fire with* water spray, foam, dry powder or vaporising liquid.

Spillage disposal Shut off all possible sources of ignition. Instruct others to keep at a safe distance. Wear breathing apparatus and gloves. Mop up with plenty of water and run to waste diluting greatly with running water. Ventilate area well to evaporate remaining liquid and dispel vapour.

Quinoline

Colourless to yellow, hygroscopic liquid; bp 238 °C; sparingly soluble in cold water.

HARMFUL IF TAKEN INTERNALLY
IRRITATING TO SKIN AND EYES

Avoid breathing vapour. Avoid contact with skin, eyes and clothing.

Toxic effects Irritates skin and mucous surfaces.

First aid *Vapour inhaled:* standard treatment (p 109).
Affected eyes: standard treatment (p 109).
Skin contact: standard treatment (p 108).

Spillage disposal Wear face-shield or goggles, and gloves. Mop up with water and detergent and run to waste, diluting greatly with running water. Ventilate area well to evaporate remaining liquid.

Quinone – *see* *p*-Benzoquinone

Resorcinol

Colourless to pink crystals; mp 110 °C. soluble in water.

IRRITATING TO SKIN AND EYES

Avoid contact with skin and eyes.

Toxic effects Absorption by the skin may result in itching and dermatitis; in severe cases, restlessness, cyanosis and convulsions may follow. The solid irritates the eyes. It may cause dizziness, drowsiness and tremors if taken by mouth.

First aid *Affected eyes:* standard treatment (p 109).
Skin contact: standard treatment (p 108).
If swallowed: standard treatment (p 109).

Spillage disposal Wear face-shield or goggles, and gloves. Mix with sand and transport to a safe, open area for burial. Site of spillage should be washed thoroughly with water and soap or detergent.

Rubidium (metal)

Hazardous properties	Ignites on exposure to air or dry oxygen; ignites on contact with fluorine or chlorine or the vapours of bromine or iodine; reaction with mercury may be violent; hydrogen is evolved vigorously on contact with cold water and ignites (B937).

Rubidium hydride

Hazardous reactions	Reacts violently with water; its reaction with acetylene is vigorous at $-60\,^{\circ}C$ (B779).

Sebacoyl dichloride (decanedioyl dichloride)

Hazardous reaction	At the end of vacuum distillation, the residue frequently decomposes spontaneously, producing voluminous black foam (B581).

Seleninyl bromide

Hazardous reactions	The liquid bromide reacts explosively with sodium and potassium, and ignites zinc dust; red phosphorus ignites and the white allotrope explodes in contact with the liquid bromide (B211).

Seleninyl chloride

Hazardous reactions	Potassium and phosphorus (white) explode on contact with the liquid, powdered antimony ignites (B682).

Selenium and compounds

Selenium is a steel grey or purplish powder, also fabricated into pellets, sticks or plates; insoluble in water. Selenium dioxide, selenous acid and the alkali-metal selenites and selenates are colourless powders or crystals, soluble in water; selenium chloride (reddish yellow), selenyl chloride (colourless or yellow) (bp 176 °C); and selenic acid (colourless) are liquids, whereas selenium tetrachloride is a cream-coloured crystalline solid. Highly poisonous hydrogen selenide (offensive smell) is generated when an acid solution of a selenium compound is reduced by metals such as tin and zinc. Selenic acid reacts vigorously with water.

SERIOUS RISK OF POISONING BY INHALATION OR SWALLOWING
IRRITATING TO SKIN, EYES AND RESPIRATORY SYSTEM
DANGER OF CUMULATIVE EFFECTS

Avoid breathing dust. Avoid contact with skin and eyes. TLV (selenium compounds as Se) (0.2 mg m^{-3}).

Toxic effects   Hydrogen selenide irritates the nose and eyes and causes inflammation of the lungs and disturbance of the digestive and nervous systems; it imparts a garlic-like odour to the breath. The chloride and solutions of the acids and salts may burn the skin—severe pain may be experienced under the finger-nails by skin absorption at the finger-tips. Selenium dioxide dust is particularly penetrating and irritates the respiratory system, eyes and skin. Assumed to be irritant and poisonous if taken by mouth. **Chronic effects** Inhalation of selenium dust over a prolonged period may cause fatigue, loss of appetite, digestive disturbance and bronchitis; dermatitis may result from prolonged exposure of skin to small amounts of selenium and its compounds. (*See* note in Ch 6, p 83.)

Hazardous reactions of selenium   Nickel, sodium and potassium interact with selenium with incandescence; the particle size of cadmium and selenium must be below a critical size to prevent explosions when making cadmium selenide—this also applies to zinc and selenium; the oxidation of recovered selenium with nitric acid is made vigorous by the presence of organic impurities; selenium may react explosively with BrF_5, ClF_3 or Na_2O_2; it ignites on contact with fluorine (B943).

First aid   *Dust or vapour inhaled:* standard treatment (p 109).
Affected eyes: standard treatment (p 109).
Skin contact: standard treatment (p 108).
If swallowed: standard treatment (p 109).

Spillage disposal   Selenium powder may be mixed with sand and treated as normal refuse, as may the disulphide. Soluble selenites and selenates can be dissolved in water and run to waste, diluting greatly with running water. Soda ash should be applied liberally to spillages of selenium dioxide, selenic and selenous acids and selenyl and selenium chlorides which may then be mopped up cautiously with plenty of water, this being run to waste, diluting greatly with running water.

Selenium tetrafluoride

Hazardous reaction   Reacts violently with water (B743).

Silane

Hazardous reactions   The very pure material ignites in air; it burns in contact with bromine, chlorine or some covalent chlorides (B815).

Silanes Review (B137).

Silicon monohydride

Hazardous reaction   The polymeric hydride reacts violently with aqueous alkali, evolving hydrogen (B779).

Silicon tetraazide

Hazardous reaction Spontaneously explosive at times (B886).

Silicon tetrachloride (silicon chloride)

Colourless fuming liquid; bp 59 °C; reacts violently with water, forming hydrochloric acid and silica.

CAUSES BURNS
IRRITATING TO SKIN, EYES AND RESPIRATORY SYSTEM

Avoid breathing vapour. Prevent contact with eyes and skin.

Toxic effects The vapour severely irritates all parts of the respiratory system. The vapour severely irritates the eyes and the liquid burns the eyes and skin. If taken by mouth there would be severe internal irritation and damage.

First aid *Vapour inhaled:* standard treatment (p 109).
Affected eyes: standard treatment (p 109).
Skin contact: standard treatment (p 108).
If swallowed: standard treatment (p 109).

Spillage disposal Instruct others to keep at a safe distance. Wear breathing apparatus and gloves. Spread soda ash liberally over the spillage and mop up cautiously with water — run to waste, diluting greatly with running water.

Silver acetylide

Hazardous reaction Touch-sensitive explosive (B280).

Silver amide

Hazardous reaction Very explosive when dry (B153).

Silver chlorite

Hazardous reactions Reacted explosively with iodomethane or iodoethane with or without solvent dilution; the salt itself is impact-sensitive, cannot be ground and explodes at 105 °C; it explodes in contact with hydrochloric acid or on rubbing with sulphur (B151).

Silver fluoride

Hazardous reactions   Reactions with calcium hydride (on grinding mixture) and titanium (at 320 °C) are incandescent; silicon reacts violently, boron explosively when ground with AgF (B152).

Silver fulminate

Hazardous reactions   A powerful detonator, it explodes violently in contact with hydrogen sulphide (B267).

Silvering solutions Review (B139).

Silver nitrate (lunar caustic)

White crystals; soluble in water.

CAUSES BURNS
IRRITATING TO EYES

Avoid contact with eyes and skin. TLV (soluble silver compounds as Ag) (0.01 mg m^{-3}).

Toxic effects   The solid and its solutions severely irritate the eyes and can cause skin burns. If taken by mouth silver nitrate can cause internal damage due to absorption in the blood followed by deposition of silver in various tissues of the body.

Hazardous reactions   Reacts with acetylene in the presence of ammonia to form silver acetylide, a powerful detonator; finely divided mixture of arsenic with excess nitrate ignited when shaken on to paper; reacts violently with chlorosulphuric acid forming nitrosulphuric acid; crystals damp with ethanol exploded when touched with spatula; intimate mixture with dry magnesium powder may ignite explosively on contact with drop of water; mixtures with phosphorus and sulphur explode under hammer blow; mixture with charcoal ignites on percussion (B153–155).

First aid   *Affected eyes :* standard treatment (p 109).
Skin contact : standard treatment (p 108).
If swallowed : standard treatment (p 109).

Spillage disposal   Wear face-shield or goggles, and gloves. Mop up with plenty of water and run to waste, diluting greatly with running water.

Silver oxalate

Hazardous reaction   Liable to explode when heated above 140 °C (B268).

Silver(I) oxide

Hazardous reactions	Slowly forms explosive silver nitride with ammonia or hydrazine; oxidises carbon monoxide exothermically and may ignite hydrogen sulphide; Se, S or P are ignited on grinding with the oxide (B156).

Silver perchlorate

White deliquescent crystals; decomposes at 486 °C; freely soluble in water.

EXPLOSIVE WHEN MIXED WITH COMBUSTIBLE MATERIAL
IRRITATING TO SKIN AND EYES

Avoid contact with skin, eyes and clothing.

Toxic effects	It irritates the skin and mucous surfaces and would be harmful if taken by mouth.
First aid	*Affected eyes:* standard treatment (p 109). *Skin contact:* standard treatment (p 108). *If swallowed:* standard treatment (p 109).
Spillage disposal	Wear face-shield or goggles, and gloves. Mop up with plenty of water and run to waste, diluting greatly with running water. Large amounts may be worth recovering.

Soda asbestos

Grey or brown granules; largely soluble in water.

HARMFUL DUST
CAUSES BURNS
IRRITATING TO SKIN, EYES AND RESPIRATORY SYSTEM

Avoid breathing dust. Prevent contact with eyes and skin.

Toxic effects	The dust severely irritates the nose and mouth. The solid and dust burn the eyes and skin. If swallowed there would be severe internal irritation and damage.
First aid	*Dust inhaled:* standard treatment (p 109). *Affected eyes:* standard treatment (p 109). *Skin contact:* standard treatment (p 108). *If swallowed:* standard treatment (p 109).
Spillage disposal	Wear face-shield or goggles, and gloves. Shovel into enamel or polythene bucket add, a little at a time with stirring, to a large volume of water. After reaction is complete, run solution to waste, diluting greatly with running water. Wash down site of spillage thoroughly with water.

Soda-lime

Hazardous reactions   Fires have been caused in waste bins into which soda-lime that has absorbed hydrogen sulphide has been thrown — considerable heat develops when this spent material is exposed to moisture and air (B139).

Sodamide – *see* Sodium amide

Sodium (metal) and sodium amalgam

Sodium — soft, silvery white sticks, pellets, wire or granules, normally coated with a grey oxide or hydroxide skin; reacts vigorously with water with formation of sodium hydroxide and hydrogen gas which may ignite. Sodium amalgam — silvery or grey spongy masses; reacts with water forming sodium hydroxide and hydrogen gas.

CONTACT WITH WATER LIBERATES HIGHLY FLAMMABLE GASES
CAUSES BURNS

Prevent contact with skin and eyes.

Toxic effects   The metal or amalgam in contact with moisture on the skin can cause thermal and caustic burns. In the same way the sodium hydroxide resulting from the reaction of the metal with water can cause burning of the skin and eyes.

Hazardous reactions   References given to special papers, booklets, *etc* on safe handling of sodium on laboratory and large scale — also the disposal of unwanted residues; anhydrous HCl, HF or sulphuric acids react slowly with sodium while the aqueous solution reacts explosively; dispersions of Na in volatile solvents become pyrophoric when solvent evaporates — serum cap closures on bottles are safest; the metal reacts with different degrees of violence with chloroform/methanol, diazomethane, *N,N*-dimethylformamide, fluorinated compounds, halocarbons, halogens, interhalogens, mercury, metal halides and oxides, non-metal halides and oxides, non-metals, oxidants, water (B887–893).

First aid   *Affected eyes:* standard treatment (p 109).
Skin contact: standard treatment (p 108).
Mouth contact: standard treatment (p 109).

Spillage disposal   (a) Sodium metal: instruct others to keep at a safe distance. Wear face-shield or goggles, and gloves. Cover with dry soda ash, shovel into dry bucket, transport to safe, open area and add, a little at a time, to a large excess of dry propan-2-ol. Leave to stand for 24 hours and run to waste, diluting greatly with running water.
(b) Sodium amalgam: cover with large volume of water in suitable vessel and allow to stand until there is no further reaction; the mercury may then be separated and recovered, and the sodium hydroxide solution run to waste, diluting greatly with running water.

Sodium acetylide

Hazardous reaction Decomposes at 150 °C evolving gas which ignites in air **(B285)**.

Sodium amide (sodamide)

Colourless or greyish white lumps or powder smelling of ammonia. Reacts violently with water with the formation of sodium hydroxide and ammonia.

> HARMFUL BY INHALATION
> CAUSES BURNS

Avoid breathing dust. Prevent contact with eyes and skin.

Toxic effects The dust severely irritates the mouth and nose. The solid and dust severely irritate or burn the eyes. The solid in contact with moisture on the skin can cause thermal and caustic burns. If taken by mouth there would be severe internal irritation and damage.

Hazardous reactions May ignite or explode on heating or grinding in air; explodes with potassium chlorate or sodium nitrite; fresh material behaves like sodium on contact with water; reference to disposal given **(B782)**.

First aid *Dust inhaled:* standard treatment (p 109).
Affected eyes: standard treatment (p 109).
Skin contact: standard treatment (p 108).
If swallowed: standard treatment (p 109).

Spillage disposal Wear face-shield or goggles, and gloves. Mix with dry sand, shovel into an enamel or polythene bucket, transport to safe, open area, and add, a little at a time with stirring, to a large volume of water; after reaction is complete, run solution to waste, diluting greatly with running water. Dispose of sand as normal refuse. Wash down site of spillage thoroughly with water.

Sodium arsenate and arsenite
– *see* Arsenic and compounds

Sodium azide

Colourless crystalline powder; soluble in water.

> TOXIC IF TAKEN INTERNALLY
> CONTACT WITH ACIDS LIBERATES A TOXIC GAS
> IRRITATING TO SKIN AND EYES

Prevent contact with skin and eyes. Avoid breathing dust. TLV 0.1 ppm.

Toxic effects The dust and solution irritate the eyes. The solid and solution irritate the skin and can cause blistering. Assumed to be very irritant and poisonous if taken by mouth.

Hazardous reactions	Decomposes somewhat explosively above its mp, particularly if heated rapidly; liable to explode with bromine, carbon disulphide or chromyl chloride; when water is added to the strongly heated azide there is a violent reaction (B881).
First aid	*Affected eyes:* standard treatment (p 109). *Skin contact:* standard treatment (p 108). *If swallowed:* standard treatment (p 109).
Spillage disposal	Wear face-shield or goggles, and gloves. Mop up with plenty of water and run to waste, diluting greatly with running water.

Sodium azidosulphate

Hazardous reaction	Weak explosive of variable sensitivity (B882).

Sodium bisulphate
– *see* Sodium hydrogensulphate

Sodium borohydride (sodium tetrahydridoborate)

White to pale grey microcrystalline powder or lumps; decomposed by water with evolution of hydrogen.

HARMFUL IF TAKEN INTERNALLY
IRRITATING TO SKIN, EYES AND RESPIRATORY SYSTEM

Avoid breathing dust. Avoid contact with skin, eyes and clothing.

Toxic effects	Dust irritates respiratory system. Irritating to skin and mucous surfaces.
Hazardous reaction	A large volume of the alkaline solution spontaneously heated and decomposed, liberating hydrogen (B185).
First aid	*Affected eyes:* standard treatment (p 109). *Skin contact:* standard treatment (p 108). *If swallowed:* standard treatment (p 109).
Spillage disposal	Wear face-shield or goggles, and gloves. Mop up with plenty of water and run to waste, diluting greatly with running water.

Sodium chlorate

Colourless crystals, soluble in water.

EXPLOSIVE WHEN MIXED WITH COMBUSTIBLE MATERIAL
HARMFUL IF TAKEN INTERNALLY

Avoid contact with combustible materials and acids. Avoid contact with skin, eyes and clothing.

Toxic effects   The dust or strong solutions may irritate the eyes and skin. The solid or solutions are damaging if taken internally, symptoms of poisoning being nausea, vomiting and abdominal pain; kidney damage may follow.

Hazardous reactions   Mixtures with ammonium salts, powdered metals, phosphorus, silicon, sulphur or sulphides are readily ignited and potentially explosive; mixture with nitrobenzene is powerful explosive; mixtures with fibrous or absorbent organic materials (charcoal, flour, shellac, sawdust, sugar) are hazardous and can be caused to explode by static, friction or shock; addition of concentrated sulphuric acid to solid chlorate causes explosion of ClO_2 generated (B662–663).

First aid   *Affected eyes:* standard treatment (p 109).
Skin contact: standard treatment (p 108).
If swallowed: standard treatment (p 109).

Fire hazard   Mixtures of sodium chlorate and combustible materials are readily ignited; mixtures with finely divided combustible materials can react explosively. *Extinguish fire with* water spray.

Spillage disposal   Shovel into bucket of water and run solution or suspension to waste, diluting greatly with running water. Site of spillage should be washed thoroughly to remove all oxidant, which is liable to render any organic matter (particularly wood, paper and textiles) with which it comes into contact, dangerously combustible when dry. Clothing wetted with the solution should be washed thoroughly.

Sodium chlorite

Hazardous reactions   Explodes on impact; intimate mixtures with finely divided or fibrous organic matter may be very sensitive to heat, impact or friction (B661).

Sodium chromate and dichromate
– *see* **Chromates and dichromates**

Sodium cyanide
– *see* **Cyanides** (water soluble)

Sodium dichromate
– *see* **Chromates and dichromates**

Sodium dihydrogenphosphide

Hazardous reaction   Ignites in air (B784).

Sodium dithionite (sodium hydrosulphite)

White or off-white crystalline powder; very soluble in water, but addition of small amount of water to the salt causes hazardous reaction.

> CONTACT WITH MOIST COMBUSTIBLE MATERIALS
> MAY CAUSE FIRE

Avoid contact with moist flammable materials including clothing.

Hazardous reaction   Addition of 10% of water caused heating and spontaneous ignition (B897).

Fire hazard   Fires resulting from contact with combustible materials are best extinguished by means of water spray.

Spillage disposal   Shovel into bucket of water and run solution or suspension to waste, diluting greatly with running water. Site of spillage should be washed thoroughly to remove all oxidant, which is liable to render any organic matter (particularly wood, paper and textiles) with which it comes into contact, dangerously combustible when dry. Clothing wetted with the solution should be washed thoroughly.

Sodium ethoxide (sodium ethylate)
and sodium methoxide (sodium methylate)

Although these compounds are sometimes used industrially as solids, they are more frequently prepared in the laboratory by reacting sodium metal with ethanol and methanol. The hazards of these solutions are dealt with here.

> HIGHLY FLAMMABLE
> CAUSE BURNS
> IRRITATING TO SKIN AND EYES

Prevent contact with skin and eyes.

Toxic effects   The solutions are extremely irritant to the eyes, causing burns. The solutions irritate the skin and may produce burns. They will cause severe internal irritation and damage if taken by mouth.

First aid   *Affected eyes:* standard treatment (p 109).
Skin contact: standard treatment (p 108).
If swallowed: standard treatment (p 109).

Spillage disposal   Shut off all possible sources of ignition. Wear face-shield or goggles, and gloves. Mop up with plenty of water and run to waste, diluting greatly with running water. Ventilate area of spillage well to evaporate remaining liquid and dispel vapour.

Sodium ethoxyacetylide

Hazardous reaction   Extremely pyrophoric; it may explode after prolonged contact with air (B389).

Sodium fluoride
– *see* **Fluoride** (water-soluble)

Sodium fluoroborate
– *see* **Tetrafluoroboric acid and salts**

Sodium fluorosilicate·
– *see* **Hexafluorosilicic acid and salts**

Sodium hydrazide

Hazardous reactions Heating to 100 °C or contact with traces of air, alcohol or water cause violent explosion (B804).

Sodium hydride

Hazardous reactions Addition to small amount of water causes explosion; reacts vigorously with acetylene in the presence of moisture even at −60 °C; the finely divided dry powder ignites in dry air; may react explosively with dimethyl sulphoxide; interaction with chlorine or fluorine is incandescent; reacts vigorously with sulphur, explosively with sulphur dioxide (B776–777).

Sodium hydrogen difluoride
– *see* **Fluorides** (water-soluble)

Sodium hydrogensulphate (sodium bisulphate)

Colourless crystals or fused masses; soluble in water.

CAUSES BURNS
IRRITATING TO SKIN AND EYES

Avoid contact with skin and eyes.

Toxic effects The solid and its strong solutions in water cause severe burns of the eyes and skin. Causes severe internal irritation and damage if taken by mouth.

First aid	*Affected eyes:* standard treatment (p 109).
	Skin contact: standard treatment (p 108).
	If swallowed: standard treatment (p 109).
Spillage disposal	Wear face-shield or goggles, and gloves. Spread soda ash liberally over the spillage and mop up cautiously with plenty of water—run to waste, diluting greatly with running water.

Sodium hydrosulphite
-*see* Sodium dithionite

Sodium hydroxide (caustic soda)

Colourless sticks, flakes, powder or pellets; soluble in water.

CAUSES SEVERE BURNS

Prevent contact with eyes and skin. TLV 2 mg m^{-3}.

Toxic effects The solid and its strong solutions cause severe burns of the eyes and skin. If taken by mouth there would be severe internal irritation and damage. Solutions as weak as 2.5N can damage eyes severely.

Hazardous reactions Very exothermic reaction with limited amounts of water; reacts vigorously with chloroform/methanol; explosion results when it is heated with zirconium; accidental contamination of metal scoop with flake NaOH, prior to its use with Zn dust, caused the latter to ignite (B777).

First aid *Affected eyes:* standard treatment (p 109).
Skin contact: standard treatment (p 108).
If swallowed: standard treatment (p 109).

Spillage disposal Wear face-shield or goggles, and gloves. Shovel into an enamel or polythene bucket, and add, a little at a time with stirring, to a large volume of water. After solution is complete run this to waste, diluting greatly with running water. Wash down site of spillage thoroughly with water.

Sodium hypochlorite solution
(containing over 5% active chlorine)

Colourless solution smelling of chlorine.

CAUSES BURNS
IRRITATING TO SKIN AND EYES

Avoid contact with skin, eyes and clothing.

Toxic effects Extremely irritating to the eyes and gives rise to burns. Bleaches and may burn the skin. Will cause internal irritation and damage if ingested.

First aid	*Affected eyes:* standard treatment (p 109).
	Skin contact: standard treatment (p 108).
	If swallowed: standard treatment (p 109).
Spillage disposal	Wear face-shield or goggles, and gloves. Mop up with plenty of water and run to waste, diluting greatly with running water.

Sodium hypophosphite
– see **Sodium phosphinate**

Sodium iodate

White crystals which decompose on heating; one part dissolves in about eleven parts of cold water.

CONTACT WITH COMBUSTIBLE MATERIAL MAY CAUSE FIRE

Avoid contact with combustible materials.

Fire hazard	Mixture of sodium iodate and combustible materials are readily ignited; mixtures with finely divided combustible materials can react explosively. *Extinguish fire with* water spray.
Spillage disposal	Shovel into bucket of water and run solution to waste, diluting greatly with running water. Site of spillage should be washed thoroughly to remove all oxidant, which is liable to render any organic matter (particularly wood, paper and textiles) with which it comes into contact, dangerously combustible when dry. Clothing wetted with the solution should be washed thoroughly.

Sodium metaperiodate
– see **Sodium periodate**

Sodium methoxide *– see* **Sodium ethoxide**

Sodium nitrate

Colourless, deliquescent crystals; soluble in water.

EXPLOSIVE WHEN MIXED WITH COMBUSTIBLE MATERIALS

Avoid contact with combustible materials.

Toxic effects	Ingestion may cause gastro-enteritis, abdominal pains, vomiting, muscular weakness, irregular pulse, convulsions and collapse; 15–30 g in one dose may be fatal.

Hazardous reactions — Reacts explosively with Al, Al_2O_3, Sb, barium thiocyanate, sodium phosphinate, sodium thiosulphate; reacts violently with acetic anhydride; may ignite fibrous organic material, charcoal, *etc* on heating; an explosive compound is formed by interaction with sodium (B868).

First aid — *If swallowed:* standard treatment (p 109).

Fire hazard — Mixtures of sodium nitrate and combustible materials are readily ignited; mixtures with finely divided, combustible materials can react explosively. *Extinguish fire with* water spray.

Spillage disposal — Shovel into bucket of water and run solution or suspension to waste, diluting greatly with running water. Site of spillage should be washed thoroughly to remove all oxidant, which is liable to render any organic matter (particularly wood, paper and textiles) with which it comes into contact, dangerously combustible when dry. Clothing wetted with the solution should be washed thoroughly.

Sodium nitrite

White or slightly yellow granules, rods or powder; hygroscopic and soluble in water. Poisonous nitrous fumes produced by the action of acids; use as food additive.

CONTACT WITH COMBUSTIBLE MATERIAL MAY CAUSE FIRE

Avoid contact with combustible materials or acids.

Hazardous reactions — Explosions are likely to occur on heating mixtures of the nitrite with ammonium salts, metal cyanides, phthalic acid or anhydride, sodium amide, sodium thiocyanate or thiosulphate; wood impregnated with solutions of the nitrite over a long period may be accidentally ignited and burn fiercely (B867).

Fire hazard — Mixtures of sodium nitrite and combustible materials are readily ignited; mixtures with finely divided combustible materials can react explosively. *Extinguish fire with* water spray.

Spillage disposal — Shovel into bucket of water and run solution to waste, diluting greatly with running water. Site of spillage should be washed thoroughly to remove all oxidant, which is liable to render any organic matter (particularly wood, paper and textiles) with which it comes into contact, dangerously combustible when dry. Clothing wetted with the solution should be washed thoroughly.

Sodium oxalate – *see* Oxalates

Sodium pentachlorophenate – *see* Pentachlorophenol

Sodium perchlorate – *see* Perchlorates

Sodium periodate (sodium metaperiodate)

White crystals, soluble in cold water.

CONTACT WITH COMBUSTIBLE MATERIALS MAY CAUSE FIRE

Avoid contact with combustible materials. Avoid contact with skin, eyes and clothing.

Toxic effects	No documentary evidence on the toxicity of this compound has been traced. However, as potassium metaperiodate is rated 'high' for its irritant characteristics and toxicity by ingestion, it must clearly be treated with respect.
First aid	*Affected eyes:* standard treatment (p 109). *Skin contact:* standard treatment (p 108). *If swallowed:* standard treatment (p 109).
Fire hazard	Mixtures with combustible materials are readily ignited; mixtures with finely divided combustible materials can react explosively. *Extinguish fire with* water spray.
Spillage disposal	Shovel into bucket of water and run solution to waste, diluting greatly with running water. Site of spillage should be washed thoroughly to remove all oxidant, which is liable to render any organic matter (particularly wood, paper and textiles) with which it comes into contact, dangerously combustible when dry. Clothing wetted with the solution should be washed thoroughly.

Sodium peroxide

Pale yellow powder; reacts violently with water forming sodium hydroxide and oxygen.

CONTACT WITH COMBUSTIBLE MATERIALS MAY CAUSE FIRE
CAUSES SEVERE BURNS

Do not breathe dust. Prevent contact with skin and eyes.

Toxic effects	The dust irritates the respiratory system and eyes. Because of the vigour of its reaction with water it may cause both thermal and caustic burns on moist skin. Would cause severe internal irritation and damage if taken by mouth.
Hazardous reactions	Admixture with acetic acid causes explosion; mixture with ammonium peroxodisulphate explodes on heating above 75 °C, grinding in mortar or exposure to drops of water or carbon dioxide; mixtures with calcium acetylide explode; contact with fibrous materials often causes ignition; mixture with hydroxy compounds usually results in fires or explosions; reacts violently or explosively with some metals, non-metals and non-metal halides; reacts vigorously or explosively with water depending on relative quantities (B894–896).
First aid	*Dust inhaled:* standard treatment (p 109). *Affected eyes:* standard treatment (p 108). *If swallowed:* standard treatment (p 109).
Fire hazard	If in contact, or mixed with organic or other oxidisable substances, ignition or explosion may take place.

Spillage disposal	Wear face-shield or goggles, and gloves. Mix with dry sand, shovel into an enamel bucket, transport to a safe, open area and add, a little at a time, to a large volume of water; after reaction is complete, run solution to waste, diluting greatly with running water. Dispose of sand as normal refuse.

Sodium peroxyacetate

Hazardous reaction	Dry salt exploded at room temperature (B301).

Sodium phenylacetylide

Hazardous reaction	Ether-moist powder ignites in air (B546).

Sodium phosphide

Hazardous reaction	Decomposed by water or moist air, evolving phosphine which often ignites (B899).

Sodium phosphinate (sodium hypophosphite)

Hazardous reaction	Evaporation of aqueous solution by heating may cause explosion, phosphine being evolved (B784).

Sodium pyrosulphate

This colourless crystalline salt is of variable composition, produced by dehydrating sodium hydrogensulphate $Na_2S_2O_7$; it is soluble in water giving an acidic solution.

CAUSES BURNS

Avoid contact with skin, eyes and clothing.

Toxic effects	The solid and its strong solutions in water cause burns to the skin and eyes. Causes severe internal irritation and damage if taken by mouth.
First aid	*Affected eyes:* standard treatment (p 109). *Skin contact:* standard treatment (p 108). *If swallowed:* standard treatment (p 109).
Spillage disposal	Wear face-shield or goggles, and gloves. Spread soda ash liberally over the spillage and mop up cautiously with plenty of water — run to waste, diluting greatly with running water.

Sodium selenate and selenite
– *see* Selenium and compounds

Sodium silicide

Hazardous reaction   Ignites in air (B893).

Sodium sulphide

The hydrated salt consists of colourless crystalline masses; the fused salt forms brownish lumps or powder; soluble in water. With acids, poisonous hydrogen sulphide gas is evolved.

CONTACT WITH ACIDS LIBERATES A TOXIC GAS
CAUSES BURNS

Prevent contact with skin and eyes.

Toxic effects   The solid or solution cause severe burns of the eyes. The solid or solution may irritate and burn the skin. Irritant and poisonous if taken by mouth.

Hazardous reactions   After exposure to moisture and air, small lumps of fused sodium sulphide are liable to spontaneous heating; mixtures with finely divided carbon react exothermically (B898).

First aid   *Affected eyes:* standard treatment (p 109).
Skin contact: standard treatment (p 108).
If swallowed: standard treatment (p 109).

Spillage disposal   Wear face-shield or goggles, and gloves. Mop up with plenty of water and run to waste, diluting greatly with running water.

Sodium tellurate and tellurite
– *see* Tellurium and compounds

Sodium tetrahydroborate
– *see* Sodium borohydride

Sodium tetrahydroaluminate
– *see* Aluminium sodium hydride

Sodium vanadate
– *see* **Vanadium compounds**

Stannic chloride – *see* Tin(IV) chloride

Stannous chloride – *see* Tin(II) chloride

Stibine (antimony trihydride)

Hazardous reactions During evaporation of the liquid at −17 °C, a weakly explosive decomposition may occur; a heated mixture with ammonia explodes; explodes with chlorine, nitric acid or ozone (B807).

Strontium acetylide

Hazardous reactions Incandesces with chlorine, bromine and iodine in the temperature range 174–197 °C (B339).

Strontium nitrate

White granules or crystalline powder; one part dissolved in about three parts of water at 20 °C·

CONTACT WITH COMBUSTIBLE MATERIALS MAY CAUSE FIRE

Avoid contact with all combustible materials.

Fire hazard Mixtures of strontium nitrate and combustible materials are readily ignited; mixtures with finely divided combustible materials can react explosively. *Extinguish fire with* water spray.

Spillage disposal Shovel into bucket of water and run solution to waste, diluting greatly with running water. Site of spillage should be washed thoroughly to remove all oxidant, which is liable to render any organic matter (particularly wood, paper and textiles) with which it comes into contact, dangerously combustible when dry. Clothing wetted with the solution should be washed thoroughly.

Styrene (phenylethylene)

Colourless liquid with penetrating disagreeable odour; bp 146 °C; immiscible with water.

FLAMMABLE
HARMFUL VAPOUR
IRRITATING TO SKIN, EYES AND RESPIRATORY SYSTEM

Avoid breathing vapour. Avoid contact with skin and eyes. TLV 100 ppm (420 mg m⁻³).

Toxic effects	The vapour irritates the eyes and respiratory system. The liquid irritates the eyes and is reported to cause severe eye injuries. Assumed to be poisonous if taken by mouth.
Hazardous reactions	Autocatalytic exothermic polymerisation becomes self-sustaining above 65 °C; on exposure to oxygen at 40–60 °C an interpolymeric peroxide was formed which exploded on gentle heating (B548).
First aid	*Vapour inhaled:* standard treatment (p 109). *Affected eyes:* standard treatment (p 109). *Skin contact:* standard treatment (p 108). *If swallowed:* standard treatment (p 109).
Fire hazard	Flash point 31 °C; explosive limits 1.1–6.1%; ignition temp. 490 °C. *Extinguish fire with* foam, dry powder, carbon dioxide or vaporising liquid.
Spillage disposal	Shut off all possible sources of ignition. Wear face-shield or goggles, and gloves. Apply non-flammable dispersing agent if available and work to an emulsion with brush and water — run this to waste, diluting greatly with running water. If dispersant not available, absorb on sand, shovel into bucket(s) and transport to safe, open area for burial. Ventilate site of spillage well to evaporate remaining liquid and dispel vapour.

Succinoyl diazide

Hazardous reaction	Exploded violently during isolation (B386).

Sulphamic acid – *see* Amidosulphuric acid

Sulphonic acids

The simpler sulphonic acids, such as benzenesulphonic, benzenedisulphonic, phenolsulphonic, phenoldisulphonic, and cresolsulphonic acids are generally supplied as solutions in water or sulphuric acid. Toluene-*p*-sulphonic acid is a colourless, hygroscopic solid.

CAUSE BURNS
IRRITATING TO SKIN AND EYES

Avoid contact with skin and eyes.

Toxic effects	The solutions irritate the skin and eyes and may cause burns. Will cause internal irritation and damage if taken by mouth.
First aid	*Affected eyes:* standard treatment (p 109). *Skin contact:* standard treatment (p 108). *If swallowed:* standard treatment (p 109).
Spillage disposal	Wear face-shield or goggles, and gloves. Spread soda ash liberally over the spillage and mop up cautiously with plenty of water — run to waste, diluting greatly with running water.

Sulphur

Hazardous reactions	Evaporation of an ethereal extract of sulphur exploded violently; ignites in fluorine and in chlorine dioxide gas (explosion possible); reacts with varying degrees of vigour with interhalogens, metal acetylides, metal oxides, metals, some non-metals, many oxidants and sodium hydride (B938).

Sulphur dichloride (sulphur chloride)

Red-brown, fuming liquid; bp 59 °C; decomposed by water with the liberation of sulphur dioxide and hydrogen chloride.

CAUSES BURNS
IRRITATING TO SKIN, EYES AND RESPIRATORY SYSTEM

Avoid breathing vapour. Prevent contact with skin and eyes. Do not put water into container.

Toxic effects	The vapour irritates the respiratory system. The vapour and liquid irritate the eyes severely and the liquid may cause burns. The liquid irritates the skin and may cause burns; its ingestion would result in severe internal irritation and damage.
Hazardous reactions	Reacts vigorously with acetone, violently or explosively with dimethyl sulphoxide; mixture with sodium is impact-sensitive, as is also the case with potassium; reacts violently with nitric acid and explosively with N_2O_4; exothermic reaction with water or steam (B686).
First aid	*Vapour inhaled:* standard treatment (p 109). *Affected eyes:* standard treatment (p 109). *Skin contact:* standard treatment (p 108). *If swallowed:* standard treatment (p 109).
Spillage disposal	Instruct others to keep at a safe distance. Wear breathing apparatus and gloves. Spread soda ash liberally over the spillage and mop up cautiously with water — run to waste, diluting greatly with running water.

Sulphur dioxide

Colourless gas with a distinctive odour; supplied in liquefied form in syphons or cylinders; somewhat soluble in water.

TOXIC BY INHALATION
IRRITATING TO SKIN, EYES AND RESPIRATORY SYSTEM

Avoid inhalation of gas. TLV 5 ppm (13 mg m^{-3}).

Toxic effects	The vapour irritates the respiratory system and may cause bronchitis and asphyxia. High concentrations of vapour irritate the eyes and may cause conjunctivitis.

Hazardous reactions   Sulphur dioxide reacts violently with BrF_5, explodes with fluorine or chlorine trifluoride; incandescence and ignition occur with some metal acetylides and metal oxides; some metals react with sulphur dioxide with varying degrees of vigour; solutions of SO_2 in ethanol or ether explode on contact with potassium chlorate; reacts explosively with sodium hydride (B914).

First aid   *Vapour inhaled:* standard treatment (p 109).
Affected eyes: standard treatment (p 109).

Spillage disposal   Surplus gas or leaking cylinder can be vented slowly into a water-fed scrubbing tower or column in a fume cupboard, or into a fume cupboard served by such a tower.

Sulphuric acid

Concentrated sulphuric acid is a colourless, viscous liquid reacting vigorously with water.

CAUSES SEVERE BURNS

Prevent contact with skin and eyes. Do not put water into container. TLV 1 mg m^{-3}.

Toxic effects   The concentrated acid burns the eyes and skin severely. The dilute acid irritates the eyes and may cause burns; it will irritate the skin and may give rise to dermatitis. The concentrated acid, if taken by mouth, will cause severe internal irritation and damage.

Hazardous reactions   Many substances and classes of substance react with concentrated sulphuric acid with varying degrees of violence — these include acetone/nitric acid, acetonitrile/SO_3, acrylonitrile, alkyl nitrates, BrF_5, copper, 2-cyanopropan-2-ol, cyclopentadiene, metal acetylides or carbides, metal chlorates and perchlorates, nitramide, nitric acid/organic matter, nitric acid/toluene, nitrobenzene, nitromethane, *p*-nitrotoluene, permanganates, phosphorus, phosphorus trioxide, potassium, sodium, sodium carbonate; the dilution of sulphuric acid with water is vigorously exothermic and must be effected by adding acid to water to avoid local boiling (B792–796).

First aid   *Affected eyes:* standard treatment (p 109).
Skin contact: standard treatment (p 108).
If swallowed: standard treatment (p 109).

Spillage disposal   Wear face-shield or goggles, and gloves. Spread soda ash liberally over the spillage and mop up cautiously with plenty of water — run to waste, diluting greatly with running water.

Sulphur tetrafluoride

Colourless gas with smell resembling that of sulphur dioxide; bp −40°C; reacts violently with water forming hydrogen fluoride and sulphur dioxide; attacks glass.

VERY TOXIC BY INHALATION
REACTS VIOLENTLY WITH WATER
CAUSES SEVERE BURNS

Prevent inhalation of gas. Prevent contact with skin and eyes. TLV 0.1 ppm (0.4 mg m^{-3}).

Toxic effects   The gas reacts with moisture on the body tissues forming highly toxic and corrosive hydrogen fluoride. It is thus extremely irritant to the eyes, skin and respiratory system, and will cause severe burns if exposure is considerable.

First aid   *Gas inhaled:* standard treatment (p 109).
Affected eyes: standard treatment (p 109).
Skin contact: standard treatment (p 108).

Spillage disposal   Surplus gas or leaking cylinder can be vented slowly into a water-fed scrubbing tower or column in a fume cupboard, or into a fume cupboard served by such a tower.

Sulphur trioxide

Hazardous reactions   Dissolution of SO_3 in dimethyl sulphoxide is very exothermic; the 1:1 addition complex with dioxan sometimes decomposes violently on storing at ambient temperature; reacts violently with diphenylmercury; reacts with barium and lead oxides with incandescence; reaction with water is vigorously exothermic, sometimes explosive (B921).

Sulphuryl chloride

Colourless or yellow, fuming liquid with a pungent odour; bp 69 °C; decomposed by water with formation of hydrochloric and sulphuric acids.

CAUSES SEVERE BURNS
IRRITATING TO SKIN, EYES AND RESPIRATORY SYSTEM

Avoid breathing vapour. Prevent contact with skin and eyes. Do not put water into container.

Toxic effects.   The vapour irritates the respiratory system. The vapour and liquid irritate the eyes and the liquid will cause burns. The liquid irritates the skin and causes burns; its ingestion would result in severe internal irritation and damage.

Hazardous reactions   Reactions with alkalies may be violently explosive; solution of sulphuryl chloride in ether decomposed vigorously; reacts violently or explosively with dimethyl sulphoxide or lead dioxide; reacts vigorously with red phosphorus on warming, explosively with N_2O_5 (B682).

First aid   *Vapour inhaled:* standard treatment (p 109).
Affected eyes: standard treatment (p 109).
Skin contact: standard treatment (p 108).
If swallowed: standard treatment (p 109).

Spillage disposal   Instruct others to keep at a safe distance. Wear breathing apparatus and gloves. Spread soda ash liberally over the spillage and mop up cautiously with water — run to waste, diluting greatly with running water.

Sulphuryl diazide

Hazardous reaction   Explodes violently when heated and often spontaneously at ambient temperatures (B885).

Tellurium and compounds

Metallic ingots breaking with a white lustrous fracture, or a greyish-black powder. Telluric acid and the alkali-metal tellurites and tellurates are colourless powders or crystals, soluble in water; tellurium oxide (TeO_2) is also colourless, but is insoluble in water. Tellurium tetrachloride is a cream-coloured solid. Poisonous hydrogen telluride (offensive smell) is generated when an acid solution of a tellurium compound is reduced by metals such as tin and zinc.

HARMFUL BY INHALATION
HARMFUL IF TAKEN INTERNALLY

Avoid breathing dust. Avoid contact with skin and eyes. TLV (tellurium) 0.1 mg m^{-3}.

Toxic effects Inhalation of dust or tellurium fume gives rise to a dry mouth, metallic taste, drowsiness, loss of appetite, excessive salivation, nausea, vomiting and a foul garlic-like odour of the breath. Skin absorption and ingestion also result in foul breath and it can be assumed that effects similar to those of inhalation will be experienced. Ingestion of soluble tellurium salts may, in addition, lead to cyanosis and unconsciousness. **Chronic effects** These are similar to the above, but arise from prolonged absorption of very small amounts of tellurium compounds.

First aid *Inhalation of dust or fume:* standard treatment (p 109).
Affected eyes: standard treatment (p 109).
Skin contact: standard treatment (p 108).
If swallowed: standard treatment (p 109).

Spillage disposal Because of its low toxicity, tellurium powder can be mixed with a large quantity of sand and dealt with as normal refuse. The soluble tellurates and tellurites and telluric acid should be dissolved in a large volume of water and run to waste, diluting greatly with running water.

Tellurium tetrachloride

Hazardous reaction Interaction with liquid ammonia at −15 °C forms an explosive nitride (B700).

Tetraaluminium tricarbide

Hazardous reactions Incandescent reaction with lead peroxide or potassium permanganate (B341).

Tetraborane

Hazardous reactions Ignites in air or oxygen and explodes with concentrated nitric acid (B191).

1,1,2,2-Tetrabromoethane (acetylene tetrabromide)

Very heavy yellowish liquid with chloroform-like odour; mp 0 °C; bp 151 °C at 54 mmHg; immiscible with water.

GIVES OFF POISONOUS VAPOUR
CAUSES IRRITATION OF SKIN AND EYES

Avoid breathing vapour. Avoid contact with skin and eyes.

Toxic effects Records of these have not been traced though it is suggested that it has narcotic properties and may cause damage to the liver. It is probably similar in action to tetrachloroethane though not so toxic.

First aid *Vapour inhaled:* standard treatment (p 109).
Affected eyes: standard treatment (p 109).
If swallowed: standard treatment (p 109).

Spillage disposal Instruct others to keep at a safe distance. Wear breathing apparatus and gloves. Apply dispersing agent if available and work to an emulsion with brush and water—run this to waste, diluting greatly with running water. If dispersant not available, absorb on sand, shovel into bucket(s) and transport to safe, open area for burial. Site of spillage should be washed thoroughly with water and soap or detergent.

Tetracarbonylnickel (nickel carbonyl)

Colourless, mobile liquid normally supplied in cylinders; bp 43 °C; practically insoluble in water.

HIGHLY FLAMMABLE
GIVES OFF VERY POISONOUS VAPOUR

Avoid breathing vapour. Avoid contact with skin, eyes and clothing. TLV 0.001 ppm (0.007 mg m^{-3}).

Toxic effects The vapour is exceedingly poisonous when inhaled. With low concentrations the initial symptoms are giddiness and slight headache. Heavier exposure causes nausea, tightness of chest, weakness of limbs, perspiring, coughing, vomiting, cold and clammy skin and shortness of breath. The toxicity is believed to be derived from both the nickel and carbon monoxide liberated in the lungs. Cases of lung cancer have occurred as a result of prolonged exposure to vapour.

Hazardous reactions Reacts explosively with liquid bromine, but smoothly in the gaseous state; a mixture with mercury will explode on vigorous shaking; on exposure to air, the carbonyl produces a deposit which becomes peroxidised and may ignite (B433).

First aid *Vapour inhaled:* standard treatment (p 109).
Affected eyes: standard treatment (p 109).
Skin contact: standard treatment (p 108).
If swallowed: standard treatment (p 109).

Fire hazard The liquid is highly flammable and is liable to explode if heated to 60 °C or above. *Extinguish fire with* dry powder, carbon dioxide or vaporising liquid.

Spillage disposal Shut off all possible sources of ignition. Instruct others to keep at a safe distance. Wear breathing apparatus and gloves. Apply dispersing agent if available and work to an emulsion with brush and water — run this to waste, diluting greatly with running water. If dispersant not available, absorb on sand, shovel into bucket(s) and transport to safe, open area for burial. Site of spillage should be washed thoroughly with water and soap or detergent.

1,2,4,5-Tetrachlorobenzene

Hazardous reaction	Explosions have occurred in the commercial production of 2,4,5-trichloro-phenol by alkaline hydrolysis of tetrachlorobenzene, leading to formation of the extremely toxic tetrachlorodibenzodioxin (TCDD) (B463).

1,1,2,2-Tetrachloroethane (acetylene tetrachloride)

Colourless, heavy liquid with chloroform-like odour; bp 146 °C; immiscible with water.

GIVES OFF VERY POISONOUS VAPOUR

Avoid breathing vapour. Avoid contact with eyes. TLV (skin) 5ppm (35 mg m^{-3}).

Toxic effects	The vapour irritates the eyes, nose and lungs and may cause drowsiness, giddiness and headache and, in high concentrations, unconsciousness. Assumed to be poisonous if taken by mouth. **Chronic effects** Continuous breathing of low concentrations of vapour over a period may cause jaundice by action on the liver; it may also affect the nervous system and blood. (*See* note in Ch 6 p 84.)
Hazardous reactions	When heated with solid NaOH, chloro- or dichloro- acetylene, which ignite in air, are formed; forms impact-sensitive mixtures with K and Na (B290).
First aid	*Vapour inhaled:* standard treatment (p 109). *Affected eyes:* standard treatment (p 109). *If swallowed:* standard treatment (p 109).
Spillage disposal	Instruct others to keep at a safe distance. Wear breathing apparatus and gloves. Apply dispersing agent if available and work to an emulsion with brush and water — run this to waste, diluting greatly with running water. If dispersant not available, absorb on sand, shovel into bucket(s) and transport to safe, open area for burial. Site of spillage should be washed thoroughly with water and soap or detergent.

Tetrachloroethylene (perchloroethylene)

Colourless liquid with faint ethereal odour; bp 121 °C; immiscible with water.

HARMFUL VAPOUR

Avoid breathing vapour. Avoid contact with skin and eyes. TLV 100 ppm (670 mg m^{-3}).

Toxic effects	The inhalation of the vapour may cause dizziness, nausea and vomiting — in high concentrations, stupor. The liquid irritates the eyes and has a degreasing action on the skin. Assumed to cause symptoms similar to those of inhalation if taken by mouth.
Hazardous reactions	Explosive mixtures are formed with N_2O_4, Ba, Li; impure material containing trichloroethylene, when treated with solid NaOH and subsequently distilled, may have a volatile, explosive fore-run (B274).

First aid	*Vapour inhaled:* standard treatment (p 109). *Affected eyes:* standard treatment (p 109). *Skin contact:* standard treatment (p 108). *If swallowed:* standard treatment (p 109).
Spillage disposal	Wear face-shield or goggles, and gloves. Apply dispersing agent if available and work to an emulsion with brush and water—run this to waste, diluting greatly with running water. If dispersant not available, absorb on sand, shovel into bucket(s) and transport to safe, open area for burial. Site of spillage should be washed thoroughly with water and soap or detergent.

Tetraethylenepentamine

Pale yellow-brown, viscous, hygroscopic liquid; bp 340 °C; miscible with water.

CAUSES BURNS
IRRITATING TO SKIN AND EYES

Avoid contact with skin, eyes and clothing.

Toxic effects	The liquid irritates the eyes and skin and may cause burns. It is irritating and harmful if taken internally.
First aid	*Affected eyes:* standard treatment (p 109). *Skin contact:* standard treatment (p 108). *If swallowed:* standard treatment (p 109).
Spillage disposal	Wear face-shield or goggles, and gloves. Mop up with plenty of water and run to waste, diluting greatly with running water.

Tetraethyl silicate

Colourless liquid; bp 166 °C; practically insoluble in water, by which it is slowly hydrolysed.

FLAMMABLE
HARMFUL IF TAKEN INTERNALLY
IRRITATING TO EYES AND RESPIRATORY SYSTEM

Avoid breathing vapour. Avoid contact with skin, eyes and clothing.

Toxic effects	The vapour irritates the eyes and respiratory system and is narcotic in high concentrations. The liquid irritates the eyes and may irritate the skin; it is irritant and damaging if taken by mouth.
First aid	*Vapour inhaled:* standard treatment (p 109). *Affected eyes:* standard treatment (p 109). *Skin contact:* standard treatment (p 108). *If swallowed:* standard treatment (p 109).
Fire hazard	Flash point 52 °C. *Extinguish fire with* foam, dry powder, carbon dioxide or vaporising liquid.

Spillage disposal	Shut off all possible sources of ignition. Wear face-shield or goggles, and gloves. Apply non-flammable dispersing agent if available and work to an emulsion with brush and water—run this to waste, diluting greatly with running water. If dispersant not available absorb on sand, shovel into bucket(s) and transport to safe, open area for burial. Ventilate site of spillage well to evaporate remaining liquid and dispel vapour.

Tetraethynyltin

Hazardous reaction	Explodes on rapid heating (B545).

Tetrafluoroboric acid and salts
(fluoroboric acid and salts)

Fluoroboric acid is a colourless fuming liquid miscible with water. The salts are generally colourless, crystalline and soluble in water.

CAUSES BURNS
IRRITATING TO SKIN, EYES AND RESPIRATORY SYSTEM

Avoid inhaling vapour or dust. Prevent contact with eyes and skin.

Toxic effects	The vapour and dust irritate all parts of the respiratory system. The acid and salts burn the eyes severely. The acid burns the skin. If taken by mouth there would be severe internal irritation and damage.
First aid	*Vapour or dust inhaled:* standard treatment (p 109). *Affected eyes:* standard treatment (p 109). *Skin contact:* standard treatment (p 108). *If swallowed:* standard treatment (p 109).
Spillage disposal	Wear face-shield or goggles, and gloves. Mop up with plenty of water and run to waste, diluting greatly with running water.

Tetrafluoroethylene

Hazardous reactions	The monomer explodes spontaneously at pressures above 2.7 bar (2.7×10^5 Pa); the inhibited monomer will explode if ignited; the liquid monomer, exposed to air, will form an explosive peroxidic polymer; explodes with iodine pentafluoride; forms explosive, polymeric peroxide with oxygen (B278).

Tetrafluorohydrazine

Hazardous reactions	Gas above −73 °C which explodes on contact with air or combustible vapours; mixtures with hydrocarbons are potentially highly explosive (B741–742).

Tetrahydrofuran

Colourless, volatile liquid with ethereal odour; bp 66 °C; liable to form explosive peroxides on exposure to light and air which should be decomposed before distilling to small volume.

> EXTREMELY FLAMMABLE
> MAY FORM EXPLOSIVE PEROXIDES
> IRRITATING TO EYES AND RESPIRATORY SYSTEM

Avoid breathing vapour. Avoid contact with eyes. TLV 200 ppm (590 mg m^{-3}).

Toxic effects The vapour irritates the eyes and respiratory system; high concentrations have a narcotic effect. It is believed that liver damage may result from skin absorption or ingestion.

Hazardous reactions Readily forms peroxides by autoxidation and reference is given to brochure on handling. Peroxidised material should not be dried with NaOH or KOH as explosions may occur (B409).

First aid *Vapour inhaled:* standard treatment (p 109).
Affected eyes: standard treatment (p 109).
Skin contact: standard treatment (p 108).
If swallowed: standard treatment (p 109).

Fire hazard Flash point −17 °C; explosive limits 2.3–11.8%; ignition temp. 321 °C. *Extinguish fire with* water spray, foam, dry powder, carbon dioxide or vaporising liquid.

Spillage disposal Shut off all possible sources of ignition. Instruct others to keep at a safe distance. Wear breathing apparatus and gloves. Mop up with plenty of water and run to waste, diluting greatly with running water. Ventilate area well to evaporate remaining liquid and dispel vapour.

Tetrahydrothiophen (thiophan; tetramethylene sulphide)

Colourless liquid with smell like that of coal gas for which it is an established odorant; bp 121 °C; immiscible with water.

> HIGHLY FLAMMABLE

Avoid breathing vapour. Avoid contact with skin, eyes and clothing.

Toxic effects The evidence available is derived from animal experiments which showed that it was not irritant to the eyes and skin of rabbits and caused no permanent damage to the eyes. Oral injection produced acute toxicity 'no worse than that of benzene'. It is thought wisest to handle with caution and avoid breathing the vapour, which advertises its presence unmistakably.

First aid *Vapour inhaled:* standard treatment (p 109).
Affected eyes: standard treatment (p 109).
Skin contact: standard treatment (p 108).
If swallowed: standard treatment (p 109).

Fire hazard Flash point 18 °C; *Extinguish fire with* foam, dry powder, carbon dioxide or vaporising liquid.

Spillage disposal Shut off all possible sources of ignition. Instruct others to keep at a safe distance. Wear breathing apparatus and gloves. Apply non-flammable dispersing agent and work to an emulsion with brush and water — run this to waste, diluting greatly with running water. If dispersant not available, absorb on sand, shovel into bucket(s) and transport to safe, open area for burial. Site of spillage should be washed thoroughly with water and soap or detergent.

Tetramethylammonium chlorite

Hazardous reaction The dry solid explodes on impact (B429).

Tetramethyllead

Hazardous reaction Liable to explode above 90 °C (B428).

Tetraselenium tetranitride

Hazardous reactions Dry material explodes on compression or on heating at 130–230 °C; contact with bromine, chlorine or fuming hydrochloric acid also causes explosion (B884).

Tetrasilane

Hazardous reactions Ignites and explodes in air or oxygen; reacts vigorously with carbon tetrachloride (B822).

Tetrasulphur tetranitride

Hazardous reactions Liable to explosive decomposition on friction, shock or heating above 100 °C (B884).

Tetrazoles Review of group (B141).

Thallium and salts

Thallium – soft, silvery sticks often stored under water. The carbonate, nitrate and sulphate are colourless, crystalline solids, soluble in water; the oxide is black and insoluble in water.

POISONOUS DUST
SERIOUS RISK OF POISONING BY INHALATION OR SWALLOWING
DANGER OF CUMULATIVE EFFECTS

Avoid breathing dust. Avoid contact with skin and eyes. TLV (skin) (soluble salts at Tl) 0.1 mg m^{-3}.

Toxic effects The dusts irritate the nose and eyes and may cause nausea and abdominal pain by absorption. The metal on contact with moist skin produces a white film of the hydroxide. Skin absorption of the soluble salts and ingestion may cause nausea, vomiting, abdominal pains, weakness of the legs and mental confusion. **Chronic effects** Exposure over a long period to small amounts of the dust or solutions may result in loss of appetite, falling out of hair, pain or weakness of limbs, insomnia and mental disturbance. (*See* note in Ch 6 p 84)

First aid *Dust inhaled:* standard treatment (p 109).
Affected eyes: standard treatment (p 109).
Skin contact: standard treatment (p 108).
If swallowed: standard treatment (p 109).

Spillage disposal Small quantities of soluble thallium salts can be dissolved in a large volume of water and the solution run to waste, diluting greatly with running water. Insoluble thallium can be thoroughly mixed with a large amount of sand and the mixture buried in a safe, open area.

Thallium azide

Hazardous reaction Relatively stable, it can be exploded on heavy impact or by heating at 350–400 °C (B883).

Thiocarbonyl tetrachloride
– *see* Trichloromethane sulphenyl chloride

Thiocyanogen

Hazardous reaction Polymerises explosively above its mp −2 °C (B336).

Thioglycollic acid – *see* Mercaptoacetic acid

Thionyl chloride (sulphinyl chloride)

Pale yellow or yellow fuming liquid with an odour like sulphur dioxide; bp 79 °C; it reacts with water to form hydrochloric acid and sulphur dioxide.

CAUSES BURNS
IRRITATING TO SKIN, EYES AND RESPIRATORY SYSTEM

Avoid breathing vapour. Prevent contact with skin and eyes. Do not put water into container.

Hazardous reactions Addition of concentrated ammonia may cause an explosion; reacts violently or explosively with dimethyl sulphoxide (B681).

First aid	*Vapour inhaled:* standard treatment (p 109).
	Affected eyes: standard treatment (p 109).
	Skin contact: standard treatment (p 108).
	If swallowed: standard treatment (p 109).
Spillage disposal	Instruct others to keep at a safe distance. Wear breathing apparatus and gloves. Spread soda ash liberally over spillage and mop up cautiously with water — run to waste, diluting greatly with running water.

Thiophan *– see* Tetrahydrothiophen

Thiophen

Clear, colourless liquid; bp 84 °C; insoluble in water.

       EXTREMELY FLAMMABLE

Avoid breathing vapour. Avoid contact with skin, eyes and clothing.

Toxic effects	Animal experiments indicate moderate toxicity.
Hazardous reaction	Reacts very violently with fuming nitric acid (B387).
First aid	*Affected eyes:* standard treatment (p 109).
	Skin contact: standard treatment (p 108).
	If swallowed: standard treatment (p 109).
Fire hazard	Flash point −6 °C *Extinguish fire with* foam, dry powder, carbon dioxide or vaporising liquid.
Spillage disposal	Shut off all possible sources of ignition. Instruct others to keep at a safe distance. Wear breathing apparatus and gloves. Apply non-flammable dispersing agent and work to an emulsion with brush and water — run this to waste, diluting greatly with running water. If dispersant not available, absorb on sand, shovel into bucket and transport to safe, open area for burial or slow evaporation.

Thiophenol *– see* Benzenethiol

Thiophosphoryl fluoride

Hazardous reactions	Ignites or explodes on contact with air; heated sodium ignites in the gas (B741).

Thorium (metal)

Hazardous reactions	Finely divided metal is pyrophoric in air; incandesces in chlorine, bromine and iodine vapour; reactions with P and S are also incandescent (B947).

Thorium hydride

Hazardous reactions   Explodes on heating in air, powdered material readily ignites in air on handling (B816).

Tin, organic compounds of

Certain alkyltin compounds — notably di-n-butyltin diacetate, dilaurate, maleate and oxide — are used quite extensively as stabilisers for pvc resins, as catalysts and biocides. These are harmful by skin absorption and some, as dust, irritate the respiratory system. TLV 0·1 mg m^{-3}.

First aid   *Affected eyes :* standard treatment (p 109).
Skin contact : standard treatment (p 108).
If swallowed : standard treatment (p 109).

Spillage disposal   Wear face-shield or goggles, and gloves. Mix with sand and transport tc safe, open area for burial. Site of spillage should be washed thoroughly with water and soap or detergent.

Tin(II) chloride (stannous chloride)

Hazardous reaction   Reaction with hydrogen peroxide is strongly exothermic, even in solution (B688).

Tin(IV) chloride (stannic chloride; tin tetrachloride)

Colourless fuming liquid; bp 114 °C; reacts with water forming hydrochloric acid.

CAUSES BURNS
IRRITATING TO SKIN, EYES AND RESPIRATORY SYSTEM

Avoid contact with skin and eyes. Do not put water into container. TLV (inorganic tin compounds as Sn) 2 mg m^{-3}.

Toxic effects   The vapour irritates the respiratory system. The vapour irritates the eyes. The liquid irritates the eyes and skin and may cause burns. Will result in internal irritation and damage if taken by mouth.

First aid   *Vapour inhaled :* standard treatment (p 109).
Affected eyes : standard treatment (p 109).
Skin contact : standard treatment (p 108).
If swallowed : standard treatment (p 109).

Spillage disposal   Instruct others to keep at a safe distance. Wear breathing apparatus and gloves. Spread soda ash liberally over the spillage and mop up cautiously with water — run to waste, diluting greatly with running water.

Titanium(II) chloride

Hazardous reaction   Ignites readily in air, particularly if moist (B688).

Titanium(IV) chloride

(titanium tetrachloride, titanic chloride)

Colourless, fuming liquid with a pungent odour; bp 136 °C; reacts with water with liberation of hydrogen chloride.

CAUSES BURNS
IRRITATING TO SKIN, EYES AND RESPIRATORY SYSTEM

Avoid breathing vapour. Prevent contact with skin and eyes. Do not put water into container.

Toxic effects   The vapour irritates the respiratory system. The vapour and liquid irritate the eyes and may cause burns. The liquid irritates the skin. Assumed to cause severe internal irritation and damage if taken by mouth.

First aid   *Vapour inhaled:* standard treatment (p 109).
Affected eyes: standard treatment (p 109).
Skin contact: standard treatment (p 108).
If swallowed: standard treatment (p 109).

Spillage disposal   Instruct others to keep at a safe distance. Wear breathing apparatus and gloves. Spread soda ash liberally over the spillage and mop up cautiously with water — run to waste, diluting greatly with running water.

o-Tolidine and *o*-tolidine dihydrochloride

(2,2′-dimethylbenzidine)

Colourless to grey or brown powder; slightly soluble in water. The use of *o*-tolidine and its salts is controlled in the United Kingdom by The Carcinogenic Substances Regulations 1967.

POISONOUS DUST
TOXIC IN CONTACT WITH SKIN
DANGER OF CUMULATIVE EFFECTS

Prevent inhalation of dust. Prevent contact with skin and eyes.

Toxic effects   The dihydrochloride and its solutions irritate the skin and eyes. **Chronic effects** There is evidence that *o*-tolidine, through continued absorption, can cause cancer of the bladder.

First aid   *Affected eyes:* standard treatment (p 109).
Skin contact: standard treatment (p 108).
If swallowed: standard treatment (p 109).

Spillage disposal   Wear breathing apparatus and gloves. Mix spillage with moist sand and shovel mixtures into glass, enamel or polythene vessels for dispersion in an excess of dilute hydrochloric acid (1 volume of concentrated acid diluted with 2 volumes of water). Allow to stand, with occasional stirring, for 24 hours and then run extract to waste, diluting greatly with running water and washing the sand. The residual sand can be treated as normal refuse. The site of the spillage should be washed with water and soap or detergent.

Toluene (toluol)

Colourless liquid with characteristic odour; bp 111 °C; immiscible with water.

HIGHLY FLAMMABLE
HARMFUL VAPOUR

Avoid breathing vapour. Avoid contact with skin and eyes. TLV 100 ppm (375 mg m^{-3}).

Toxic effects Inhalation of the vapour may cause dizziness, headache, nausea and mental confusion. The vapour and liquid irritate the eyes and mucous membranes. Absorption through the skin and ingestion would cause poisoning. **Chronic effects** If the toluene contains benzene as an impurity, breathing of vapour over long periods may cause blood disease. Prolonged skin contact may cause dermatitis.

Hazardous reactions Reacts violently with BrF_3 at $-80\,°C$; liable to explode when mixed with N_2O_4; inadequate control in nitration of toluene with mixed acids may lead to runaway or explosive reaction (B534).

First aid *Vapour inhaled:* standard treatment (p 109).
Affected eyes: standard treatment (p 109).
Skin contact: standard treatment (p 108).
If swallowed: standard treatment (p 109).

Fire hazard Flash point $4.4\,°C$; explosive limits $1.4–6.7\%$; ignition temp. $536\,°C$. *Extinguish fire with* foam, dry powder, carbon dioxide or vaporising liquid.

Spillage disposal Shut off all possible sources of ignition. Wear face-shield or goggles, and gloves. Apply non-flammable dispersing agent if available and work to an emulsion with brush and water — run this to waste, diluting greatly with running water. If dispersant not available, absorb on sand, shovel into bucket(s) and transport to safe, open area for atmospheric evaporation. Ventilate site of spillage well to evaporate remaining liquid and dispel vapour.

Toluene-2-diazonium perchlorate

Hazardous reaction Explosive when wet (B528).

Toluene-2-, -3- and -4-diazonium salts

Covers hazardous reactions with ammonium sulphide, hydrogen sulphide or potassium iodide (B532–533).

Toluene di-isocyanate
- see 2,4-Di-isocyanatotoluene

Toluene-4-sulphonic acid
- see Sulphonic acids

Toluene-4-sulphonyl azide

Hazardous reaction Distillation residue containing this will explode if temperature exceeds $120\,°C$ (B533).

Toluidines

o-Toluidine and *m*-toluidine are red to dark brown liquids; *p*-toluidine consists of pale brown crystals. Insoluble in water.

HARMFUL VAPOUR
TOXIC IN CONTACT WITH SKIN
GIVES OFF POISONOUS VAPOUR

Avoid breathing vapour. Avoid contact with skin and eyes. TLV (skin) (-*o*) 5 ppm (22 mg m^{-3}).

Toxic effects	Excessive breathing of the vapour or absorption through the skin may cause headache, drowsiness, cyanosis, mental confusion and, in severe cases, convulsions. They are dangerous to the eyes and the above effects may also be experienced if they are taken by mouth.
First aid	*Vapour inhaled:* standard treatment (p 109). *Affected eyes:* standard treatment (p 109). *Skin contact:* standard treatment (p 108). *If swallowed:* standard treatment (p 109).
Spillage disposal	Wear face-shield or goggles, and gloves. Mix with sand and shovel mixture into glass, enamel or polythene vessel for dispersion in an excess of dilute hydrochloric acid (1 volume of concentrated acid diluted with 2 volumes of water). Allow to stand, with occasional stirring, for 24 hours and then run acid extract to waste, diluting greatly with running water and washing the sand. The sand can be disposed of as normal refuse.

Tolylcoppers

Hazardous reaction	*o*-, *m*- and *p*-isomers usually explode strongly on exposure to oxygen at 0 °C, or weakly above 100 °C in vacuo (B529).

Trialkyl-aluminiums and -bismuths Reviews (B143).

Triallyl cyanurate (2,4,6-trisallyloxy-*s*-triazine)

Colourless liquid or solid; mp 27 °C; bp 162 °C at 2mmHg; hydrolysed by water forming allyl alcohol which is irritant.

HARMFUL IF TAKEN INTERNALLY

Avoid contact with skin, eyes and clothing.

Toxic effects	Arises because of the allyl alcohol formed on hydrolysis.
First aid	*Affected eyes:* standard treatment (p 109). *Skin contact:* standard treatment (p 108). *If swallowed:* standard treatment (p 109).
Spillage disposal	Wear face-shield or goggles, and gloves. Clear up with dust-pan and brush. May be disposed of, after mixing with sand, as normal refuse or flushed away to waste with water.

Triallyl phosphate

Hazardous reaction Distillation residue exploded (B571).

Triazenes Review of group (B144).

2,4,6-Triazido-1,3,5-triazine

Hazardous reaction Explodes on impact, shock or rapid heating to 170–180 °C (B380).

Tribenzylarsine

Hazardous reaction Oxidises slowly at first in air but becomes violent (B614).

Tribromomethane – *see* Bromoform

Tribromosilane

Hazardous reaction Usually ignites when poured in air (B213).

Tributylamine – *see* Butylamines

Tributylbismuth

Hazardous reactions Explodes in oxygen and ignites in air (B597).

Tributylborane

Hazardous reaction Mixture of n- and iso- isomers ignited on exposure to air (B597).

Tributylphosphine

Colourless liquid with garlic-like odour; bp 240 °C; practically insoluble in water; usually packed under nitrogen to avoid oxidation.

> FLAMMABLE
> HARMFUL VAPOUR

Avoid breathing vapour. Avoid contact with skin, eyes and clothing.

Toxic effects	Irritates respiratory system and skin; very irritant to eyes; moderately toxic by ingestion.
First aid	*Vapour inhaled:* standard treatment (p 109). *Affected eyes:* standard treatment (p 109). *Skin contact:* standard treatment (p 108). *If swallowed:* standard treatment (p 109).
Fire hazard	Flash point 40 °C; ignition temperature 200 °C. *Extinguish fire with* foam, dry powder, carbon dioxide or vaporising liquid.
Spillage disposal	Shut off all possible sources of ignition. Wear face-shield or goggles and gloves. Apply non-flammable dispersing agent if available and work to an emulsion with brush and water — run this to waste, diluting greatly with running water. If dispersant not available, absorb on sand, shovel into bucket(s) and transport to safe open area for atmospheric evaporation. Ventilate site of spillage well to evaporate remaining liquid and dispel vapour.

Tricadmium dinitride

Hazardous reactions	Explodes on heating and on contact with water (B628).

Tricalcium dinitride

Hazardous reactions	Spontaneously flammable in air; incandesces in chlorine or bromine vapour (B626).

Tricalcium diphosphide

Hazardous reaction	Liberates phosphine with water — this ignites spontaneously (B626).

Trichloroacetaldehyde – *see* Chloral

Trichloroacetic acid

Colourless, hygroscopic crystals; mp 58 °C; soluble in water.

> SERIOUS RISK OF POISONING BY INHALATION, SWALLOWING
> OR SKIN CONTACT
> CAUSES SEVERE BURNS

Prevent contact with skin and eyes.

Toxic effects   Severely irritates the eyes and skin, producing blisters after a latent period. Assumed to cause severe irritation and damage if taken by mouth.

First aid   *Affected eyes:* standard treatment (p 109).
Skin contact: standard treatment (p 108).
If swallowed: standard treatment (p 109).

Spillage disposal   Wear face-shield or goggles, and gloves. Spread soda ash liberally over the spillage and mop up cautiously with plenty of water—run to waste, diluting greatly with running water.

Trichloroacetyl chloride

Colourless liquid with acrid, pungent odour; bp 118 °C; reacts with water forming trichloro-acetic and hydrochloric acids.

> SERIOUS RISK OF POISONING BY INHALATION, SWALLOWING
> OR SKIN CONTACT
> CAUSES SEVERE BURNS

Prevent inhalation of vapour. Prevent contact with skin and eyes.

Toxic effects   The vapour irritates the eyes and respiratory system. The liquid burns the eyes and skin. Assumed to cause severe internal irritation and damage if taken by mouth.

First aid   *Vapour inhaled:* standard treatment (p 109).
Affected eyes: standard treatment (p 109).
Skin contact: standard treatment (p 108).
If swallowed: standard treatment (p 109).

Spillage disposal   Instruct others to keep at a safe distance. Wear breathing apparatus and gloves. Spread soda ash liberally over the spillage and mop up cautiously with water — run to waste, diluting greatly with running water.

1,2,4-Trichlorobenzene

Colourless liquid or solid; mp 17 °C; insoluble in water.

> HARMFUL IF TAKEN INTERNALLY
> HARMFUL VAPOUR

Avoid breathing vapour. Avoid contact with skin, eyes and clothing.

Toxic effects	These are not well documented, but it has been ascribed moderate acute and chronic systemic toxicity when it is ingested or the vapour inhaled.
First aid	*Vapour inhaled :* standard treatment (p 109). *Affected eyes :* standard treatment (p 109). *Skin contact :* standard treatment (p 108). *If swallowed :* standard treatment (p 109).
Spillage disposal	Wear face-shield or goggles, and gloves. Mix with sand and transport to a safe, open area for burial. The site of the spillage should be washed thoroughly with water and soap or detergent.

1,1,1,-Trichloroethane (methylchloroform)

Colourless liquid; bp 74 °C; insoluble in water.

HARMFUL VAPOUR

Avoid breathing vapour. Avoid contact with skin, eyes and clothing. TLV 350 ppm (1900 mg m^{-3}).

Toxic effects	The vapour may irritate the eyes and respiratory system; it is narcotic in high concentrations. The liquid may irritate the skin; it irritates the eyes without causing serious damage and must be assumed to be harmful if taken by mouth.
Hazardous reactions	Mixture with potassium may explode on light impact; violent decomposition, with evolution of HCl, may occur when it comes into contact with aluminium, magnesium, or their alloys (B295).
First aid	*Vapour inhaled :* standard treatment (p 109). *Affected eyes :* standard treatment (p 109). *Skin contact :* standard treatment (p 108). *If swallowed :* standard treatment (p 109).
Spillage disposal	Wear face-shield or goggles, and gloves. Apply dispersing agent if available and work to an emulsion with brush and water—run this to waste, diluting greatly with running water. If dispersant not available, absorb on sand, shovel into bucket(s) and transport to safe, open area for atmospheric evaporation. Site of spillage should be washed thoroughly with water and soap or detergent.

Trichloroethylene

Colourless liquid with sweetish chloroform-like odour; bp 87 °C; immiscible with water.

HARMFUL VAPOUR

Avoid breathing vapour. Avoid contact with skin and eyes. TLV 100 ppm (535 mg m^{-3}).

Toxic effects	Inhalation of the vapour may cause headache, dizziness and nausea—with high concentrations, unconsciousness. The vapour and liquid irritate the eyes. Ingestion produces similar effects to inhalation of the vapour.

Hazardous reactions Decomposes with strong alkalies with evolution of spontaneously flammable dichloroacetylene; reacts violently with many metals, *eg* Al, Ba, Be, Li, Mg, Ti; reacts violently with anhydrous perchloric acid, explosively with N_2O_4 and with liquid oxygen (when initiated) (B281).

First aid *Vapour inhaled:* standard treatment (p 109).
Affected eyes: standard treatment (p 109).
If swallowed: standard treatment (p 109).

Spillage disposal Wear face-shield or goggles, and gloves. Apply dispersing agent if available and work to an emulsion with brush and water — run this to waste, diluting greatly with running water. If dispersant not available, absorb on sand, shovel into bucket(s) and transport to safe, open area for atmospheric evaporation. Site of spillage should be washed thoroughly with water and soap or detergent.

Trichloromethane sulphenyl chloride
(perchloromethyl mercaptan; thiocarbonyl tetrachloride)

Oily, yellow liquid; bp 149 °C with slight decomposition; insoluble in water — hydrolysis is slow.

VERY TOXIC IF TAKEN INTERNALLY
IRRITATING TO SKIN, EYES AND RESPIRATORY SYSTEM

Toxic effects The vapour irritates the eyes and respiratory system. The liquid irritates the skin and eyes and is highly toxic if taken by mouth.

First aid *Vapour inhaled:* standard treatment (p 109).
Affected eyes: standard treatment (p 109).
Skin contact: standard treatment (p 108).
If swallowed: standard treatment (p 109).

Spillage disposal Instruct others to keep at a safe distance. Wear breathing apparatus and gloves. Apply dispersing agent if available and work to an emulsion with brush and water — run this to waste, diluting greatly with running water. If dispersant not available, absorb on sand, shovel into bucket(s) and transport to safe, open area for burial. Wash site of spillage thoroughly with water and soap or detergent.

Trichloro(methyl)silane
– *see* **Methyltrichlorosilane**

Trichloronitromethane (chloropicrin)

Colourless liquid with intense penetrating odour; bp 112 °C; almost insoluble in water.

SERIOUS RISK OF POISONING BY INHALATION, SWALLOWING
OR SKIN CONTACT
GIVES OFF VERY POISONOUS VAPOUR
IRRITATING TO SKIN, EYES AND RESPIRATORY SYSTEM

Do not breath vapour. Prevent contact with skin and eyes. TLV 0.1 ppm (0.7 mg m^{-3}).

Toxic effects   The vapour irritates the respiratory system severely leading in severe cases to bronchitis and recurrent asthmatic attacks through lung damage; it causes nausea and vomiting. The vapour irritates the eyes severely, causing intense lachrymation. The vapour and liquid irritate the skin. If swallowed, the liquid causes vomiting and diarrhoea.

Hazardous reactions   Above a critical volume, bulk containers can be shocked into detonation; reacts violently with aniline at 145 °C, and with alcoholic sodium hydroxide (B218).

First aid
Vapour inhaled: standard treatment (p 109).
Affected eyes: standard treatment (p 109).
Skin contact: standard treatment (p 108).
If swallowed: standard treatment (p 109).

Spillage disposal   Instruct others to keep at a safe distance. Wear breathing apparatus and gloves. Apply dispersing agent if available and work to an emulsion with brush and water — run this to waste diluting greatly with running water. If dispersant not available, absorb on sand, shovel into bucket(s) and transport to safe open area for burial. Site of spillage should be washed thoroughly with water and soap or detergent.

2,4,5-Trichlorophenol

Colourless crystals or grey flakes with strong, phenolic odour; mp 67 °C; practically insoluble in water.

SERIOUS RISK OF POISONING BY INHALATION, SWALLOWING OR SKIN CONTACT
IRRITATING TO SKIN, EYES AND RESPIRATORY SYSTEM

Avoid breathing dust. Avoid contact with skin, eyes and clothing.

Toxic effects   Inhalation, ingestion or skin absorption of the dust or solid may result in lung, liver or kidney damage; symptoms of poisoning are an increase followed by a decrease in respiratory rate and urinary output, fever, increased bowel action, weakness of movement, collapse and convulsions. Skin contact may cause dermatitis.

First aid
Affected eyes: standard treatment (p 109).
Skin contact: standard treatment (p 108). But see note on phenol p 107.
If swallowed: standard treatment (p 109).

Spillage disposal   Wear face-shield or goggles, and gloves. Mix with sand and transport to safe, open area for burial. Site of spillage should be washed thoroughly with water and soap or detergent.

ααα-Trichlorotoluene
– *see* Benzylidene chloride

2,4,6-Trichloro-s-triazine
(cyanuric chloride; cyanuryl chloride)

Colourless crystals with pungent odour; hydrolyses in presence of water, forming hydrochloric acid.

HARMFUL BY INHALATION
IRRITATING TO SKIN, EYES AND RESPIRATORY SYSTEM

Avoid breathing dust. Avoid contact with skin and eyes.

Toxic effects	The dust irritates the respiratory system, eyes and skin. Assumed to be irritating and damaging to the alimentary system if taken by mouth.
First aid	*Dust inhaled:* standard treatment (p 109). *Affected eyes:* standard treatment (p 109). *Skin contact:* standard treatment (p 108). *If swallowed:* standard treatment (p 109).
Spillage disposal	Wear face-shield or goggles, and gloves. Mop up with plenty of water and run to waste, diluting greatly with running water.

Tricopper diphosphide

Hazardous reactions	Powder burns in chlorine; mixtures with potassium chlorate explode on impact, and with potassium nitrate on heating; CuP and CuP_2 behave similarly (B723).

Tricresyl phosphate – *see* Tritolyl phosphate

Tricyclo[3.1.0.0^{2,6}]hex-3-ene – *see* Benzvalene

Triethylaluminium

Hazardous reactions	Reacts explosively with methanol, ethanol, propan-2-ol, vigorously with 2-methylpropan-2-ol; a mixture with *N,N*-dimethylformamide explodes when heated (B512).

Triethylamine

Colourless liquid with strong ammoniacal odour; bp 89 °C; miscible with water.

EXTREMELY FLAMMABLE
IRRITATING TO SKIN, EYES AND RESPIRATORY SYSTEM

Avoid breathing vapour. Avoid contact with skin and eyes. TLV 25 ppm (100 mg m^{-3}).

Toxic effects	The vapour irritates the eyes and respiratory system. The liquid irritates the eyes. Assumed to be irritant and poisonous if taken by mouth.
Hazardous reaction	Complex with N_2O_4, containing excess of latter, exploded below 0 °C when free of solvent (B514).
First aid	*Vapour inhaled:* standard treatment (p 109). *Affected eyes:* standard treatment (p 109). *Skin contact:* standard treatment (p 108). *If swallowed:* standard treatment (p 109).
Fire hazard	Flash point −7 °C; explosive limits 1.2–8.0%. *Extinguish fire with* water spray, dry powder, carbon dioxide or vaporising liquid.
Spillage disposal	Shut off all possible sources of ignition. Instruct others to keep at a safe distance. Wear breathing apparatus and gloves. Mop up with plenty of water and run to waste, diluting greatly with running water. Ventilate area well to evaporate remaining liquid and dispel vapour.

Triethylantimony

Hazardous reaction	Inflames in air (B515).

Triethylarsine

Hazardous reaction	Inflames in air (B513).

Triethylbismuth

Hazardous reactions	Ignites in air, and explodes at about 150 °C (B513).

Triethylborane

Hazardous reaction	Ignites in air (B513).

Triethylenetetramine

Colourless to yellowish liquid; bp 278 °C; miscible with water.

CAUSES BURNS
IRRITATING TO SKIN, EYES AND RESPIRATORY SYSTEM

Avoid breathing vapour. Avoid contact with skin and eyes.

Toxic effects	The vapour irritates the eyes and respiratory system. The liquid burns the skin and eyes. Assumed to be irritant and poisonous if taken by mouth.

First aid   *Vapour inhaled:* standard treatment (p 109).
                           Affected eyes: standard treatment (p 109).
                           Skin contact: standard treatment (p 108).
                           If swallowed: standard treatment (p 109).

Spillage   Wear face-shield or goggles, and gloves. Mop up with plenty of water and
disposal   run to waste, diluting greatly with running water.

Triethylgallium

Hazardous   Ignites in air (B513).
reaction

Triethylphosphine

Hazardous   Explosive product by reaction of oxygen at low temperature (B515).
reaction

Triethynylaluminium

Hazardous   Residue from sublimation of dioxan complex is explosive; trimethylamine
reactions   complex may also explode on sublimation (B464).

Triethynylantimony

Hazardous   Explodes on strong friction (B467).
reaction

Triethynylarsine

Hazardous   Explodes on strong friction (B464).
reaction

1,3,5-Triethynylbenzene

Hazardous   Exploded on rapid heating and compression (B588).
reactions

Triethynylphosphine

Hazardous   Explodes on strong friction and may explode spontaneously on standing
reactions   (B467).

Trifluoroacetic acid and anhydride

Colourless liquids with pungent odour; bp 72 °C and 40 °C respectively; the acid is miscible with water, and the anhydride reacts with water forming the acid.

> SERIOUS RISK OF POISONING BY INHALATION, SWALLOWING OR SKIN CONTACT
> CAUSES SEVERE BURNS

Avoid breathing vapour. Prevent contact with skin and eyes.

Toxic effects The vapour irritates the eyes and respiratory system. The liquid burns the eyes and quickly penetrates the skin to cause deep-seated burns. Assumed to cause severe burning and damage if taken by mouth. There is no reported toxic effect due to the presence of fluorine as in the case of highly toxic fluoroacetic acid.

First aid *Vapour inhaled:* standard treatment (p 109).
Affected eyes: standard treatment (p 109).
Skin contact: standard treatment (p 108).
If swallowed: standard treatment (p 109).

Spillage disposal Instruct others to keep at a safe distance. Wear breathing apparatus and gloves. Spread soda ash liberally over the spillage and mop up cautiously with water — run to waste, diluting greatly with running water.

Trifluoroperoxyacetic acid (peroxytrifluoroacetic acid)

Hazardous reactions An extremely powerful oxidising agent that must be used with great care (B283).

3,3,3-Trifluoropropyne

Hazardous reaction Liable to explode (B344).

Tri-isobutylaluminium

Hazardous reaction Powerful reductant supplied in hydrocarbon solvent; undiluted material ignites in air (B596).

Tri-isopropylphosphine

Hazardous reactions Reacts rather vigorously with most peroxides, ozonides, *N*-oxides, and chloroform (B575).

Trilead dinitride

**Hazardous
reaction**
Decomposes explosively during vacuum degassing (B881).

Trimagnesium diphosphide

**Hazardous
reactions**
Ignites on heating in chlorine, or in bromine or iodine vapours; reaction with nitric acid causes incandescence; with water, phosphine is evolved and may ignite (B862).

Trimercury dinitride

**Hazardous
reactions**
Explodes on friction, impact or in contact with sulphuric acid (B830).

Trimercury tetraphosphide

**Hazardous
reactions**
Ignites in chlorine or when warmed in air; mixture with potassium chlorate explodes on impact (B830).

Trimethylaluminium

**Hazardous
reaction**
Extremely pyrophoric (B373).

Trimethylamine and solutions

Colourless gas with fishy odour that clings to clothes; bp 3 °C; available in liquefied form in cylinders and also in aqueous and ethanolic solutions.

EXTREMELY FLAMMABLE
IRRITATING TO SKIN, EYES AND RESPIRATORY SYSTEM

Avoid breathing vapour. Avoid contact with skin and eyes.

Toxic effects
The vapour irritates the mucous membranes and respiratory system; in high concentrations it may affect the nervous system. The vapour and solutions irritate the eyes. The solutions may irritate the skin. Assumed to be poisonous if taken by mouth.

First aid
Vapour inhaled: standard treatment (p 109).
Affected eyes: standard treatment (p 109).
Skin contact: standard treatment (p 108).
If swallowed: standard treatment (p 109).

Fire hazard (a) Gas: explosive limits 2.0–11.6 % ignition temp. 190 °C. Since the gas is supplied in a cylinder, turning off the valve will reduce any fire involving it; if possible, cylinders should be removed quickly from an area in which fire has developed. (b) Solutions in water and ethanol: *extinguish fire with spray, foam, dry powder, carbon dioxide or vaporising liquid.*

Disposal Surplus gas or leaking cylinder can be vented slowly into a water-fed scrubbing tower or column in a fume cupboard, or into a cupboard served by such a tower. Solutions may be run to waste, diluting greatly with running water.

Trimethylamine oxide

Hazardous preparation The oxide exploded during concentration (B376).

Trimethylarsine

Hazardous reactions Inflames in air; reaction with halogens is violent (B374).

Trimethylbismuth

Hazardous reaction Ignites in air (B374).

Trimethylborane

Hazardous reaction Ignites in air (B374).

Trimethylchlorosilane (chloro(trimethyl)silane)

Colourless, volatile, fuming liquid with pungent odour; bp 57 °C; reacts violently with water.

EXTREMELY FLAMMABLE
CAUSES BURNS
IRRITATING TO SKIN, EYES AND RESPIRATORY SYSTEM

Prevent inhalation of vapour. Prevent contact with skin, eyes and clothing.

Toxic effects The vapour and fumes are strongly irritant to the eyes, skin and respiratory system. The liquid burns the skin and eyes and will cause severe damage if taken internally.

First aid *Vapour inhaled:* standard treatment (p 109).
Affected eyes: standard treatment (p 109).
Skin contact: standard treatment (p 108).
If swallowed: standard treatment (p 109).

Fire hazard   Flash point −18 °C. *Extinguish fire with* dry sand, dry powder or carbon dioxide.

Spillage   Instruct others to keep at a safe distance. Wear breathing apparatus and
disposal   gloves. Spread soda ash liberally over the spillage and mop up cautiously with water — run to waste, diluting greatly with running water.

Trimethylgallium

Hazardous   Ignites in air and reacts violently with water (B375).
reactions

2,4,4-Trimethylpentene (di-sobutylene)

Colourless liquid; bp 102 °C; insoluble in water.

               HIGHLY FLAMMABLE

Avoid breathing vapour. Avoid contact with skin and eyes.

Toxic effects   The vapour is slightly irritant at low concentrations, more so and narcotic at high concentrations. The liquid irritates the eyes and may irritate the skin.

First aid   *Vapour inhaled:* standard treatment (p 109).
                    Affected eyes: standard treatment (p 109).
                    Skin contact: standard treatment (p 108).
                    If swallowed: standard treatment (p 109).

Fire hazard   Flash point below 22 °C. *Extinguish fire with* foam, dry powder, carbon dioxide or vaporising liquids.

Spillage   Shut off all possible sources of ignition. Wear face-shield or goggles, and
disposal   gloves. Apply non-flammable dispersing agent if available and work to an emulsion with brush and water — run this to waste, diluting greatly with running water. If dispersant not available, absorb on sand, shovel into bucket(s) and transport to safe open area for atmospheric evaporation. Ventilate site of spillage well to evaporate remaining liquid and dispel vapour.

Trimethyl phosphate

Hazardous   Distillation residue exploded (B376).
reaction

Trimethylphosphine

Hazardous   May ignite in air (B376).
reaction

Trimethylthallium

Hazardous reactions Liable to explode above 90 °C; ignites in air (B377).

2,4,6-Trimethyl-1,3,5-trioxan
-*see* Paraldehyde

Trinitroacetonitrile

Hazardous reaction Explodes if heated quickly to 220 °C (B337).

2,2,2-Trinitroethanol

Hazardous reaction Shock-sensitive explosive which has exploded during distillation (B300).

Trinitromethane (nitroform)

Hazardous reactions Exploded during distillation; exploded in mixture with an impure ketone; frozen mixtures of trinitromethane and propan-2-ol exploded during thawing (B229).

2,4,6-Trinitrotoluene (TNT)

Hazardous reactions The explosion temperature of TNT was reduced by addition of 1% red lead, sodium carbonate or potassium hydroxide (B519).

Trioxygen difluoride

Hazardous reactions Powerful oxidant which causes ignition or explosions even at −183 °C with some oxidisable materials (B738).

Triphenylaluminium

Hazardous reaction Evolves heat and sparks on contact with water (B609).

Triphenyltin hydroperoxide

Hazardous reaction   Explodes reproducibly at 75 °C (B610).

Trisilane

Hazardous reaction   Ignites or explodes in air or oxygen (B821).

Trisilver nitride

Hazardous reaction   Sensitive explosive when dry (B158).

Trisilylamine

Hazardous reaction   Ignites in air (B821).

Tritellurium tetranitride

Hazardous reaction   Two explosive forms described (B884).

Trithorium tetranitride

Hazardous reaction   Burns incandescently in air, vividly in oxygen (B884).

Tritolyl phosphate (tricresyl phosphate)

Pale brown, almost colourless, liquid; bp 410 °C; immiscible with water.

**SERIOUS RISK OF POISONING BY
INHALATION, SWALLOWING OR SKIN CONTACT**

Avoid contact with skin and eyes.

Toxic effects   When absorbed through the skin or ingested, tritolyl phosphate may cause serious damage to the nervous and digestive systems. Poisoning may show itself in degrees of muscular pain and paralysis.

First aid   *Affected eyes:* standard treatment (p 109).
Skin contact: standard treatment (p 108).
If swallowed: standard treatment (p 109).

Spillage disposal	Wear face-shield or goggles, and gloves. Apply dispersing agent if available and work to an emulsion with brush and water—run this to waste, diluting greatly with running water. If dispersant not available, absorb on sand, shovel into bucket(s) and transport to safe, open area for burial. Site of spillage should be washed thoroughly with water and soap or detergent.

Trivinylbismuth

Hazardous reaction	Ignites in air (B493).

Uranium (metal)

Hazardous reactions	Storage of foil in closed containers in presence of air and moisture may produce a pyrophoric surface; the metal incandesces in ammonia at dull red heat; ignites in fluorine at ambient temperature, in chlorine at 150–180 °C, in bromine vapour at 210–240 °C and in iodine vapour at 260 °C; incandesces in sulphur vapour and with selenium (B950–951).

Uranium compounds

The commonest uranium compounds encountered in the laboratory are the acetate, nitrate and double zinc and magnesium acetates, all of which are yellow crystalline salts soluble in water. Uranium hexafluoride is a colourless or pale yellow crystalline solid which sublimes readily at about 56 °C. Users in the United Kingdom who stock appreciable quantities of uranium compounds are advised to ascertain their responsibilities under The Ionising Radiation (Unsealed Radioactive Substances) Regulations, 1968.

> SERIOUS RISK OF POISONING BY
> INHALATION OR SWALLOWING
> DANGER OF CUMULATIVE EFFECTS

Avoid breathing dust or vapour. TLV (as uranium) (0.2 mg m^{-3}). (*See* Ch 6, p 82, 87.)

Toxic effects	The dust may irritate the lungs and cause retention of uranium in the body with subsequent damage to the kidneys. The vapour of the hexachloride irritates the respiratory system and may injure the kidneys. Assumed to cause internal damage if taken by mouth.
First aid	*Dust or vapour inhaled:* standard treatment (p 109). *Affected eyes:* standard treatment (p 109). *If swallowed:* standard treatment (p 109).
Spillage disposal	Small quantities of soluble uranium salts can be dissolved in a large volume of water and run to waste, diluting greatly with running water. Small amounts of insoluble compounds can be mixed with a large excess of sand and buried in a safe, open area.

Uranium dicarbide

Hazardous reactions Ignites on grinding in a mortar or on heating in air to 400 °C; incandescent reactions with fluorine, chlorine and bromine; violent reaction with hot water (B340).

Uranium hexafluoride

Hazardous reactions Reacts very vigorously with benzene, toluene or xylene, violently with water or ethanol (B746).

Uranium(III) hydride

Hazardous reaction Dry powdered hydride ignites in air (B808).

Uranium(IV) hydride

Hazardous reaction Finely divided hydride ignites on contact with oxygen (B816).

Uranyl diperchlorate

Hazardous reaction Attempted recrystallisation from ethanol caused explosion (B685).

Vanadium compounds

Vanadium pentoxide is a red-brown to dark brown powder. The other compounds most commonly encountered are the sodium and ammonium vanadates, which are colourless, crystalline, and soluble in water.

**HARMFUL BY INHALATION
IRRITATING TO SKIN, EYES AND RESPIRATORY SYSTEM**

Avoid breathing dust. TLV (pentoxide dust) (0.5 mg m^{-3}), (pentoxide fume) (0.05 mg m^{-3}).

Toxic effects The dust or fume of vanadium pentoxide causes irritation of the respiratory system, chest constriction, coughing, and the tongue assumes a blackish-green colour. The dust or fume irritates the eyes and may cause conjunctivitis. If taken by mouth vanadium compounds cause vomiting, excessive salivation and diarrhoea; large doses may damage the nervous system, causing drowsiness, convulsions and unconsciousness.

First aid *Dust or fume inhaled:* standard treatment (p 109).
Affected eyes: standard treatment (p 109).
If swallowed: standard treatment (p 109).

Spillage disposal Soluble vanadium compounds can be dissolved in water and run to waste, diluting greatly with running water. Insoluble compounds can be mixed with moist sand and buried in a safe, open area.

Vanadium trichloride

Hazardous reaction   Reaction with Grignard reagents is almost explosively violent under some conditions (B697).

Vanadyl triperchlorate

Hazardous reactions   Explodes above 80 °C, and ignites many organic solvents on contact (B695).

Vinyl acetate

Colourless liquid; bp 73 °C; 1 g dissolves in 50 cm³ water at 20 °C.

EXTREMELY FLAMMABLE
HARMFUL VAPOUR

Avoid breathing vapour. Avoid contact with skin and eyes.

Toxic effects   The vapour may be narcotic when inhaled in high concentrations. The liquid irritates the eyes and may irritate the skin by its defatting action; it is assumed to be harmful if taken by mouth.

Hazardous reactions   Polymerisation may accelerate to dangerous extent; the vapour reacts vigorously in contact with silica gel or alumina; unstabilised polymer exposed to oxygen at 50 °C gave interpolymeric peroxide which was explosive (B397).

First aid   *Vapour inhaled in high concentrations:* standard treatment (p 109).
Affected eyes: standard treatment (p 109).
Skin contact: standard treatment (p 108).
If swallowed: standard treatment (p 109).

Fire hazard   Flash point −8 °C; explosive limits 2.6–13.4%; ignition temp 427 °C. *Extinguish fire with* foam, dry powder, carbon dioxide or vaporising liquid.

Spillage disposal   Shut off all possible sources of ignition. Instruct others to keep at a safe distance. Wear breathing apparatus and gloves. Apply non-flammable dispersing agent if available and work to an emulsion with brush and water — run this to waste, diluting greatly with running water. If dispersant not available, absorb on sand, shovel into bucket(s) and transport to safe, open area for atmospheric evaporation. Site of spillage should be washed thoroughly with water and soap or detergent.

Vinyl acetate ozonide

Hazardous reaction   Explosive when dry (B401).

Vinyl bromide (bromoethylene)

Colourless liquid or gas; bp 16 °C; insoluble in water.

> EXTREMELY FLAMMABLE
> VERY TOXIC BY INHALATION

Prevent inhalation of vapour. Prevent contact with skin and eyes.

Toxic effects Inhalation of vapour in high concentrations may produce dizziness and narcosis. The liquid irritates the eyes and may irritate the skin by its defatting action; it is assumed to be harmful if taken by mouth. In view of the recent observation that vinyl chloride (*see* below) can cause a cancer of the liver, it must be assumed that vinyl bromide is likely to behave in a similar manner.

First aid *Vapour inhaled in high concentrations:* standard treatment (p 109).
Affected eyes: standard treatment (p 109).
Skin contact: standard treatment (p 108).
If swallowed: standard treatment (p 109).

Fire hazard Flash point below −8 °C. *Extinguish fire with* water spray, foam, dry powder, carbon dioxide or vaporising liquid.

Spillage disposal Shut off all possible sources of ignition. Instruct others to keep at a safe distance. Wear breathing apparatus and gloves. Apply non-flammable dispersing agent if available and work to an emulsion with brush and water—run this to waste, diluting greatly with running water. If dispersant not available, absorb on sand, shovel into bucket(s) and transport to safe, open area for atmospheric evaporation. Site of spillage should be ventilated thoroughly.

Vinyl chloride (chloroethylene)

Colourless gas with pleasant, sweet odour; bp −14 °C; slightly soluble in water.

> EXTREMELY FLAMMABLE
> VERY TOXIC BY INHALATION
> DANGER OF VERY SERIOUS IRREVERSIBLE EFFECTS

Prevent inhalation of gas. Prevent contact with liquid.

Toxic effects Inhalation of vapour in high concentrations produces dizziness and narcosis. The liquid may irritate and burn the skin, the latter due to its freezing action. As a result of observations made in the US (1974), exposure to working atmospheres of vinyl chloride monomer (VCM) has been shown to cause a rare liver cancer, angiosarcoma. This may not manifest itself until over 20 years after initial exposure. The UK Health and Safety Executive has set an interim hygiene standard (Vinyl Chloride Code of Practice for Health Precautions, February 1975) stating that the average concentration of the monomer over a whole working shift must not exceed 25 ppm, and providing that wherever practicable exposure should be brought as near as possible to zero concentration.

Hazardous reaction Formation of unstable polyperoxide may occur (B293).

First aid	*Vapour inhaled in high concentrations:* standard treatment (p 109).
	Skin contact with liquid: standard treatment (p 108).
Fire hazard	Flash point $-78\,°C$; explosive limits 4–22%; ignition temp. 472 °C. Since the gas is supplied in a cylinder, turning off the valve will reduce any fire involving it; if possible, cylinders should be removed quickly from an area in which a fire has developed.
Spillage disposal	Surplus gas or leaking cylinder can be vented slowly to air in a safe, open area or gas burnt off through a suitable burner in a fume cupboard.

Vinylethylene – *see* Buta-1,3-diene

Vinyl fluoride (fluoroethylene)

Hazardous reaction	Explosive when mixed with air (explosive limits 2.6–22%).

Vinylidene chloride (1,1-dichloroethylene)

Colourless, volatile liquid with chloroform-like odour; bp 32 °C; almost insoluble in water.

EXTREMELY FLAMMABLE
MAY FORM EXPLOSIVE PEROXIDES
HARMFUL VAPOUR

Avoid breathing vapour. Avoid contact with skin and eyes.

Toxic effects	Inhalation of vapour may cause drowsiness and anaesthesia; maximum safe working concentration is about 25 ppm. The frequent inhalation of small quantities can result in chronic effects which take the form of liver and kidney damage. The liquid irritates the skin and eyes and must be assumed to be poisonous if taken by mouth.
Hazardous reactions	Rapidly absorbs oxygen from air forming explosive peroxide; reaction products formed with ozone are particularly dangerous (B288).
First aid	*Vapour inhaled:* standard treatment (p 109).
	Affected eyes: standard treatment (p 109).
	Skin contact: standard treatment (p 108).
	If swallowed: standard treatment (p 109).
Fire hazard	Flash point $-15\,°C$ (open cup); explosive limits 5.6–11.4%; ignition temp. 458 °C. *Extinguish fire with* water spray, foam, dry powder, carbon dioxide or vaporising liquid.
Spillage disposal	Shut off all possible sources of ignition. Instruct others to keep at a safe distance. Wear breathing apparatus and gloves. Apply non-flammable dispersing agent if available and work to an emulsion with brush and water—run this to waste, diluting greatly with running water. If dispersant not available, absorb on sand, shovel into bucket(s) and transport to safe, open area for atmospheric evaporation. Site of spillage should be ventilated thoroughly.

Vinyllithium

Hazardous reaction
Violently pyrophoric when freshly prepared (B297).

2-Vinylpyridine

Colourless liquid rapidly darkening to red-brown due to polymerisation; bp 158 °C; tert. butylcatechol is commonly added to minimise polymer formation; 2.7 g dissolves in 100 g water at 20 °C.

IRRITANT VAPOUR
HARMFUL BY SKIN CONTACT
FLAMMABLE

Avoid breathing vapour. Avoid contact with skin, eyes and clothing.

Toxic effects
These are not well documented, but animal experiments suggest high toxicity. The vapour irritates the skin, eyes and respiratory system. The liquid irritates the skin and eyes and must be assumed to be irritant and injurious if taken by mouth.

First aid
Vapour inhaled: standard treatment (p 109).
Affected eyes: standard treatment (p 109).
Skin contact: standard treatment (p 108).
If swallowed: standard treatment (p 109).

Fire hazard
Flash point 42 °C. *Extinguish fire with* foam, dry powder, carbon dioxide or vaporising liquid.

Spillage disposal
Shut off all possible sources of ignition. Instruct others to keep at a safe distance. Wear breathing apparatus and gloves. Apply non-flammable dispersing agent if available and work to an emulsion with brush and water — run this to waste, diluting greatly with running water. If dispersant not available, absorb on sand, shovel into bucket(s) and transport to safe, open area for burial. Site of spillage should be washed thoroughly with water and soap or detergent.

Xenon compounds Review (B145).

Xylenes (xylols)

Colourless liquids; bps 144 °C (*o*-), 139 °C (*m*-) and 138 °C (*p*-); immiscible with water.

HARMFUL VAPOUR
HARMFUL BY SKIN ABSORPTION
FLAMMABLE

Avoid breathing vapour. Avoid contact with skin and eyes. TLV 100 ppm (435 mg m^{-3}).

Toxic effects Inhalation of the vapour may cause dizziness, headache, nausea and mental confusion. The vapour and liquid irritate the eyes and mucous membranes. Absorption through the skin and ingestion would cause poisoning. **Chronic effects** If the xylene contains benzene as an impurity, repeated breathing of vapour over long periods may cause blood disease. Prolonged skin contact may cause dermatitis.

Hazardous reactions Aerobic and nitric acid oxidations of *p*-xylene to terephthalic acid both carry special hazards (B550).

First aid *Vapour inhaled:* standard treatment (p 109).
Affected eyes: standard treatment (p 109).
Skin contact: standard treatment (p 108).
If swallowed: standard treatment (p 109).

Fire hazard Flash points 17 °C (*o*-) and 25 °C (*m*- and *p*-); explosive limits approximately 1–7%; ignition temperatures 464 °C (*o*-), 528 °C (*m*-) and 529 °C (*p*-). *Extinguish fire with* foam, dry powder, carbon dioxide or vaporising liquid.

Spillage disposal Shut off all possible sources of ignition. Wear face-shield or goggles, and gloves. Apply non-flammable dispersing agent if available and work to an emulsion with brush and water—run this to waste, diluting greatly with running water. If dispersant not available, absorb on sand, shovel into bucket(s) and transport to safe, open area for atmospheric evaporation or burial. Ventilate site of spillage well to evaporate remaining liquid and dispel vapour.

Xylenols

With the exception of 2,4-xylenol, which is often encountered as a yellow-brown liquid, the commoner xylenols are colourless crystalline solids, slightly soluble in water.

HARMFUL BY SKIN ABSORPTION
CAUSES BURNS

Avoid breathing vapour. Avoid contact with skin and eyes.

Toxic effects The vapour of heated xylenols is irritant to the respiratory system. They irritate or burn the eyes and skin severely. Considerable absorption through the skin or ingestion may cause headache, dizziness, nausea, vomiting, stomach pain, exhaustion and coma. **Chronic effects** Repeated inhalation or absorption of small amounts may result in damage to the liver or kidneys.

First aid *Vapour inhaled:* standard treatment (p 109).
Affected eyes: standard treatment (p 109).
Skin contact: standard treatment (p 108). But see note on phenol p 107.
If swallowed: standard treatment (p 109).

Spillage disposal Wear face-shield or goggles, and gloves. Mix with sand and transport to a safe, open area for burial. Site of spillage should be washed thoroughly with water and soap or detergent.

Xylidines

Most of the common xylidines are red to dark-brown liquids (3,4-xylidine consists of pale brown crystals); insoluble in water.

TOXIC IN CONTACT WITH SKIN
GIVES OFF POISONOUS VAPOUR

Avoid breathing vapour. Avoid contact with skin and eyes. TLV (skin) 5 ppm (25 mg m^{-3}).

Toxic effects   Excessive breathing of the vapour or absorption through the skin may cause headache, drowsiness, cyanosis, mental confusion and, in severe cases, convulsions. The xylidines are dangerous to the eyes and the above effects may also be experienced if they are taken by mouth. **Chronic effects** Prolonged exposure to the vapour or slight skin exposures over a period may affect the nervous system and the blood, causing fatigue, loss of appetite, headache and dizziness.

First aid   *Vapour inhaled:* standard treatment (p 109).
Affected eyes: standard treatment (p 109).
Skin contact: standard treatment (p 108).
If swallowed: standard treatment (p 109).

Spillage disposal   Wear face-shield or goggles, and gloves. Mix with sand and shovel mixture into glass, enamel or polythene vessel for dispersion in an excess of dilute hydrochloric acid (1 volume of concentrated acid diluted with 2 volumes of water). Allow to stand, with occasional stirring, for 24 hours and then run acid extract to waste, diluting greatly with running water and washing the sand. The sand can be disposed of as normal waste.

Zinc (metal)

Hazardous reactions   A Zn dust explosion occurred during sieving of hot dry material; a mixture of As_2O_3 with an excess of Zn filings exploded on heating; Zn powder reacts with CS_2 with incandescence; a paste of Zn powder and CCl_4 will burn after ignition; zinc forms pyrophoric Grignard-type compounds with bromoethane and is expected to react explosively with chloromethane; warm Zn powder incandesces in F_2 and Zn foil ignites in cold chlorine if traces of moisture are present; the metal reacts explosively when heated with anhydrous $MnCl_2$, and violently with KO_2, TiO_2 and ZnO_2; zinc dust residues from reduction of nitrobenzene are often pyrophoric when dry; As, Se and Te all react on heating with Zn powder with incandescence; interaction of Zn powder and sulphur on heating considered too violent as a school experiment; mixtures of Zn dust with potassium chlorate or ammonium nitrate are liable to explode on impact, *etc*; a scoop contaminated with flake NaOH ignited Zn dust; in contact with air and limited amounts of water, zinc dust will generate heat and become incandescent (B952–5).

Zinc dihydrazide

Hazardous reaction   Explodes at 70 °C (B819).

Zinc peroxide

**Hazardous
reactions**

Hydrated peroxide explodes at 212 °C; mixtures with Al or Zn powders burn brilliantly (B917).

Zirconium (metal)

**Hazardous
reactions**

Pyrophoric hazards of Zr powder are well documented; mixture of Zr powder and CCl_4 exploded on heating; reacts vigorously with potassium chlorate or nitrate or with CuO or lead oxides; reacts explosively with alkali metal hydroxides or carbonates on heating; hydrated sodium tetra-borate and the alkali metal chromates, dichromates, molybdates, sulphates or tungstates react violently or explosively with the metal; Zr powder, damp with 5–10% water, may ignite; although water is used to prevent ignition, the powder, once ignited, will burn under water more violently than in air (B955).

Zirconium dicarbide

**Hazardous
reactions**

Ignites in cold fluorine, in chlorine at 250 °C, bromine at 300 °C and iodine at 400 °C (B341).

Zirconium dichloride

**Hazardous
reaction**

Ignites in air when warm (B688).

Zirconium tetrachloride

**Hazardous
reaction**

Ignited lithium metal strip (B700).

Chapter 9

Safety in Hospital Biochemistry Laboratories

Hazards not found in other chemical laboratories

Chemical laboratories which deal with biological specimens are found in a variety of institutions; the most obvious are hospitals, but similar work is carried out in veterinary establishments, forensic departments, university clinical departments and the laboratories of pharmaceutical companies. Such laboratories share the common hazards of all chemistry laboratories; however, the human and animal tissue specimens which they analyse (blood, urine, gastric contents, bile, cerebrospinal fluid and faeces) come, in many cases, from sources which are potentially infected with a variety of pathogens. Attention has been increasingly drawn in recent years to the risks of acquiring infection in hospital laboratories. One or two decades ago tuberculosis was one of the main hazards, and, in spite of the great decline in the number of new tuberculosis cases notified, the risk is still sufficiently great to lead to the issue of a revised edition of a Public Health Laboratory Service report on this topic.[1] Further, a working party under the chairmanship of Sir James Howie is currently considering the formulation of a Code of Practice for hospital laboratories to minimise the risk of laboratory-acquired infection. A report is expected in late 1976.

Legislation

Under the Health and Safety at Work etc Act 1974, the Health and Safety Executive is responsible for enforcing statutory requirements relating to occupational health and safety. With minor exceptions, the Act imposes statutory duties on all people at work; no one, so far as is reasonably practicable, must cause harm to anyone as a result of his activities at work. Until detailed new regulations are introduced under the Act regulations already made under the Factory Acts remain in force. The extension of statutory provisions to cover the vastly increased number of people and types of organisations envisaged in the Health and Safety Act has brought many problems in its wake but, as the work of the Health and Safety Executive develops, many of the problems considered in this chapter will become the subject of regulations.

Hepatitis

It is important to note that one of the major hazards of work in hospital laboratories, *ie* the risk of contracting viral hepatitis, has now been recognised by the listing in 1976 of this condition as an industrial disease for the particular classes of worker who are liable to it. The risk of this infection is particularly associated with patients with acute renal failure who are undergoing haemodialysis. Large volumes of blood may be used in their treatment, both to combat the associated anaemia by transfusion and to prime the dialysers. Some of this blood could come from symptomless carriers of infection. The risk to the patient may be seen from the statement that the amount of hepatitis virus associated with the HB_sAg antigen can be such that if 1 ml of a carrier's blood were added to Loch Ness and mixed, then each drop from the loch would contain virus.[2] A case has been reported[3] in which a renal transplant patient was found to be a symptomless carrier of hepatitis; three months after first detection of the HB_sAg antigen in her blood, its concentration had reached 3.8×11^{11} particles/ml of serum. There is thus a possibility that a patient in a dialysis unit may become infected with hepatitis, and that specimens received from such units may bring a high risk of infection into the laboratory.

High-risk specimens

While such 'high-risk' specimens frequently come from dialysis units, it should be borne in mind that 'high-risk' specimens can also come from any ward or clinic, and can be contaminated with a variety of infective agents. For example, a faecal specimen for occult blood examination may be found to be heavily contaminated with poliomyelitis virus. A blood specimen from a patient with undiagnosed jaundice may contain the virus of infective hepatitis or of serum hepatitis, which may also be present in blood from patients in the convalescent stage of these diseases, or even from people who are completely symptom-free, as pointed out above. An outbreak of hepatitis in a dialysis unit in the US was caused by a urine specimen from a patient with post-transfusion hepatitis (in another part of the hospital) being kept for a short time in a dialysis unit refrigerator in which were being stored blood for transfusion and food for patients and staff.

A Public Health Laboratory Service working party reported[4] in 1968 that members 'who have scrutinized haematology and biochemistry laboratory procedures from the standpoint of infection control have been struck by such lapses from ordinary hygiene as the communal use of unplugged mouth pipettes, even by the technicians with a solid background of bacteriological training'. They suggested practical instruction with the object of demonstrating 'the application to the special context of the dialysis unit or the clinical laboratory of routine precautions used every day by the bacteriologists when he handles infected cultures', and also said that 'most biochemistry and haematology laboratory technicians, certainly most of the senior ones, will have had some bacteriological teaching and experience during their training. They will simply need a reminder that by and large they should observe basic microbiological precautions when doing biochemical or haematological tests on dialysis unit specimens. Reminded that blood from a dialysis unit patient may be more dangerous than a stool from a typhoid patient, they will realise what precautions are desirable. Science graduates in biochemistry may be more vulnerable in this regard'.

Later studies have confirmed the susceptibility of non-medical graduates to infection.[5]

It may be possible in some hospitals to set aside a special area for dealing with 'high-risk' specimens, but since many routine biochemistry laboratories are housed in very cramped quarters, and are dealing with a heavy and increasing work load, this will rarely happen. As a result it will be difficult to apply adequate precautions, and the PHLS working party concluded that 'the best compromise would seem to be to concentrate on a few fundamental precautions based on probably major risks rather than to lay down highly elaborate rituals that may not be observed'.

General precautions

The major risks to be guarded against are the ingestion or inhalation of infected matter, whether in aerosol form or as dust particles from dried material, and also its entry to the body through a broken skin surface. Clearly, precautions must be taken but it must be emphasised that, if they are to be observed, they must be understood and accepted by laboratory workers and be seen to be relevant to the risk involved. General precautions should include the following points:

1. No eating, drinking or smoking in the laboratory, except in rooms set aside for this purpose. Such rooms may have to be provided by altering existing accommodation.

2. No food should be kept in laboratory refrigerators, nor should refrigerators used specifically for food be positioned in the laboratory.

3. The gum on envelopes, labels or forms should not be licked.

4. Gloves should be worn when dealing with 'high-risk' specimens. Surgeons' gloves may be more generally suitable than the lighter polythene type which are a very loose fit and puncture easily.

5. Laboratory coats should be fully buttoned-up; other protective clothing should be worn when appropriate. For example, it has been suggested that when processing specimens staff should wear not only gloves, but also plastic aprons; these may be of the type generally worn by surgeons or a smaller domestic type. Sleeved gowns, fastening at the back, may be preferred. A number of new designs of laboratory coat have come on to the market, but resistance by staff still has to be overcome. Wearing of safety spectacles has also been advised when handling 'high-risk' specimens or homogenising faeces.

6. Broken glass should not be picked up with unprotected fingers.

7. Adequate locker facilities should be available, so that staff do not have to bring any personal belongings into the working areas. In some laboratories it may be necessary to compromise by setting aside cupboard or drawer space in the working areas for this purpose and to insist that handbags and shopping, for example, are not allowed to touch bench surfaces.

8. Staff should remove gloves and laboratory coats and wash their hands thoroughly before going for coffee, lunch or tea, and before going home. Use of hand-cream after washing is advisable in order to maintain the condition of the skin.

9. All direct pipetting by mouth should be forbidden. For general pipetting procedures a rubber aspirator bulb should be used. Small volumes (*eg* 0.1ml or less) are difficult to control with an aspirator bulb and a rubber tube and mouthpiece should be used. Care should be taken to prevent contamination of the tube and mouthpiece, and a cotton wool plug should be placed in either the pipette or the rubber tube.

10. If a specimen is spilled, the area should be swabbed down with a suitable disinfectant (1 per cent hypochlorite solution); *ie* a 1:10 dilution of commercial hypochlorite containing 10g available chlorine per 100ml and a suitable indicator—an example is 'Chloros'.

11. The hands should be kept away from the face and neck.

12. Broken glassware should be placed in covered tins for disposal.

13. Any cuts or abrasions should be kept covered when in the laboratory.

Record keeping

It is important that the compulsory record of all accidents, even trivial ones, should be kept; this could be useful in the event of an outbreak of infection among staff.

Reception of specimens

The possibility of setting aside a special area for the initial processing of specimens from patients known to be, or suspected of, suffering from infectious conditions has already been mentioned. The PHLS working party suggested that 'most pathologists would prefer to designate special senior technical staff in their own laboratories to deal with specimens thought to present a hazard and otherwise simply to review biochemical and haematological techniques so as to include the kinds of precaution routinely practised in microbiological laboratories'.

It will be necessary in most laboratories to treat all specimens as potentially dangerous, and it is thus desirable to organise the reception system so that specimens are handled by a limited number of designated staff. The problem will be eased in those laboratories where there is a separate specimen reception and separation area. In such an area it should be possible to set aside bench space for dealing with request forms only, thus reducing the risk of infection to clerical staff, although one has to realise that many forms will already have been contaminated by contact with specimen containers in the wards. There should be a settled policy on how to deal with contaminated forms and specimens. Forms with visible contamination should be copied and then disposed of. The ideal approach, practised in at least one laboratory, is that each 'high-risk' specimen and the relevant request form are received in a plastic bag; after removal of the specimen, the form is heated inside the bag at 120–150 °C for 30 minutes, then photocopied and the original destroyed. Some specimens may have to be discarded at once if contamination is gross. If such 'high-risk' specimens (from *eg* patients undergoing haemodialysis, receiving multiple blood transfusions or suffering from hepatitis or unexplained jaundice) are to be given special treatment, they must be identified at ward level by some distinguishing mark rather than by supplying a list of names to the laboratory.

Centrifuging

The best situation for a centrifuge is in a well ventilated fume-cupboard the outlet from which is filtered to reduce contamination of the atmosphere outside the laboratory. Some centrifuges will have to be redesigned to localise contamination from broken specimens and to allow easy cleaning. Some blood specimens are sent to the laboratory in tubes with push-in plastic stoppers. Ideally the stoppers should not be removed before centrifuging; this will avoid production of an aerosol in the centrifuge bowl. Unfortunately, however, red cells may still be held around the stopper and may contaminate the clear plasma, so this approach is rarely practicable. Because of the retention of blood around the bottom of the stopper, even gentle removal is likely to cause production of droplets, contaminating hands and atmosphere. The British Standard Specification for Medical Specimen Containers for Haematology and Biochemistry (BS 4851: 1972 and AMD 1656) provides for the supply of tubes with screw caps; these tubes are now widely used and are much safer than older types. They can safely be centrifuged before removing the cap provided that during centrifuging the cap remains clear of the bucket. Stoppers are best put into bags which can later go for autoclaving and/or incineration. Centrifuging in an angle-head is known to lead to contamination of surrounding air and bench area; swing out heads should be used.[6] Unnecessarily fast centrifuging should be avoided, although this restriction poses a problem in some cases; for some analyses serum is preferred as specimen, and it is not always easy to obtain enough serum if centrifugation is carried out slowly. If there is breakage in the centrifuge the entire contents, and possibly even the air in the room, will be contaminated. The remaining tubes should be removed and handled carefully; the bucket and broken tube should be kept overnight in a 2 per cent solution of glutaraldehyde before removing tube with forceps. The centrifuge bowl should be swabbed down with 2 per cent glutaraldehyde avoiding contact with skin and eyes.

Pipetting

If the serum or plasma must be transferred to a second tube before analysis disposable tubes should be used. Transfer of specimen by (plugged) Pasteur pipettes should be done gently, avoiding bubble formation and bursting, to reduce aerosol production; the teat should not be contaminated with specimen. Used pipettes of all types should be discarded, tip downwards, into 1 per cent hypochlorite solution and left totally immersed overnight before disposal.

Tubes containing reaction mixtures should be carefully drained into a sink and the empty tubes put into 1 per cent hypochlorite for cleaning.

Tubes containing unwanted specimen or red cells should be put into waxed paper bags containing cellulose wadding or into plastic bags which are supported in a special holder which incorporates a heat-sealing device. All the material can be incinerated. If an autoclave is available, non-disposable tubes and disposable material can be kept separately in standard autoclave containers; disposable material can then be burnt after autoclaving.

Ideally, used pipettes, even after overnight soaking in hypochlorite solution should be autoclaved before final rinsing and drying.

The amount of pipetting by mouth has already been much reduced in many laboratories as a result of the increased use of automatic analytical devices. It has been pointed out,[4] however, that more knowledge is needed on the microbial contamination of such devices by infected specimens. Where automatic analysis is being carried out, the wastes should go directly into the sink outlet together with a good flow of water in such a way as to eliminate any risk of aerosol formation.

Mechanical analysis

Increasingly, laboratories of this type employ various types of mechanised analytical equipment; the most ubiquitous is the continuous-flow instrument marketed as the AutoAnalyzer, but other mechanised analysers are constantly being introduced and the special problems of each with respect to safety must be considered by the user.

The unit of the AutoAnalyzer system with greatest potential danger for infectious hazard is the dialyser. The membrane, which is in contact with all specimens, has to be changed regularly, and is a potential source of infection; so also are the nipples on the dialyser plates and elsewhere in the apparatus, especially if a small clot is present. Gloves must be worn when changing the membrane and indeed when changing tubing anywhere in the system. If a spare dialyser unit is available, this can be put into use while the original unit is disinfected with 2 per cent glutaraldehyde. However, disinfection will rarely be possible if a membrane has to be changed during a run on multiple analysers.

AutoAnalyzer sample cups and their contents can be carefully removed by gloved hands either into a bag containing wadding for incineration or into hypochlorite for later autoclaving and incineration. Sample plates should be kept overnight in 1 per cent hypochlorite before re-use.

Any AutoAnalyzer procedure which leads to the production of an aerosol (*eg* mixing of blood for haemoglobin determination) should be modified.

'Discrete' analysers (Vickers, Greiner, Centrifichem, *etc*) treat each specimen in a separate tube or cuvette. There may be risks from aerosol or droplet formation at sampling or mixing stages. Non-metal rotors should be cleaned at least monthly in warm 1 per cent hypochlorite solution[7] or in 2 per cent glutaraldehyde; cuvettes themselves should be cleaned weekly[7] with a similar cold solution; in each case thorough washing with water should follow. Water-baths should be checked periodically for the presence of pathogenic organisms. Waste from these analysers (and from continuous-flow systems) enters the laboratory drainage system, which requires regular (daily where the work load is high) treatment with 10 per cent hypochlorite solution. Disposal of specimen containers and reagent/reaction vessels should be similar to that of unwanted specimen and red cells (above).

Faeces and urine

Faeces can be a potent source of infection. It is common to have to determine the fat or nitrogen content of a series of specimens; part of the process is the homogenising of the specimen in water using a blender. New types of blender are now available which remove the risks of contamination of surroundings associated with earlier types. Disposal of large amounts of faeces should be by autoclaving followed by incineration. Specimens sent for occult blood examination should be put into a waxed or sealable plastic bag and burnt.

Urine should be transported to the laboratory in an unbreakable container. Some laboratories have used Winchesters in wicker baskets; others have used two-litre plastic bottles.

It is known that organic solvents will disrupt and destroy some viruses; however, their effect on many viruses, including those of hepatitis, is unknown. It is obvious, therefore, that staff handling large volumes of urine sometimes are exposed to a high degree of risk, especially from aerosols released from separating funnels. Use of well-ventilated hoods, and careful technique, might reduce this risk. It has been suggested that before disposal of urine concentrated hypochlorite solution (2ml/100ml of urine) should be added and allowed to stand overnight; the urine should then be poured off preferably into a sluice (which few laboratories have) without splashing and the container sterilised. There may be troublesome foaming on addition of the hypochlorite solution; autoclaving may be a better procedure.

Other precautions

Benches and sinks should be cleaned every evening with 1 per cent hypochlorite; this should also be used to wipe up any spilled biological material.

Other measures may include the use of plastic specimen racks which can be easily disinfected, and improvements to washing facilities, such as the installation of elbow or foot-operated taps at handbasins which should be reached without staff having to touch door handles or other objects. Use of disposable paper towels should be compulsory. Pens or pencils may be hung by the side of recorders or intercom sets so that they may be used by staff wearing gloves.

Finally, a senior member of the laboratory staff must be designated Laboratory Safety Officer, with the responsibility of supervising safety precautions, ensuring their observance, and searching out any loopholes in the scheme.

Acknowledgements

In preparation of this material, we have had the great benefit of discussion with and advice from Dr J. Connolly, Consultant Virologist, Royal Victoria Hospital. We are also indebted to the following colleagues who have allowed us to consult their respective lists of safety precautions: Professor L. G. Whitby, Department of Clinical Chemistry, Edinburgh University. Professor T. P. Whitehead, Department of Clinical Chemistry, Birmingham University. P. M. G. Broughton, Esq., Department of Chemical Pathology, Leeds University.

References

1. Department of Health & Social Security, *Precautions against tuberculosis infection in the diagnostic laboratory*, HM(70)60, October, 1970.
2. *Lancet*, 1969, **2**, 989 leading article.
3. J. Naginton, *Brit. Med. J.*, 1970, **4**, 409.
4. *Brit. Med. J.* 1968, **3**, 454 Public Health Laboratory Services, Working Party on Haemodialysis Units.
5. N.R. Grist, Personal Communication, 1976.
6. F. Whitwell, *et al., J. clin. Path.*, 1957, **10**, 88.
7. A. Clarke, Personal Communication.
See also: Department of Health & Social Security & Welsh Office, *Safety in pathology laboratories*, 1972.
C.B. Phillips 'Prevention of Laboratory-Acquired Infections', in *Handbook of laboratory safety*, 384 Cleveland, Ohio: Chemical Rubber Co.

Chapter 10

Precautions against Radiations

This Chapter deals with safety during work with the ionising radiations that are most likely to be encountered in chemical laboratories. Precautions against x-rays are covered briefly, but those against radioactive materials are dealt with more fully. No attempt is made to discuss the precautions necessary where highly radioactive materials are handled, as this gives rise to many special problems, and also may well involve the construction of special laboratories, on which expert advice should be sought. The attention of anyone responsible for work with ionising radiations in research laboratories should be drawn to the Department of Employment and Productivity Code of Practice,[1] and also that for the Universities.[2] Premises to which the Factories Acts are applicable are also subject to regulations under these acts.[3] Hospital laboratories are subject to the Department of Health and Social Security Code of Practice.[4] All such Codes of Practice spring ultimately from the Recommendations of the International Commission on Radiological Protection, ICRP.[5]

The Codes of Practice referred to above are the documents laying down details of the radiation safety measures to be followed in the UK, but many other countries have similar rules, some as Codes of Practice, some as official

Handbooks,[6] some as laws. In addition, some helpful publications have been issued by the World Health Organisation[7] and by the International Atomic Energy Agency.[8]

At the time of writing, the Health and Safety at Work etc Act 1974 has been enacted to cover all hazards at work or arising from work, but it is too early to indicate its effects in detail or to give references to regulations, codes, *etc*, resulting from it. Similarly, an EEC Directive on Radiological Protection is expected to be incorporated into British legislation by 1978.

Types of radiation

The characteristics of the main types of ionising radiation will now be described.

X-Rays. These are electromagnetic rays, like light, but each photon has a much shorter wavelength, *ie* a higher energy. They are produced outside atomic nuclei whenever moving electrons, or other charged particles, are stopped, or when in an atom their energy changes sufficiently. In practice, they are generated by accelerating electrons in a vacuum tube (the x-ray tube) with a high voltage, and then stopping them suddenly at the 'target', when most of their energy is lost as heat, but some appears as x-rays. X-rays are physically identical with γ-rays (*see below*) except in their mode of production, but because of the great range of voltage that can be used for accelerating the electrons, they may be softer (*ie* of lower energy) or harder than any known γ-rays. X-rays in the laboratory are usually generated at between 10 kV and 100 kV.

α-*Rays.* These are streams of α-particles, each the nucleus of a helium atom travelling at a high speed. An α-particle is emitted by certain radioactive nuclides, especially among the heavy elements, occasionally accompanied by a γ-ray. The energies of α-rays usually reach several MeV, and they ionise so intensely that they are easily stopped, even by a sheet of paper.

β-*Rays.* These are electrons emitted by many radioactive nuclides, often in conjunction with γ-radiation. A typical β-emitter is radioactive phosphorus (^{32}P) which emits β-particles with a range of energies from zero to 1.7 MeV, but no γ-radiation. These β-rays are much more penetrating than α-rays, and at 1.7 MeV will require some 8 mm of water or tissue to stop them completely.

γ-*Rays.* These are physically the same as x-rays, but emitted by the nuclei of certain radioactive nuclides, each yielding γ-rays of characteristic energy or energies. They may be soft (*eg* ^{125}I, at 0.027 and 0.035 MeV). Most γ-emitters also emit β-radiation or α-radiation, but because these are relatively soft they can easily be removed by filtration, leaving only the γ-rays. The γ-radiation can be reduced in intensity rather than stopped by absorbing material; that from ^{60}Co γ-rays, for instance, is reduced by about 50 per cent on passing through a slab of lead some 15 mm thick.

Bremsstrahlung. This is a form of x-radiation generated when β-rays are stopped. It is of significance only for intense sources which emit β-rays but not γ-rays; when γ-rays are emitted, these are much more intense than the *Bremsstrahlung.*

Neutrons. Some radioactive materials, such as plutonium, produce neutrons spontaneously, but not copiously. In the laboratory, neutron sources are usually composed of a mixture of an α-emitter, such as radium-226 or polonium-210, or a high-energy γ-emitter like antimony-124, with beryllium. Atoms of beryllium, when bombarded in this way, undergo nuclear transformation that results in the emission of a neutron. Commercial neutron generators are also available; in these, atoms of deuterium or tritium are accelerated by at least 100 kV and strike a target containing deuterium. Under these conditions, the D–D nuclear reaction generates neutrons of about 5 MeV energy, and the D–T reaction neutrons of about 14 MeV. Such neutrons, when slowed down, can be used, for example, for activation analysis.

Positrons. Positrons are similar to electrons, but have a positive charge. They are emitted by certain radioactive nuclides, such as ^{11}C or ^{64}Cu, and are very short-lived. After being slowed down by ionisation, a positron and an electron combine to annihilate one another, but in the process two photons are emitted in exactly opposite directions. This is called annihilation radiation, and both photons have an energy of 0.51 MeV.

Quantities and units

Radiation quantities and units are defined by the International Commission on Radiations Units and Measurements. This Commission's recommendations cover a much wider field than concerns us here, and a fuller discussion of the units, quantities and symbols *etc* will be found in the Commission's report.[9]

The *activity* of a quantity of a radioactive material is normally stated in *curies*. The activity of the quantity of a radioactive nuclide such that 3.700×10^{10} disintegrations per second occur in it is called one curie. This is approximately the number of disintegrations per second in 1 g of radium in equilibrium with its daughter products. Since the curie is a large unit, however, the millicurie (mCi) and the microcurie (μCi) are also used. The specific activity of a radioactive material is usually expressed in units such as curies per gramme.

The rate of disintegration of a radioactive nuclide varies over a wide range, but for any single nuclide it is constant, unaffected by temperature or any other factor. It is usually expressed in terms of the half-life, *ie* the time taken for the activity of a specimen to decrease to one-half its initial value. Half-lives as short as 10^{-9} s, and as long as 10^9 years, are known.

The unit of *exposure* to x-rays or γ-rays is the *röntgen*, which has been defined as an exposure to 'x- or γ-radiation such that the associated corpuscular emission per 0.001293 g of air produces, in air, ions carrying one electrostatic unit of quantity of electricity of either sign'. The 'associated corpuscular emission' refers to the ionisation produced; 0.001293 g of air is approximately the mass of 1 cm^3 of dry atmospheric air at NTP. The definition implies that the only ionisation measured is that arising from the absorption of the x-rays and γ-rays by atoms of the air, and in practice this is not easy to accomplish. The amount of energy absorbed in an exposure of 1 röntgen depends on many factors, such as the absorbing material and the energy of the radiation, so that *absorbed dose* is measured by a different unit, the *rad*; one rad of absorbed dose equals an energy absorption of 100 erg/g (10^{-2} J/kg) of the absorbing material. The dose rate of an irradiation may be expressed in such units as rad/hour.

The röntgen is about equal to 83 erg/g of air for moderately hard x-radiation and is sufficiently similar to the rad to be treated as approximately equal for many purposes when air or soft tissues are irradiated with moderately hard x-rays or γ-rays. The rad may, however, be used for all ionising radiations, whereas the röntgen is applicable only to x-rays and γ-rays.

The biological effect of a given dose of radiation depends on a number of factors, such as the total time during which the various doses of radiation are received, the dose rate during irradiation, and the type of radiation. Heavy particles such as neutrons or protons produce greater biological damage for a given dose than do x, γ or β-rays. For this reason, for protection purposes the term *dose equivalent* (DE) is introduced, where DE = D $\times$ QF, D being the dose in rad, and QF the Quality Factor expressing the relative biological effect of the radiation considered (QF = 1 for x, γ and β-rays, 10 for fast neutrons and α-rays, and so on). This dose equivalent is expressed in *rem* (rad-equivalent-man), and maximum permissible doses are specified in this unit. For all x, γ and β-ray work, however, the dose equivalent in rem is numerically equal to the dose in rad.

The SI units for radiation measurements are coming into use more slowly than other SI units. They are:

Absorbed Dose. 1 gray (gy) = 1 J kg⁻¹ (= 100 rad).

Radioactivity. 1 becquerel (bq) = 1 disintegration per second (approx. 27.03×10^{-12} Ci)

Effects of radiation

When radiation falls on atoms of matter, energy is generally lost by ionisation. If a detached electron has sufficient energy, it is likely to lose it by causing secondary ionisation. Many such ionisation processes take place in irradiated matter, and even small doses may cause some chemical changes; special techniques are necessary to demonstrate them unless the doses delivered are very high. Visible changes can take place when the dose is high enough, occurring in plastics for example at 10^6 to 10^7 rad, and doses in this range are also used commercially for sterilisation purposes.

The effects of radiation on living matter are essentially damaging and are smaller if the dose rate is decreased. It is still not known whether there is a threshold dose of radiation below which there is no apparent effect on an organism, or whether all radiation is deleterious even at very low dose rates. Small organisms such as bacteria are much more resistant to the effects of radiation than animals, and usually survive doses of radiation that would be fatal to animals.

Cells which are actively dividing are particularly susceptible to damage by radiation. For this reason, cells in the blood-forming tissues are said to be especially sensitive to radiation, and are regarded as requiring special protection. The gonads similarly require special protection, but for a different reason. It is thought that radiation damage to reproductive cells is cumulative, *ie* related to total dose accumulated, and depends little on dose rate. Changes induced in these cells will include alteration of the genetic material, or mutations, and when these changes occur they will be faithfully reproduced in any offspring which are conceived from a changed cell. Mutations are generally harmful, though frequently only slightly so, and generally recessive, *ie* they cannot be recognised for several generations, until someone inherits a particular mutation through both parents. The Medical Research Council[10] concludes that if an individual in a population receives a radiation exposure of between 30 and 80 röntgen to the gonads, 'no noticeable effect would be produced whether on their immediate offspring or upon their descendants', provided that this radiation exposure was received by only a small proportion of the whole population. The information we have on the genetic effects of radiation comes from experience with plants and animals, and man is assumed to be substantially similar in this respect. Other organs of the body may be affected by radiation when the doses involved are greater.

Permissible doses of radiation

It is no longer supposed that there is a 'tolerance dose' of radiation which can be received with impunity by a radiation worker. Instead, the ICRP defines a 'permissible dose' in these terms:[5e] 'Any exposure to radiation is assumed to entail a risk of deleterious effects. However, unless man wishes to dispense with activities involving exposure to ionising radiations, he must recognise that there is a degree of risk and must limit the radiation dose to a level at which the assumed risk is deemed to be acceptable to the individual and to society in view of the benefits derived from such activities. . . . The term Maximum Permissible Dose has become established to describe the doses that are regarded as being the maximum that should be permitted under particular conditions. The Commission recognises that the term is not an entirely satisfactory one to describe values that must necessarily, at present and in the foreseeable future, involve a considerable element of judgement. Nevertheless, until increased knowledge of the risks of radiation is available to make a more quantitative assessment of acceptable doses, the Commission proposes to retain the term Maximum Permissible Dose for exposures of radiation workers to controllable sources. . . . Long experience in the use of x-rays, radium and other radioactive materials, together with information on radiation injuries in man and other organisms, has indicated that values for Maximum Permissible Doses can be set such that there is a low probability of radiation injury without undue restriction of the uses and benefits of ionising radiations. These facts form the present basis of the Commission's recommended values.'

In selecting the doses appropriate as permissible doses for particular conditions, the Commission is clearly committed to considering the risks involved. It would be unreasonable, for instance, to specify permissible doses of radiation for laboratory workers such that the consequential radiation hazard was smaller or larger by several orders of magnitude than the ordinary hazards of the laboratory.

The values for permissible doses are listed in Table 10.1. There are two categories of dose for different groups of people:

(*a*) Adults exposed to radiation in the course of their work.

(*b*) Members of the public.

Group *b* is limited to the smaller radiation doses, partly because it includes pregnant women and children. In addition, the genetic dose to the whole population is largely determined by the gonad doses received by this group, and these are therefore restricted to low doses.

Group *a* comprises those adults exposed to radiation in the course of their work. The Codes of Practice operative in the UK require that any individual in this group who could receive as much as three-tenths of the doses listed in Table 10.1 must be 'designated' as a radiation worker, while anyone who

Table 10.1. Summary of permissible doses recommended by the ICRP (1966).

Organ or tissue	MPD for adults exposed in the course of their work[a]	Dose levels for members of the public
Gonads, red bone marrow[b]	5 rem in a year[c]	0.5 rem in a year
Skin, bone, thyroid	30 rem in a year	3 rem in a year[d]
Hand, forearms, feet, ankles	75 rem in a year	7.5 rem in a year
Other single organs	15 rem in a year	1.5 rem in a year

a. One half of the year's permissible dose (rounded up to the next whole number) may be received in one-quarter of the year except that the abdomen of women of reproductive capacity may not receive more than 1.3 rem in one-quarter of the year, nor may a woman in whom pregnancy has been established receive more than 1 rem to the foetus during the remaining period of the pregnancy.
b. Also for whole-body uniform irradiation.
c. In work of an exceptional nature, it is permissible to exceed a whole-body dose of 5 rem in a year provided the second and third provisions of a are observed and provided the total cumulative dose at age N year does not exceed 5(N-18) rem. Planned special exposures in emergency situations are permitted provided that in any single event twice the annual dose limit is not exceeded, and in a lifetime five times this limit.
d. The dose limit to the thyroid of children up to 16 years of age is 1.5 rem per year.

could not receive as much radiation as this need not be designated.

The maximum permissible whole-body dose of 5 rem in a year corresponds to 100 mrem per week, or 2.5 mrem per hour, at an even rate. For work in which radiation is likely to be received at a fairly constant rate, the scheme of work, shielding, and so forth, should be designed so that these values are not exceeded.

Where a person under 18 years of age is engaged in work involving ionising radiation, the permissible doses of Table 10.1 apply, but not the relaxations permitted by c. Further, the cumulative gonad dose received by any individual at age 30 should not exceed 60 rem.

In some schools and colleges, the teaching programme includes work with x-ray machines or radioactive substances, occasionally involving pupils of under 16 years of age. In addition to International Recommendations,[5f] official guidance is available to all concerned with this in the UK.[11] It includes the recommendation that no pupil should receive more than 50 millirem from external radiation in a year, and restricts the use of unsealed sources so that the possibility of exposure to internal radiation is similarly limited.

When practicable, the radiation received by each individual radiation worker must be measured, usually by means of a film-badge.[8f] This is a special photographic film in its envelope mounted in a holder which is worn on the person. The holder contains several metal foils, so that the blackening of the different areas of the film will yield information on the kind of radiation and its energy, as well as the dose. The films in such badges are usually changed every two or four weeks, and detailed information may be obtained from the National Radiological Protection Board, Harwell, Didcot, OX11 0RQ.

Where unsealed radioactive materials are in use, there are also the hazards of contamination and inhalation or ingestion. When taken into the body, many radioactive materials are absorbed or concentrated in some tissues more than in others and, as Table 10.1 makes clear, some tissues are associated with smaller permissible doses than others. The permissible intake of any radioactive material is based on these considerations, and is often small—orders of magnitude less than the amount of the same material that would give significant hazard for external radiation. Particular care must be exercised during manipulations involving radioactive gases. For dusts or a wide range of nuclides, the ICRP[5a] gives the permissible levels for radioactivity in air and drinking water, and also the maximum permissible body burden.

Contamination of surfaces may give rise to these hazards, as well as the external exposure to hands *etc* from these surfaces, and permissible levels of contamination in research laboratories are given in the Department of Employment and Productivity Code.[1]

In the event of ingestion or inhalation of a radioactive material, it may be possible to estimate the body burden by radiation measurements. With a material like iodine, selectively concentrated in one organ, this is done easily. A GM tube or scintillation counter is used to estimate the radioactivity of the thyroid, or whatever other organ is involved, calibrated against a sample of the same radioactive material in a suitable phantom, *ie* a container constructed to represent the part of the body involved. People who are regularly working with large quantities of radioiodine should undergo this test from time to time. For most γ-emitting materials, an estimate can be made by the technique of 'whole body counting'; the equipment for this is expensive and not widely available, but when this kind of measurement is necessary, arrangements can be made for it to be carried out. For some other radioactive materials, such as uranium, analysis of excreta can give a useful indication of body content. Table 10.2 gives the main properties of a few commonly used radioactive nuclides.

Administrative arrangements

Where ionising radiation is used, arrangements for radiation safety must be made by the responsible authority, and this is not simple. Up to the time of writing, the relevant recommendations have no legal force, although many pressures are brought to bear. In particular, if the radiations should cause

Table 10.2. Main properties of some radioactive nuclides.

Nuclide	Half-life	Max. energy of main β-rays (MeV)	Main γ-ray energy (MeV)	Critical organ	Body burden† (μCi)	Maximum permissible‡ concentration (μCi/cm³) Air	Water	K-factor‖
^{3}H	12.3a	0.018	..	Whole body	10^3	5×10^{-6}	0.1	..
^{14}C	5760a	0.159	..	Fat	300	4×10^{-6}	0.02	..
^{22}Na	2.6a	0.54	{ 0.51, 1.28	Whole body	10	2×10^{-7}	10^{-3}	12.0
^{24}Na	15h	1.39	{ 1.37, 2.75	GI tract	7	10^{-6}	6×10^{-3}	18.4
^{32}P	14.3d	1.7	..	Bone	6	7×10^{-8}	5×10^{-4}	..
^{35}S	87d	0.167	..	Testis	90	3×10^{-7}	2×10^{-3}	..
^{42}K	12.4h	3.6	1.52	Stomach	10	2×10^{-6}	9×10^{-3}	1.4
^{45}Ca	165d	0.25	..	Bone	30	3×10^{-8}	3×10^{-4}	..
^{47}Ca	4.7d	2.0	1.31	Bone	5	2×10^{-7}	10^{-3}	5.7
^{51}Cr	27.8d	..	0.323	Whole body	800	10^{-5}	0.05	0.16
^{55}Fe	2.7a	..	0.0059	Spleen	10^3	9×10^{-7}	0.02	..
^{59}Fe	45d	0.46	{ 1.10, 1.29	Spleen	20	10^{-7}	2×10^{-3}	6.4
^{58}Co	71d	0.485	{ 0.51, 0.81	Whole body	30	8×10^{-7}	10^{-3}	5.5
^{60}Co	5.3a	0.31	{ 1.17, 1.33	Whole body	10	3×10^{-7}	10^{-3}	13.2
^{64}Co	12.8h	0.66	0.51	Spleen	10	2×10^{-6}	0.01	1.2
^{90}Sr	28a	0.54	..	Bone	2	10^{-9}	10^{-5}	..
^{125}I	60d	..	0.035	Thyroid	10	2×10^{-8}	10^{-4}	..
^{131}I	8d	0.61	0.36	Thyroid	0.7	9×10^{-9}	6×10^{-5}	2.2
^{132}I	2.3h	2.14	0.95	Thyroid	0.3	2×10^{-7}	2×10^{-3}	11.8
^{137}Cs	30a	0.51	0.66	Whole body	30	6×10^{-8}	4×10^{-4}	3.3
^{226}Ra*	1620a	3.15	2.2	Bone	0.1	3×10^{-11}	4×10^{-7}	8.2

* Together with daughter products.
‡ For 40 hours exposure per week.
† Maximum permissible continuous body-burden for radiation workers.
‖ Exposure-rate at 1 cm from a point source of 1 mCi in r/h.

harm to anyone, or if anyone should think he has been harmed by them, a legal action for damages would be much more difficult to resist if precautions officially recommended had not been taken. The full recommendations[1-4] should of course be consulted by anyone concerned, as only the main outlines can be presented here. The situation is now altered by the Health and Safety at Work etc Act under which the Health and Safety Executive have rights of inspection and control, and may also promulgate statutory Codes of Practice.

In setting up a scheme for radiation protection, it is necessary to preserve a balance between two extreme points of view. One will urge that it is most important that the work is done, and no safety precautions can be tolerated that interfere with this, and the other is afraid of damage from the slightest exposure to radiation and insists that the work must be reorganised accordingly. While the ICRP has specified permissible doses of radiation, which are given in Table 10.1, it also recommends that in practice the dose received should be reduced below these levels where practicable. This can usually be done without much difficulty, but people may with confidence be subject to doses approaching the permissible values where the levels cannot practicably be reduced.

In radiation safety work, one is sometimes hampered by the great sensitivity of the measuring equipment available, which can measure doses far smaller than those known to produce any effect at all on living matter. In Britain, the 'natural background' of cosmic radiation, radiation from our own natural radioactivity, and so forth, totals about 0.1 rem per year, 2 per cent of the smallest permissible dose for a radiation worker. Yet small changes in the background radiation can be detected with accuracy, and a natural background could still be measured even if it were much smaller still. This is in sharp contrast with some other hazards, where measurement is difficult at levels lower than those known to produce physiological changes. No one must conclude that because a Geiger counter clicks, there is a serious radiation dose.

Each person in a laboratory must clearly take all proper steps to protect himself and others from hazards arising in his work, whether from ionising radiations or not. This requires knowledge and training, and it is the responsibility of the governing body to see that this has been provided. Each organisation requires Radiological Safety Officers, probably on a part-time basis in each department, to ensure that proper precautions are observed in the laboratories, and such people will have the assistance when necessary of an expert in the subject, probably a radiological physicist or health physicist. Local rules should be drafted to relate general Codes of Practice to local circumstances, and these should be handed out to all the people concerned. Special sets of records are required, such as a register of sealed sources,* a

---

* A *sealed source* is defined as 'Any radioactive substance sealed in a container or bonded wholly within material (otherwise than for the purpose of storage, transport or disposal or as a nuclear fuel element) including the immediate container or the bonding'.

stock record for unsealed sources,† and records of radiation doses. When new radiation work is to start, attention should be given to safety problems which arise, in particular as to the suitability of apparatus or premises, and whether the work is planned in the safest practicable way.

Unless it can be shown that the radiation to be received by a particular person is most unlikely to exceed three-tenths of the maximum permissible dose (Table 10.1), the person concerned must be 'designated'. He will be required to undergo a medical examination (including a blood examination) before starting such work,[7a] and again annually if in fact the radiation received does exceed this level. The preliminary examination is intended partly to ascertain whether such a person is suffering from defects which might be aggravated by exposure to radiation, and partly to record the normal blood-count values *etc* for the individual; the results of subsequent examinations will be compared with this. In addition, a 'designated' individual is required to read the appropriate rules, and to be subject to personnel monitoring, usually by wearing a film-badge.

Conditions and places of storage of radioactive materials need to be supervised to ensure safety, especially as some radioactive materials, such as radium, can emit radioactive gases. Transport of such materials requires careful thought, as there are severe restrictions, for instance on the transport of the more hazardous radioactive materials by road. The relevant UK Code of Practice, and Statutory Instruments,[12] should be consulted for details of permissible quantities, labelling *etc*. The trivial quantities that will be involved in most laboratories can be taken by road without legal control provided that the conditions of p 51 of the Code of Practice are met.

Staff must be trained in advance to deal with a radiation emergency in the laboratory should it arise. The most usual emergency is likely to be a spill of radioactive material, possibly including splashing on to a person who may be cut or bruised. A 'spill-pack', containing items most likely to be needed in such an emergency, should be immediately available together with detailed instructions. Suggestions for contents will be found in the Codes of Practice.[1,4] Thought ought also to be given to radiation hazards in the event of fire, since in a fierce fire many radioactive materials would become airborne and could be inhaled by firemen fighting the blaze. The local fire brigade should be consulted about such matters.

Laboratory organisation

It is essential that radioactive materials are used only by those who are equipped and trained to do so, and it is therefore advisable to arrange that orders for the supply of these materials cannot be issued by others. The Radiological Safety Officer needs to know in advance of all supplies of these

† An *unsealed source* is 'Any radioactive substance which is not a sealed source'.

materials, so that appropriate safety arrangements can be made, and he should, therefore, receive copies of all such orders. On receipt, the materials must be stored in a place with adequate protection against tampering. If there is any possibility of volatilisation of radioactive material, forced ventilation may be necessary. Access to such a store must be restricted to the laboratory workers authorised for this kind of work, and it is of particular importance to restrict access when the radioactive material is to be used for human subjects, as for instance in blood volume or red-cell life studies. In the UK the application of radioactive materials to a human subject requires prior authorisation from the Department of Health and Social Security.

For radioactive materials which would become a hazard in an emergency, appropriate signs are required on doors (and sometimes, for the fire brigade, inside windows) together with the name of the responsible person to be informed. Emergency equipment should be immediately available, such as protective clothing, absorbent materials, barriers for roping off affected areas, warning notices, monitoring equipment. To reduce the area of contamination in a serious emergency, it may be necessary to evacuate a laboratory, fans should be turned off, doors closed and contaminated clothing should be discarded as close as practicable to the site of the emergency. If contamination is on a person, steps should be taken to reduce or remove this without allowing it to enter the blood stream; the area should be washed with saline, but away from cuts, eyes or other sensitive areas; if scrubbing is necessary, it should not be so vigorous that the skin becomes inflamed or broken. If the contamination proves difficult to remove, it may be advisable to wash with an inactive solution of the same element. Medical advice should be sought as early as possible.

In dealing with an emergency, work involving relatively high doses may sometimes need to be carried out. This should only be done in accordance with a pre-determined plan, so drawn up that no one individual receives more than the 'planned special exposure' mentioned in *c* of Table 10.1.

The disposal of radioactive waste is, in the UK, subject to the provisions of the Radioactive Substances Act 1960, and the regulations made under that Act by the Ministry of Housing and Local Government.[13] With certain exceptions, it is necessary for each user of radioactive materials to be registered with the Radiochemical Inspectorate of the Department of the Environment, and for a Certificate to be issued prescribing where the materials may be stored and used, and in what quantity, under what conditions and so on. The Ministry also issues authorisations for waste disposal, laying down the limits of radioactive waste for the institution and the methods of disposal to be adopted. These are designed so that liquid waste in sewers shall not be a hazard either to the sewage workers or to those who may subsequently be concerned with the sludge or the purified water. The limits for dustbin disposal are set by the need not only to protect the refuse collectors, but also those who may be associated with the rubbish tips. In incineration, it is necessary to restrict the air contamination to a safe level, and also the radioactivity of the ash. In practice, it is necessary for all radioactive waste disposal for an institution to be coordinated by one Radiological Safety Officer, to ensure that the specified limits for the whole institution are observed.

All radiation workers should understand that 'adequate safety' is primarily a matter of restricting the radiation dose received to less than the recommended limits, and, so far as reasonably practicable, to as low a level as can be managed. It is not primarily a matter of restricting the hours of work, or of having long holidays with pay. These used to be part of the official recommendations but are not so today. Neither is safety primarily a matter of massive shielding, although this is sometimes necessary. Occasionally work with intensely radioactive sources is best carried out with no shielding at all, but under conditions controlled very carefully indeed.

There are in fact three factors to be taken into account:

1. Distance from the source. For practical purposes, the radiation intensity may be assumed to be reduced inversely as the square of the distance from the source.

2. Time for an operation. When other factors are unchanged, the dose received is proportional to the time taken for an operation.

3. Use of shielding. The reduction of radiation by shielding depends on the material used, its thickness, and the radiation energy from the particular radioactive material considered.

For a particular procedure, it may be necessary to evaluate the optimum combination of distance, time and shielding, but this will depend on so many local factors that no general rules are applicable. It is likely, however, that the use of remote tools will enable distances to be increased without corresponding increase in the time of operation. Such tools are available commercially (or can be made locally) for opening ampoules or screw-top bottles, operating burette taps or syringes, picking up glass vessels and so forth. Indeed, for high-activity work, manipulators are available which will operate over thick protective walls and still give great precision of manipulation.

Rooms or laboratories where x-rays are used, or where there are more than trivial quantities of radioactive material, should have the door marked with the international trefoil sign in black and yellow.[4, 5b, 14] This is not to be interpreted as a danger sign, or even labelled as such, but as a warning that caution is needed since ionising radiation *may* be present. This sign, surmounted by the word **FIRE** in red, is sometimes used to indicate a room (such as the room for a radiotherapy cobalt treatment unit) containing enough radioactive material to be a hazard to firemen in the event of a fire. Because of its specific concentration in the thyroid gland, radio-iodine is particularly dangerous in this context, as also are alpha-ray emitters, and the radiostrontiums. Certain types of laboratory equipment are known to contain a radioactive source in this category simply to produce ionisation in gases.

Precautions in work with x-rays

X-rays are normally produced in a special tube so shielded that no appreciable emission of radiation occurs outside the 'useful beam', but the intensity inside this beam may be very great, so that due precautions need to be taken even against the radiation scattered from the beam by any material irradiated. It is of particular importance that the useful beam of x-rays should at no time irradiate directly any part of any individual; when necessary, protective clothing should be worn.

X-rays are reduced in intensity by any material, but the greater the density, and the greater the atomic number, the greater the attenuation. Thus a given attenuation can be achieved with either iron or lead, but if iron is used a greater mass, as well as a greater thickness, will be required although the cost may in some circumstances be less.

X-rays are frequently used for crystallographic purposes, the diffraction of the radiation in different directions giving an indication of the internal structure of a crystal. For this purpose, a small but intense beam of radiation is required, often of very soft (*ie* low photon energy) radiation. It is particularly important to ensure that the x-ray tube is not energised whenever adjustments need to be made, and that hands cannot come near the x-ray beam. Automatic shutters can be used, or warning lights wired into the same circuit as the high-tension transformer but, since familiarity breeds contempt, constant vigilance is necessary.[15]

Small mobile x-ray machines are sometimes used in research laboratories. The useful beam should never be larger than is necessary for the required irradiation; restricting cones of different sizes, or adjustable diaphragms, should be available. The useful beam should be directed away from people and adjacent occupied areas. When it is necessary for someone to be close to, or even in, the beam, protective shields or clothing should always be used, but efforts should be made to use clamps, or remote control devices whenever possible.

Protection in work with sealed sources

Radiation from sealed radioactive sources is much less intense than from an x-ray machine. They may be used, for example, to ionise gases, for the removal of electrostatic charges or in certain types of analytical apparatus; for measurements of scatter or absorption, as in thickness gauges; or for irradiating solutions or living organisms in order to produce radiation changes. It is necessary to ensure that the amount of radiation from such sources reaching working positions does not exceed permissible levels.

Often a source is incorporated in a piece of equipment so arranged that the radiation can be turned on and off as required, usually by moving the source by remote control into and out of a protected housing. Any equipment of this kind must be designed to 'fail safe'. If it is electrically operated, the equipment should automatically return the source to the 'safe' position in the event of a power failure, possibly by a spring, but it is usually desirable to make provision so that in any case this operation can be carried out manually—by a long tool, if necessary.

All such sources must be entered in the appropriate register; they should therefore have identifying numbers or marks and be audited at set intervals. Checks need to be made periodically to ascertain that the sealing of each source remains satisfactory. Different methods are used for different types of source. A radium source is sealed with absorbent cotton wool for a day or two, and the cotton wool then put close to the electrode of an ionisation chamber. If the source is leaking, radon will have escaped and this radioactive gas will be trapped in the cotton wool and give off α-rays. If the source contains almost any other radioactive material, a moist filter-paper or cloth held in forceps is wiped over the surface of the source, and then assessed for radioactivity. If a source is found to be leaking it should be immediately sealed in a leakproof container and returned to the supplier for repair. Checks will be necessary to find out how far contamination has spread, followed by appropriate decontamination.

The loss of a source may be serious, especially if it contains one of the more hazardous radioactive materials, such as ^{90}Sr or ^{226}Ra. Where such sources are used, appropriate detecting equipment should be available so that if one is lost it can then more easily be detected and recovered. Radium needles have been recovered intact in this way from incinerators, drains and demolished buildings; one such needle exploded and contaminated a section of a hospital so severely that several rooms had to be demolished and rebuilt.

Precautions in work with unsealed radioactive materials

When radioactive material is unsealed, more care must be exercised, because of the greatly increased risk of contamination, inhalation and ingestion. The degree of care is related not only to the amount of activity and toxicity of the material, but also to the form in which it is manipulated. For many years it has been the practice to classify laboratories according to the scheme laid out in Table 10.3. For a Grade C laboratory, it would suffice to modify a first-class chemical laboratory to ensure, for instance, that floor and bench coverings will not absorb radioactive solutions. A Grade B laboratory needs to be thought out in detail in relation to its functions, and should be reserved

Table 10.3. Classification of laboratories.

Class of nuclide	Grade of laboratory necessary for activity of		
	C	B	A
1	$<10\mu$Ci	<1mCi	>1mCi
2	<1mCi	<100mCi	>100mCi
3	<100mCi	<10Ci	>10Ci
4	<10Ci	<1000Ci	>1000Ci

Multiplying factors for different procedures

Procedure	Multiply permissible activity by
Storage of stock solutions	$\times 100$
Very simple wet operations	$\times 10$
Normal chemical operations	$\times 1$
Complex wet operations with risk of spill, or simple dry operations	$\times 0.1$
Dry and dusty operations	$\times 0.01$

Classification of radionuclides

Class	Examples
1 (high toxicity)...................	^{226}Ra
2 (medium toxicity, upper subgroup)	^{22}Na, ^{45}Ca, ^{60}Co, ^{90}Sr, ^{125}I, ^{131}I
3 (medium toxicity, lower subgroup)	^{14}C, ^{24}Na, ^{32}P, ^{35}S, ^{42}K, ^{51}Cr, ^{55}Fe ^{59}Fe, ^{58}Co, ^{64}Cu, ^{132}I
4 (low toxicity)	^{3}H, ^{85}Kr, ^{133}Xe

for such work. Compared with a normal chemical laboratory, more space generally is necessary per worker, especially in fume cupboards, and the benches and floors should be designed to take heavy shielding when required. There must be provision for each worker to have rubber gloves, and foot-, knee- or elbow-operated taps are desirable. Benches, floors and other surfaces should be easily decontaminable. A Grade A laboratory usually has to be specially designed and equipped for a particular purpose, possibly with changing rooms, independent filtered-air ventilation and large effluent tanks for the reception and storage of radioactive waste. It is not intended that the scheme in Table 10.3 should be enforced rigidly, but that it be used as a general guide to the kind of laboratory appropriate to particular work. Appendix 3 of the Department of Employment and Productivity Code[1] gives the full IAEA classification of radionuclides into the four classes for toxicity, but does not use a classification for laboratories, relying on the experience and discretion of Radiation Safety Officers to provide appropriate laboratory facilities for the work to be done.

Cleanliness is essential in all radioisotope laboratories, even more so than in many other types of laboratory. Care should be taken to avoid spills (for example, by conducting certain operations over a large drip-tray) and to clear up at once any spillage that does occur; if the spillage dries, active material can become airborne. When possible laboratories should be used solely for radioisotope work, but even when this is not possible specified sections of laboratories should be so reserved. To reduce the risk of internal deposition of radioactivity, no smoking, drinking, eating or application of cosmetics should be allowed. Work with radioactive materials at widely different levels of activity should be so segregated as to reduce the possibility of cross-contamination affecting experimental results. The counting of radioactive samples should not be carried out in the same room as the manipulation of even modest quantities of radioactivity.

When appreciable amounts of γ-active materials are used, shielding will be necessary. This can take many forms, depending on the work to be done, but in general it is more economical in material to put the protective shielding as close as possible to the radioactive material. On occasion a few lead pots will be almost as effective as a bench fully-protected with hundreds of pounds of interlocking lead bricks, but sometimes there is no alternative to the latter. Concrete or barytes blocks are often used, being cheaper than lead but more bulky. Such protective devices need to be planned and commissioned in full knowledge of the local circumstances. Much information on the attenuation of different radiations in different materials can be found in a number of publications[5b, c, d] and [8b, 16]

Frequently, contaminated articles need to be transported for disposal. Some of these, such as animal excreta, contaminated filter-papers or other absorbent paper, can be suitably sealed for the purpose in polythene bags. Polythene sheeting may also be used to line drip-trays and the like, so that if a spill does occur absorbent paper may be used to mop it up and then folded into the polythene sheet for disposal.

Glassware used to hold radioactive solutions frequently becomes contaminated, and occasionally this contamination may be very difficult to remove completely. Where recovery of the articles is desired, they should be cleaned in a chromic acid bath reserved for this purpose. It is in any event an excellent precaution against cross-contamination to mark glassware so that one article is used only for one radioisotope, different colours can be used for indicating different radioisotopes used in the laboratory. For glassware used to contain solutions of high radioactivity (such as a pipette for making up dilution of a stock solution on arrival) *two* marks can be used, so that this article is not used subsequently for very dilute tracer solutions.

In working with radioactive dusts or vapours, it is obviously necessary to use an adequate fume cupboard so vented that any radioactivity discharged will not enter the same or any other premises in appreciable amount. This may require the use of a filter, or scrubbing, in the air exhaust, together with a powerful fan and a discharge well clear of the roof of the building.[17] In certain circumstances, it may be necessary to check that even under gale conditions no blow-back of activity into the laboratory can occur, and in general the principle to be established is that air-flow is always from low-activity areas to areas of high activity, although this may need to be modified when sterile radiopharmaceuticals are being prepared.[18]

Newcomers to radioisotope work need, of course, to familiarise themselves with radiation safety, as outlined here. Before they embark on work with radioactive materials, especially in the more toxic categories, they should first of all carry out the sequence of operations they intend to use with inactive material, and then possibly with a relatively non-toxic material such as ^{32}P or ^{24}Na. They should manipulate radioactive materials in much the same way as they would dangerous bacteriological materials (*see* Chapter 9). Indeed, the techniques used to prevent radioactive contamination are similar to those used to maintain sterility in a hospital operating theatre.

Contaminated surfaces which cannot be discarded must be cleaned, and the Codes of Practice contain official recommendations for this. The magnitude of the problems depends on the radioactive material concerned and its chemical form. Simple washing with water or a detergent may suffice. If, however, the radioactive material is of high specific activity, minute traces may be absorbed on the surface, which may be removed most easily by washing with an inactive 'carrier' solution of the same substance. Sometimes rubbing, or scrubbing, is necessary, but if the radioactive material has soaked into a surface it may be impossible to remove it; a choice must be made between sealing it in (which may be practicable with small quantities) and dismantling the laboratory. For this reason, plain wooden benches, even if waxed, are to be avoided and one surfaced with stainless steel, or a plastic such as Formica, is much to be preferred.

Measuring instruments

Most of the many measuring instruments for radiation protection are among the following types.

Film-badges. In the UK, different types of monitoring film are available from Kodak and Ilford. One type of film-holder is described in a British Standard[19] but other types are in use. It is important that the holder be designed to match the characteristics of the particular film to be used. Holders can be bought for less than £1 each, and individual films for a few pence. It is, however, grossly uneconomic as well as unsatisfactory to run a film-badge service for small numbers of people, say less than 200, except in special circumstances. The services and advice of the National Radiological Protection Board should be sought.

Thermoluminescent dosimeters (TLD). Certain crystals, such as LiF, have the property that irradiation with ionising radiation causes some electrons to change to an excited state where they remain until the material is heated; they then return to the ground state with the emission of a photon of light. Such material therefore can be used for radiation measurement when properly calibrated. TLD is particularly useful when a dosimeter must be small, or in an awkward position, and they are used, for instance, to measure finger-tip doses. They can also be used for personnel dosimetry generally to replace film-badges, and they have the advantage that the system can be automated to a greater extent.

Personnel pocket dosimeters. These are available from more than one manufacturer, but different types are designed for different radiations; thus it is important not to use for x-ray work a dosimeter designed for measuring hard γ-rays. Each dosimeter may cost £15 or more, and it is necessary to have a recharging unit.

Ionisation-chamber dosimeters. These will measure the dose rate at a given place with good accuracy. If the amount of radiation is small, the accuracy may be poor, and the instrument may have a long time-constant. Such dosimeters usually have several ranges of sensitivity, and the ionisation chamber is sometimes separated from the rest of the instrument by a cable, to facilitate measurements in corners and so on. Such instruments may cost up to several hundred pounds, depending on the accuracy, flexibility and speed of response. Dosimeters may be mains-operated, or battery-operated to facilitate portable use.

Geiger-counter ratemeters. These instruments are much more sensitive to radiation than those mentioned above, but it is difficult to ensure that a given reading corresponds to a given dose rate when, for instance, the energy of the radiation changes. Such an instrument can only give quantitative measurements for the radiations for which it was calibrated, but some similar instruments are calibrated arbitrarily in 'counts per second'. They all have the advantage that they can detect very small quantities of radiation, except that it is not easy to detect very soft α- or β-radiation. Many instruments have—sometimes as an optional extra—a geiger-counter tube in a probe cable-connected to the rest of the instrument in such a way that any contamination of hands, surfaces, drains and so forth can be easily checked. Some ratemeters are battery-operated, but many are mains-operated for laboratory use. Such a mains-operated instrument may well be installed in a laboratory used for dispensing radioactive materials, and left on during occupation of the laboratory for ease of checking contamination. Several portable instruments are likely to be needed for other laboratories and other parts of the building. These instruments may cost from £100 or so to several hundred pounds. Some simple versions are useful only for assessing whether, for instance, a particular dustbin does or does not contain an appreciable amount of γ-active material and are of no use for quantitative measurements.

Scintillation counters. Certain crystals, and a few other materials absorb a photon or charged particle and emit light proportional to the radiation energy absorbed. This light is detected by a photo-multiplier and converted to a pulse of current which can then be processed electronically. A crystal scintillation counter is likely to be much more sensitive to radiation than a Geiger counter, though more expensive. Where a liquid scintillator is used, it is possible to mix with it radioactive material emitting very soft radiations, and thus to measure them. One of the few ways of detecting bench contamination with tritium, for example, is to wipe the bench with a tissue using a standard routine, and to incorporate the tissue in a liquid scintillation counter for measurement. The simplest scintillation counters cost about £200, a sophisticated automatic system about £10 000.

Non-ionising radiations. Safety precautions for non-ionising radiations have received much less attention than those for ionising radiations, and no internationally-agreed standards have been laid down which compare with the ICRP Recommendations. Nevertheless, some advice can be given.

Ultraviolet radiation. Ultraviolet light may give rise to superficial eye damage and to burning of the skin, not unlike that produced by over-exposure to the sun. A few seconds' exposure of the eye to an unscreened arc at a distance of several yards may cause an extremely painful condition generally known as 'eye flash'. The characteristic symptoms are a feeling as of sand in the eyes accompanied by intense pain, intolerance to light, watering of the eyes and possibly temporary loss of vision. In a laboratory where ultraviolet radiation is produced in a confined space, there is a possible danger to persons in the area of excessive concentration of ozone, which has an irritant action on the upper respiratory system. The American Medical Association has recommended that individuals should not be exposed to intensities of ultraviolet light greater than 0.5 μW/cm^2 (7-hour day), or 0.1 μW/cm^2 (24-hour day). A monitoring instrument suitable for the purpose of measurement has been described.[20]

Precautions. Protective cabinets or screens should be placed around the source of emission and a screen of ultraviolet-absorbing glass should be interposed between the worker and the source of radiation. It is easier to do this if the area of ultraviolet radiation is limited to the minimum necessary. Approved goggles should be worn and, when necessary, hands and forearms should be protected by cotton material or suitable barrier creams. To guard against the danger from ozone already mentioned, the room should be well ventilated.

Infrared radiation. Since infrared radiation is readily absorbed by surface tissues it does not inflict any deep injuries. In the case of the eyes, the heat absorbed by the lens of the eye from infrared radiation is not readily dispersed, and a cataract may be produced. This condition is often known as 'chain-maker's eye', but it is also found in glassblowers and furnace attendants.

Precautions. All sources of intense infrared radiation should be shielded as near to the source as practicable by heat-absorbing screens, to prevent the radiation entering the eyes of the workers. Approved goggles or eye-shields should be worn, and since this may result in less light entering the eye, the general illumination of the laboratory should be increased appropriately.

Laser beams. There are many laboratory uses of lasers, and these can be dangerous as the beam energy is concentrated into a very small area. Even at low intensity, damage to small areas of the retina of the eye can all too easily result from a laser beam, especially if it is viewed in an optical instrument without proper attenuation, and recovery may be slow or non-existent. At higher powers, a laser beam can cause severe and rapid damage to skin. A comprehensive *Guide for control of laser hazards*[21] has recently been published. Safe levels are said to depend on wavelength, length and frequency of repetition of exposure, *etc*, and are complicated by the fact that some lasers generate radiation in the ultraviolet or infrared regions as well as visible light. This guide should be consulted, as it is impracticable even to summarise its conclusions here.

Ultrasonics. Ultrasonic radiation is being increasingly used for many purposes, including cleaning in special baths. Where this radiation is used at high intensity, biological results can certainly be expected; indeed, it is used in certain medical procedures to destroy tissue. The energy intensity capable of causing damage is said to vary with frequency and with duration and frequency of exposure.[22] In certain circumstances, it is alleged[23] that intensities as low as 4 mW cm^{-2} can cause a decrease in DNA synthesis, but whether this is the limiting effect is open to question. Certainly intensities above the cavitation threshold can be expected to produce tissue damage, and this level is said to vary from 0.08 to 6 W cm^{-2} depending on frequency.

Microwave radiation. The microwave region of the electromagnetic spectrum is increasingly used for a variety of purposes, among them rapid heating. In particular, microwave ovens have been known to leak dangerous amounts of microwave energy when the doors are not properly shut. The health hazards of this form of radiation were discussed at an International Symposium sponsored by the World Health Organisation and the US Department of Health, Education and Welfare, the proceedings of which have been published.[24] Thermal effects are considered to be unlikely at intensities below 1 mW cm^{-2}, and methods of measurement are discussed. Much more research is thought necessary before any conclusions are reached on cumulative or delayed effects.

References

1. Department of Employment and Productivity, *Code of practice for the protection of persons exposed to ionising radiation in research and teaching.* London: HMSO, 1968, (revised 1974).
2. *Radiological protection in universities.* London: Committee of Vice-Chancellors and Principals of the Universities of the United Kingdom, 36 Gordon Square, WC1, 1966.
3. a. Ministry of Labour, *Ionizing radiation precautions for industrial users*, Safety, Health and Welfare New Series No. 13. London: HMSO, 1961.
   b. The Ionizing Radiations (Sealed Sources) Regulations, 1969, (SI 1969 No. 808). London: HMSO.
   c. The Ionizing Radiations (Unsealed Radioactive Substances) Regulations, 1968, (SI 1968 No. 780). London: HMSO.
4. Ministry of Health, *Code of practice for the protection of persons exposed to ionizing radiations arising from medical and dental use.* London: HMSO, 1972.
5. *Recommendations of the International Commission on Radiological Protection.* London and New York: Pergamon.
   a. Publication 2 — Report of Committee II on permissible dose for internal radiation, 1959.
   b. Publication 15 — Report of Committee III on Protection against ionizing radiations from external sources, 1969.
   c. Publication 4 — Report of Committee IV on protection against electromagnetic radiation above 3 MeV and electrons, neutrons and protons, 1963.
   d. Publication 5 — Report of Committee V on the handling and disposal of radioactive materials in hospitals and medical research establishments, 1964.
   e. Publication 9 — Recommendations of the International Commission on Radiological Protection adopted September 17, 1965.
   f. Publication 13 — Radiation protection in schools for pupils up to the age of 18 years, 1970.
6. *Eg* Reports of the National Council on Radiation Protection and Measurements (US).
   a. No. 8. Control and removal of radioactive contamination in laboratories.
   b. No. 10. Radiological monitoring methods and instruments.
   c. No. 24. Protection against radiations from sealed gamma sources.
   d. No. 28. Manual of radioactivity procedures.
   e. No. 30. Safe handling of radioactive isotopes.
   f. No. 32. Radiation protection in educational institutions.
7. Publications of the World Health Organisation, Geneva.
   a. Medical supervision in radiation work. Technical report series No. 196, 1960.
   b. Radiation hazards in perspective. Technical report series No. 248, 1962.
   c. Public health responsibilities in radiation protection. Technical report series No. 254, 1963.
   d. Public Health and the medical use of ionizing radiations. Technical report series No. 306, 1965.
8. 'Safety Series' of the International Atomic Energy Agency, Vienna.
   a. No. 1. Safe handling of radioisotopes.
   b. No. 2. Health physics addendum to No. 1.
   c. No. 3. Medical addendum to No. 1.
   d. No. 6. Regulations for the safe transport of radioactive materials.
   e. No. 7. Notes on certain aspects of No. 6.
   f. No. 8. Use of film badges for personnel monitoring.
   g. No. 9. Basic safety standards for radiation protection.

9. Recommendations of the International Commission on Radiation Units and Measurements, esp. No. 11. *Radiation quantities and Units*, 1968, ICRU Publications, PO Box 4869, Washington DC, 20008, US.

10. *The hazards to man of nuclear and allied radiations*, Cmnd 1225, Report to the Medical Research Council, London 1960.

11. *The use of ionizing radiations in educational establishments and notes for the guidance of schools establishments of further education and colleges of education in the use of radioactive substances and equipment producing x-rays.* Administrative Memorandum 2/76. Available free of charge from the Department of Education and Science, Elizabeth House, York Rd, London, SE1 7PH.

12. a. Department of the Environment, *Code of practice for the carriage of radioactive materials by road.* London: HMSO, 1975.
    b. The Radioactive Substances (Carriage by Road) (Great Britain) Regulations, 1974, (S.I. 1974 No. 1735). London: HMSO.
    c. The Radioactive Substances (Road Transport Workers) (Great Britain) Regulations, 1970 (S.I. 1970 No. 1827). London: HMSO.

13. a. The Radioactive Substances Act, 1960. and various exemption orders. London: HMSO.
    b. The Control of Radioactive Waste, Cmnd 884, London: HMSO, reprinted 1962.

14. a. *A basic symbol to denote the actual or potential presence of ionizing radiation*, BS 3510: 1968, London: BSI.
    b. *Colours for specific purposes*, BS 3810: 1964. London: BSI.

15. B.E. Stern, *Health Phys.* 1970, **19**, 133.

16. Handbook of radiological protection. Part I — Data. London: HMSO, 1971.

17. K. Everett and D. Hughes, *A guide to laboratory design.* London: Butterworth, 1975.

18. *Guidelines for the preparation of radiopharmaceuticals in hospitals*, British Institute of Radiology, Special Report No. 11, London, 1975.

19. *Film badges for personnel radiation monitoring.* BS 3664: 1963. London: British Standards Institution, 2 Park Street, W1 4AA.

20. H.L. Andrews, *Rev. Sci. Instrum*, 1945, **16**, 253.

21. a. *A guide for control of laser hazards.* (1973). Issued by The American Conference of Governmental Industrial Hygienists, PO Box 1937, Cincinnati, Ohio, 45201.
    b. A.J.H. Goddard, *Phys. Bull.* 1976, **27**, 245.

22. P.N.T. Wells, in *Cardiovascular applications of ultrasound*, R. S. Reneman, (Ed). Amsterdam and London: North Holland, 1974.

23. N. Prasad, R. Prasad, S.C. Bushong, L.B. North, E. Rhea, *Lancet*, 1976, **i**, 1181.

24. *Biologic effects and health hazards of microwave radiation.* Warsaw: Polish Medical Publishers, 1974.

Index

This index is designed to help readers make quick reference to topics in the book, with the exception of the coloured alphabetically-arranged monograph section of Chapter 8. Certain chemicals of particular importance referred to in the other nine chapters are indexed and in these cases the Chapter 8 page number has been included in bold type.

Index